AF894710

Topics in Current Chemistry · Vol. 146

Physical Organic Chemistry

Physical Organic Chemistry

With 21 Figures and 27 Tables

AKADEMIE-VERLAG BERLIN
1988

Die Originalausgabe erscheint im Springer-Verlag
Berlin · Heidelberg · New York · London · Paris · Tokyo
in der Schriftenreihe „Topics in Current Chemistry“,
Volume 146

Vertriebsrechte für die sozialistischen Länder:
Akademie-Verlag Berlin
Vertriebsrechte für alle Staaten mit Ausnahme der sozialistischen Länder:
Springer-Verlag Berlin · Heidelberg · New York · London · Paris · Tokyo

Alle Rechte vorbehalten
© Springer-Verlag Berlin · Heidelberg 1987

ISBN 3-540-18541-0 Springer-Verlag Berlin · Heidelberg · New York
ISBN 0-387-18541-0 Springer-Verlag New York · Heidelberg · Berlin

ISBN 3-05-500514-7 Akademie-Verlag Berlin

Erschienen im Akademie-Verlag Berlin, DDR-1086 Berlin, Leipziger Straße 3—4
Lizenznummer: 202 · 100/510/88
Printed in the German Democratic Republic
Gesamtherstellung: VEB Druckerei „Thomas Müntzer“, 5820 Bad Langensalza
LSV 1275
Bestellnummer: 763 845 8 (3072/146)
14800

Editorial Board

Prof. Dr. *Michael J. S. Dewar*	Department of Chemistry, The University of Texas Austin, TX 78712, USA
Prof. Dr. *Jack D. Dunitz*	Laboratorium für Organische Chemie der Eidgenössischen Hochschule Universitätsstraße 6/8, CH-8006 Zürich
Prof. Dr. *Klaus Hafner*	Institut für Organische Chemie der TH Petersenstraße 15, D-6100 Darmstadt
Prof. Dr. *Edgar Heilbronner*	Physikalisch-Chemisches Institut der Universität Klingelbergstraße 80, CH-4000 Basel
Prof. Dr. *Ŝhô Itô*	Department of Chemistry, Tohoku University, Sendai, Japan 980
Prof. Dr. *Jean-Marie Lehn*	Institut de Chimie, Université de Strasbourg, 1, rue Blaise Pascal, B. P. Z 296/R8, F-67008 Strasbourg-Cedex
Prof. Dr. *Kurt Niedenzu*	University of Kentucky, College of Arts and Sciences Department of Chemistry, Lexington, KY 40506, USA
Prof. Dr. *Kenneth N. Raymond*	Department of Chemistry, University of California, Berkeley, California 94720, USA
Prof. Dr. *Charles W. Rees*	Hofmann Professor of Organic Chemistry, Department of Chemistry, Imperial College of Science and Technology, South Kensington, London SW7 2AY, England
Prof. Dr. *Fritz Vögtle*	Institut für Organische Chemie und Biochemie der Universität, Gerhard-Domagk-Str. 1, D-5300 Bonn 1

Table of Contents

Rearrangements of "Carbanions"

Gernot Boche

Fachbereich Chemie der Philipps-Universität Marburg, Hans-Meerwein-Straße,
D-3550 Marburg, FRG

Table of Contents

After a brief introduction "carbanion" and "carbanion" radical rearrangements are reviewed with special concern to the literature of the last ten years. Radical "anion" systems are included because of the increasing importance of electron transfer reactions.

The first topic deals with the cyclopropyl-allyl and the intimately related allyl "anion" rearrangement. Next, electron transfer induced rearrangements of cyclopropanes (stereoisomerization and formation of ring-opened derivatives via trimethylene radical "anions") are reviewed. In relation with this subject the question is discussed whether the rearrangement of cyclopropanes in the presence of a base is initiated by proton or electron transfer. Section 3 deals with electron transfer induced rearrangements of the generalized cyclobutene-butadiene radical "anion" system, followed by the cyclization of 5-hexenyl "anions". Because of the widespread use of 5-hexenyl radical cyclizations as a mechanistic tool it is important to separate the radical from the "anion" rearrangement. Section 5 deals with the isomerization of α-substituted vinyllithium compounds. Section 6 describes how the structure of a "carbanion" is strongly influenced by a variation of the gegenion, the solvent, complexing agents and the temperature. Electron transfer valence isomerizations are treated next. The complexity of "carbanion" structures in solution, as shown in sections 6 and 7, is also observed in the solid state as demonstrated by the polymorphism of organolithium compounds. Section 9 is devoted to some results concerning rearrangements of and within alkyllithium aggregates, and the final topic refers to "carbanion"-accelerated rearrangements.

1 Introduction

The literature on rearrangements of "carbanions" published up to 1979 has already been reviewed [1-8]. Sigmatropic rearrangements and other migrations of saturated groups (hydrogen, alkyl groups, hetero-atoms), unsaturated carbon (olefins, acetylenes, aromatic compounds) and doubly bonded oxygen species (homoenolization, Favorskii rearrangement, Ramberg-Bäcklund reaction) are treated especially in the articles of Grovenstein [6] and Hunter, Stothers and Warnhoff [8]. Electrocyclic reactions have been extensively reviewed by Staley [5], Buncel [3] and Hunter [4].

Most of the literature on rearrangements of "carbanions" reviewed in this article has been published since 1979. Since "carbanions" normally neither exist in solution (nor in the solid state), and since the gegenion, the solvent and chelating agents strongly influence the structure and the reactivity of a certain (alkali or alkali earth) organometallic compound, we stress those influences whenever there is information in the literature. Furthermore, since it has been possible in recent years to characterize different aggregates of organolithium species, their *intra-* and *inter-*"molecular" rearrangements and their different reactivities we included these results into this article, too. Thus, we would like to emphasize the importance of non-skeletal rearrangements of "carbanions". Similarly, we have summarized rearrangements of radical "anions" induced by electron transfer reactions because of the increasing importance of this type of reaction lately.

2 Rearrangements Involving Three-Membered Rings

2.1 Cyclopropyl-Allyl-"Anion" Rearrangements

It was only after Woodward and Hoffmann had predicted a conrotatory mode for the thermal cyclopropyl-allyl anion transformation in 1965 [9] that new interest developed in this reaction [10]. But it was first shown by Huisgen and coworkers [11] by means of the iso-π-electronic uncharged aziridine *1* which gives the azomethine ylids *2* and *3*, respectively, that the stereochemistry of the thermal and the photochemical reactions agrees with the prediction.

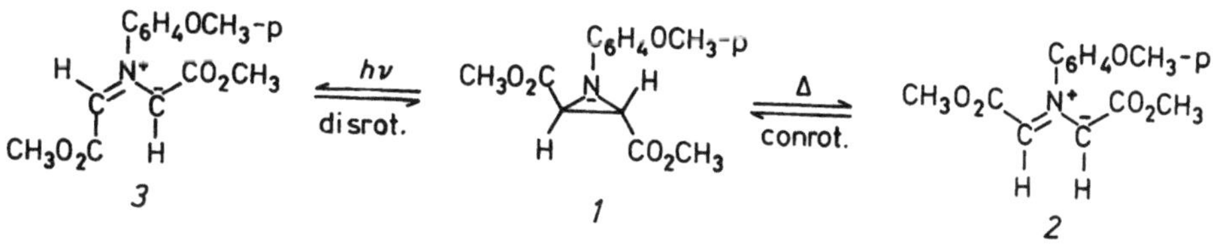

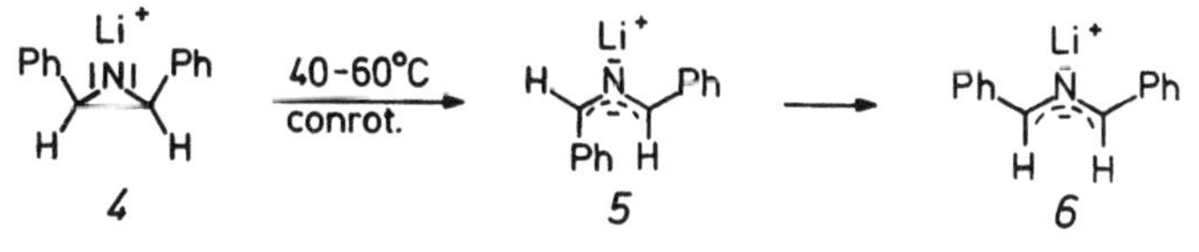

Kauffmann and coworkers [12] studied another hetero-analogous but in this case an "anionic" system, N-lithio-cis-2,3-diphenylaziridine (*4*). *4* transforms thermally into *endo,exo*-1,3-diphenyl-2-azallyllithium (*5*).

The rearrangement of *5* to *6* competes successfully with the trapping of *5* with *trans*-stilbene which established the stereochemistry of *5*. The stereochemistry of *2* and *3* has similarly been determined by 1,3-dipolar cycloaddition reactions.

Why is it that the predicted modes of rearrangement were first confirmed by means of these two systems and not with a "real" cyclopropyl-allyl "anion" system? The long history of the cyclopropyl-allyl "anion" rearrangement shows that this is because of several problems [10b]. The first of these, which is also indicated from the results with *1* and *4*, and which is confirmed by MO calculations, is the *slow* thermal conrotation of a cyclopropyl "anion" to give the corresponding allyl "anion", as compared to the *fast* isomerization of the allyl "anion" to give the most stable isomer.

Table 1 gives the results of several MO calculations for the conrotatory ring-opening of the cyclopropyl anion to give the allyl anion, and of the isomerization of the allyl anion [13].

Table 1. Relative energies ($\Delta\Delta H_f$, kcal/mol) with respect to the cyclopropyl anion of the transition state of the conrotatory cyclopropyl-allyl anion rearrangement A, the allyl anion B, and the transition state of the allyl anion isomerization C.

	A	B	C

MINDO/3	31.2	−9.8	—
STO-3G	66.7	−9.7	25.3
4-31G	38.0	−40.0	27.6

It is evident from Table 1 that the "activation energy" for the allyl anion isomerization is much lower than for the conrotation of the cyclopropyl anion to give the allyl anion. Consequently, in order to verify the predicted conrotatory mode one has to trap the first formed allyl "anion" before it isomerizes to give the thermodynamically most stable isomer, e.g., in a cycloaddition reaction. Exactly this was possible with *2*, *3* and *5*. So far, however, a similarly fast reaction has not been found for allyl "anions" [13].

The second problem is concerned with the question of which organometallic compounds behave like cyclopropyl "anions" and undergo such a ring-opening reaction. Interestingly, *no* ring-opening reaction has been observed, e.g., with cyclopropyllithium *7* [14] and 1-cyano-2,2-diphenylcyclopropyllithium *8* [15]. On the other hand, 1,2,3-triphenylcyclopropane reacts with n-butyllithium/tetramethylethylene diamine (TMEDA) to give a mixture of isomers of 1,2,3-triphenylallyllithium which indicates that the intermediately formed 1,2,3-triphenylcyclopropyllithium *9* is able to undergo a ring-opening reaction [16].

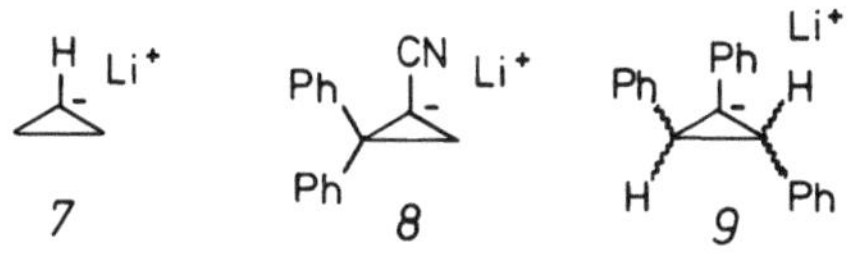

If one compares *7*, *8* and *9* (and many more cyclopropyl "anions") [10b] it turns out that electron acceptor substituents at *both* carbon atoms, which become terminal centers of the allyl "anion", facilitate the ring scission. In addition it is necessary that the negative charge of the cyclopropyl "anion" is stabilized by electron withdrawing (!) groups; hydrogen or alkyl groups at C-1 render the ring-opening reaction more difficult or impossible. This, of course, is quite the opposite to what one would expect if a real cyclopropyl *anion* were the ring-opening species. Thus, *ion pair effects* play an essential role in the rearrangement of cyclopropyl "anions".

In spite of these difficulties a kinetical criterium has been elaborated for the thermal conrotation of a cyclopropyl "anion". The result of this study has recently been confirmed by a special cyclopropyl-allyl "anion" rearrangement allowing trapping reactions of the allyl "anion".

In their kinetic studies, Boche and coworkers [13,17] first examined the rates of the ring-opening reactions of the isomeric 1-cyano-2,3-diphenylallyllithium compounds *10* and *14*, and of the isomerization reactions of the 1,3-diphenyl-2-cyanoallyllithium species (*11* ⇌ *12* ⇌ *13*).

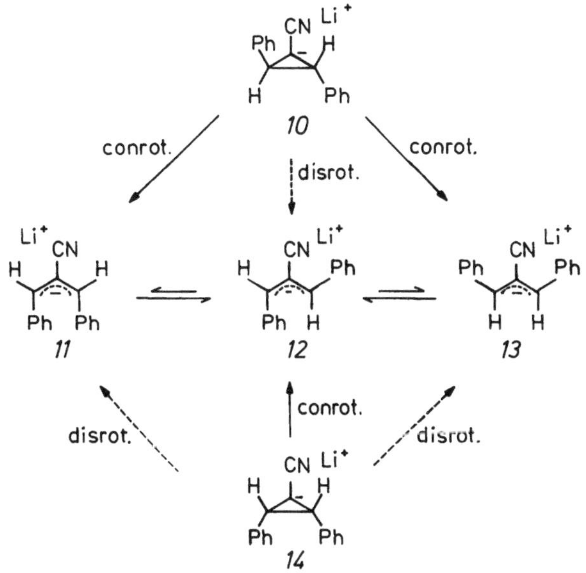

It was shown that the ring-opening reaction e.g. of *10* to give *13* (and/or *11* which is not clear from the available data) is about 1500 times *slower* than the isomerization of the allyl "anion" *13* (and/or *11*) to give the more stable *12*. A similar situation occurs when one starts from the cyclopropyl "anion" *14* [18]. These experimental data thus nicely confirm the results of the theoretical calculations discussed in connection with Table 1.

The rates of the ring-opening reactions of the cyclopropyl "anions" *10* and *14* have also been compared with the rate of the "forbidden" disrotatory ring-opening reaction of the structurally related cyclopropyl "anion" *15* (to give *16*) which was introduced by Wittig and coworkers [19, 13].

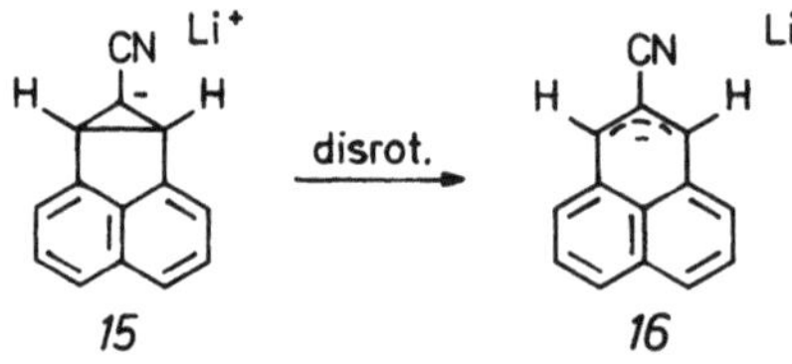

At 20 °C the following ratios of rate constants are measured:

$$\frac{k_{10}}{k_{15}} = 5500\ ; \qquad \frac{k_{14}}{k_{15}} = 740$$

These data demonstrate that the "forbidden" disrotatory ring opening of *15* is much slower than the ring-opening reactions of the cyclopropyl "anions" *10* and *14* which are not prevented from occurring in the predicted conrotatory mode. This kinetical criterium for the thermal conrotatory cyclopropylallyl "anion" transformation has been published independently by Ford and coworkers [20].

Similarly, a thermal conrotation has been observed in the case of the tricyclic cyclopropyllithium species *17* [21].

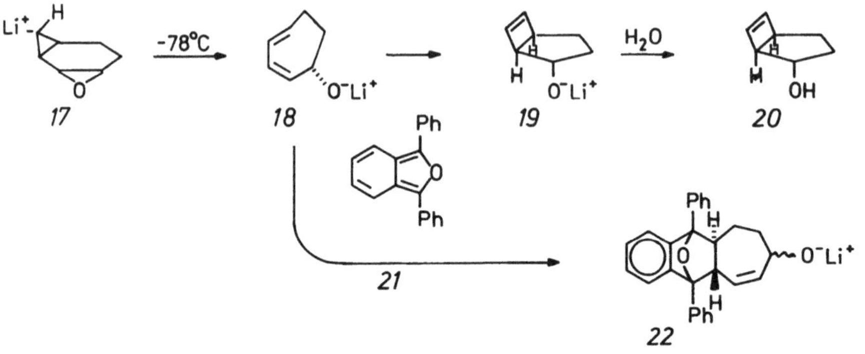

The conrotatory mode of ring opening of the β-lithiocyclopropyloxirane *17* is suggested first by the isolation of the *cis*-fused cyclobutene *20*; conclusive evidence for the intermediate *18* (the precursor of *19* and *20*) is provided by trapping the diene *18* in a Diels-Alder reaction with *21* to give the *trans*-fused adduct *22*.

The ring opening of *17* is in strong contrast to the normally unreactive cyclopropyllithium compounds with hydrogen at C-1 and alkyl substituents at C-2 and C-3 as mentioned earlier. Therefore, in this rather special cyclopropyl "anion", electrocyclic transformation combined with a Grob-type fragmentation the ring-opening reactions of *both* three-membered rings must be *concerned.* Furthermore the reaction should profit from the formation of a lithium-oxygen bond in *18*.

It is thus unambigously proven that the thermal cyclopropylallyl "anion" rearrangement follows the predicted conrotatory mode.

Photochemical cyclopropyl-allyl "anion" transformations have been observed by Newcomb and Ford [20c)] and by Fox [22)]; for example, the photochemical disrotation of *10* to give *12*.

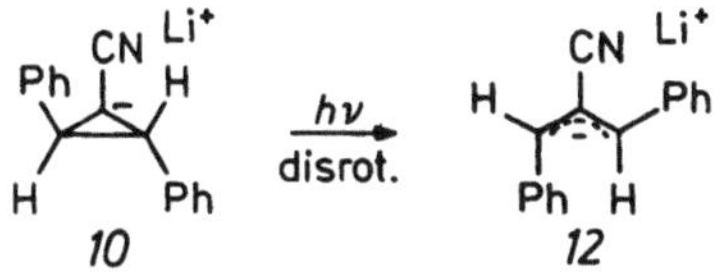

10 *12*

It is not clear, however, whether a photochemical "anion" or a thermal cyclopropyl radical ring-opening — the latter caused by photochemical electron ejection — took place.

2.2 Allyl-"Anion" Isomerizations

Thompson and Ford determined the rotational barriers of allyllithium *23a*, allylpotassium *23c*, allylcesium *23d*, (Z)-1-methylallylpotassium *24*, (Z)-1-isopropylallylpotassium *25* and (E)-1-isopropylallylpotassium *26* [23)].

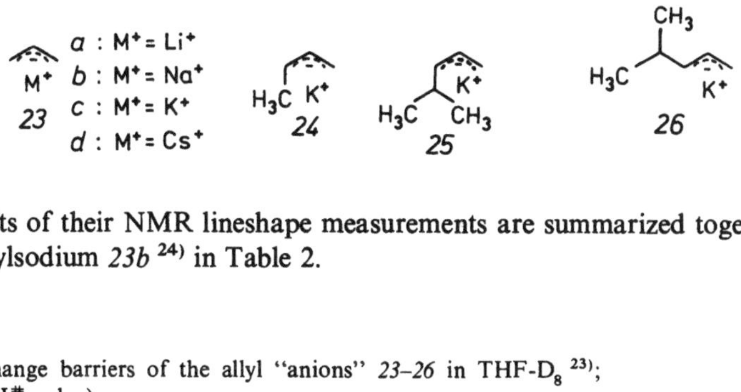

23 *24* *25* *26*

The results of their NMR lineshape measurements are summarized together with those of allylsodium *23b* [24)] in Table 2.

Table 2. Exchange barriers of the allyl "anions" *23–26* in THF-D_8 [23)]; a: Ref. 24 ($\Delta H^{\#}$ value)

	$\Delta G_T^{\#}$ [kcal/mol]	T [°C]
allyllithium *23a*	10.7 ± 0.2	−51
allylsodium *23b*	11.5	—[a]
allylpotassium *23c*	16.7 ± 0.2	68
allylcesium *23d*	18.0 ± 0.3	68
(Z)-1-methylallylpotassium *24*		
C_1–C_2	18–22	
C_2–C_3	17.0 ± 0.3	68
(Z)-1-isopropylallylpotassium *25*		
C_1–C_2	> 19.3	68
C_2–C_3	17.0 ± 0.3	47
(E)-1-isopropylallylpotassium *26*		
C_2–C_3	< 14.0	28

The result with allyllithium *23a* differs only insignificantly from that of an earlier report (10.5 kcal/mol) [25]. Complexation of *23a* with TMEDA does not influence the rate of exchange. Hexamethylphosphoric triamide (HMPT), 15-crown-5 ether and [2.1.1]cryptand in tetrahydrofuran (THF) led to rapid decomposition of *23a*. Addition of n-butyllithium had essentially no effect on the barrier. Since the ^{13}C NMR chemical shifts of *23a* are independent of the solvent, it is assumed that *23a* exists as a contact ion pair or higher aggregate in the NMR experiments. (The other alkali metals should also form contact ion pairs with the allyl anion because of their well-known tendency to form contact ion pairs even more readily than the lithium cation [26]).

Concerning the aggregation, it has been reported that allyllithium *23a* in *THF* is predominantly monomeric possibly with some dimers [27]. Since *23a*, on the other hand, is highly aggregated in *diethyl ether* [25,28], but the barriers are the same in both solvents [25], it is believed [23] that the state of aggregation of allyllithium *23a* does not affect its rotational barrier.

The earlier determinations of the aggregation state and the discussions of its influence on the rotational barrier have recently been questioned [29]. Cryoscopic measurements show conclusively that allyl*lithium* is a *dimer* in THF at −108 °C. Furthermore, the *ab initio* rotational barrier computed for isolated allyl*sodium* *23b* (11.5 kcal/mol) [30] is the same as the experimentally observed value (11.5 kcal/mol) [24] though the latter value must refer to a *solvated* species. It is hence concluded that the discrepancy between the *ab initio* calculated barrier for isolated allyl*lithium* (17.7 kcal/mol) and the experimental value in THF solution (10.7 kcal/mol, see Table 2) is due to *dimerization*. MNDO calculations indeed indicate an asymmetric dimer structure of allyl*lithium*. The asymmetric structure is consistent both with earlier [32] and more recent [29] NMR results.

The rotational barriers of allylpotassium *23c* (16.7 kcal/mol) and allylcesium *23d* (18.0 kcal/mol) are much higher than those of *23a* and *23b*. This observation and the question of whether aggregates are involved in the rotational process, clearly emphasize once more the important influence of the gegenion in "carbanion" reactions. The experimental results with *23a–d* thus lead only to a lower limit (18.0 kcal/mol) for the rotational barrier of the allyl *anion* in solution. Incidently, this value comes close to the lowest calculated (MP2/4-31 + G//4-31 + G) value of the allyl *anion* rotational barrier (22.2 kcal/mol) [33].

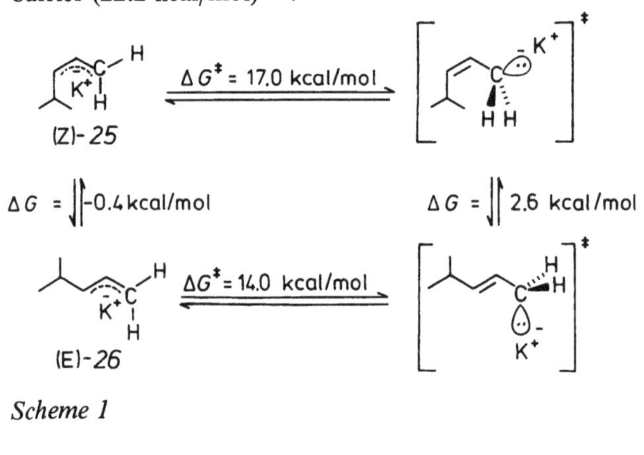

Scheme 1

In the case of the *alkyl*-substituted allylpotassium compounds *24*, *25* and *26* the anticipated effects of a C-1 alkyl group are to increase electron density at C-3 of the allyl "anion", to increase the C-1–C-2 bond order, and to decrease the C-2–C-3 bond order. Exactly this is found in the NMR experiments (Table 2). Interestingly, the rotational barrier of the CH_2-group in (Z)-*25* is higher than in (E)-*26*. This is not due to lowering of the ground-state energy of (Z)-*25* relative to (E)-*26* because the isomer ratio (Z)-*25*:(E)-*26* = 65:35 accounts for an energy difference of only 0.4 kcal/mol. The difference of 3.0 kcal/mol in the free enthalpies of activation consequently means that the (Z)-*25* transition state is 2.6 kcal/mol *less* stable than the (E)-*26* transition state (Scheme 1).

It has been proposed [23] that the CH_2 group of the isomer (Z)-*25* rotates in such a way that the cation migrates through the *anti*-face of the allyl plane, away from the isopropyl group (Scheme 1). On the other hand, the (E)-1-isopropyl substituent in (E)-*26* permits the *syn*-location of the cation in the transition state (Scheme 1). Calculations show a significant preference for *syn*-transition states in such rearrangements [30].

Rotational barriers and conformational equilibria of 1,3-diphenylallyl lithium and a series of 2-substituted 1,3-diphenylallyl lithium compounds have been determined by Boche and coworkers [18]. Table 3 summarizes the amounts of *endo,endo*-, *endo,exo*- and *exo,exo*-conformers *27* and the free energies of activation of their mutual transformations.

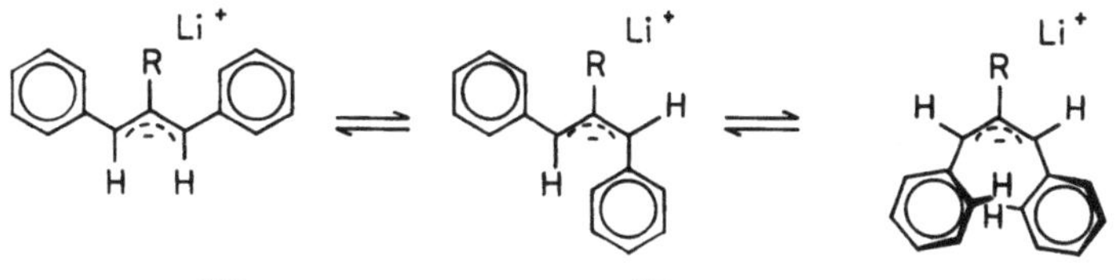

exo,exo-27 *endo,exo-27* *endo,endo-27*

The different equilibria essentially result from the steric interactions of the phenyl group(s) and the substituent R in the *endo,exo*- and *exo,exo*-conformations, and the two phenyl groups in the *endo,endo*-conformation. The *endo,endo*-conformer is favored as the ν-value [35] of R increases. Thus, the decrease of the $\Delta G^{\neq}$ values on going

Table 3. *Exo,exo*-, *endo,exo*- and *endo,endo*-conformers (%) and free energies of activation ($\Delta G^{\neq}_{273\,°C}$ [kcal/mol]) for rearrangements in THF starting from the *endo,exo*-isomers of the 2-substituted 1,3-diphenylallyl lithium compounds *27a–g* [18]

27	R	*exo, exo-27*	← $\Delta G^{\neq}$	*endo, exo-27*	$\Delta G^{\neq}$ →	*endo, endo-27*
a	H	92	17.8	8		–
b	CH_3	8	14.2	92		–
c	CN	4.5	16.4	91	16.4	4.5
d	C_2H_5	15	13.8	68	13.8	17
e	C_6H_5	–		56	14.3	44
f	$CH(CH_3)_2$	–		38	12.5	62
g	$C(CH_3)_3$	–		–		100

from *27a* to *27f* should clearly result from an increasing destabilization of the ground states. Transition-state effects, if at all, should play a minor role.

Ion-pair effects, as in the case of the unsubstituted allyl alkali metal compounds *23a–d*, do not markedly influence the rotational barriers of the 1,3-diphenylallyllithium species *27a–f*, although the reason is different: *27a–f* are solvent-separated ion pairs [36]. Addition of HMPT to the THF solution of *27a* raises the $\Delta G^{\#}_{38\,°C}$ value by 0.9 kcal/mol which corresponds to a rate retardation of 5–6 times. In the case of the methyl-substituted allyl "anion" *27b* HMPT slows down the rate by a factor of only 2–3. With the 2-cyano "anion" *27c* the $\Delta G^{\#}$ values of the Li^+ compound in THF and the Li^+, Na^+ and K^+ species in dimethyl sulfoxide (DMSO) are the same [18]. In the case of *27e* and *27f* the rate of the rearrangement is not affected if HMPT or TMEDA are added to the THF solutions. The observations of some solvent dependence in the case of the sterically less hindered *27a* and *b*, but of no effect with the more crowded *27e* and *f*, are in line with the general observation that solvent-separated ion pairs are favored with respect to contact ions pairs by increasing steric hindrance [25]. Hence, these experimental results could be interpreted to mean that in the case of *27a* and *b* contact ion pairs participate in the allyl "anion" rearrangement reaction.

2.3 Electron Transfer Induced Rearrangements of Cyclopropanes and Consecutive Reactions [37]

Electron transfer to cyclopropane *28* should lead to the cyclopropane radical "anion" *29*, which, in principle, can rearrange to give the ring-opened trimethylene radical "anion" *30*. Further reduction of the trimethylene radical "anion" should give a 1,3-"dianion" (*31*). A less likely pathway to give *31* is conceivable via the cyclopropane "dianion" *32* (Scheme 2).

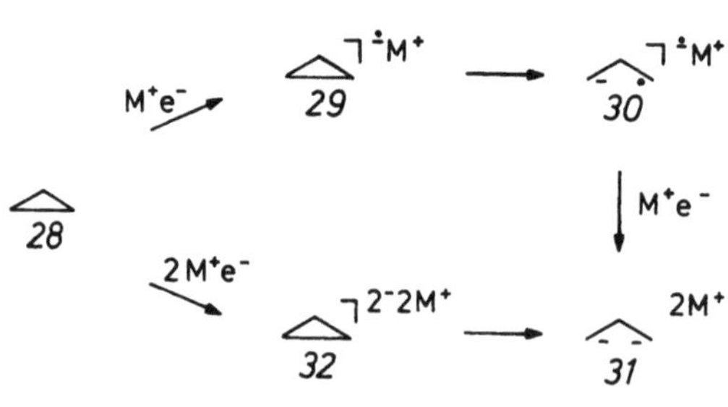

Scheme 2

The preparation of *29* was reported in 1963 [38]. Three years later, however, these results were shown to be false [39,40]. Cyclopropanes with *electrophoric* substituents (e.g. π-electron systems like carbonyl or aromatic groups), on the other hand, easily accept electrons. In many cases this leads to skeletal rearrangements and further reactions.

2.3.1 Cyclopropanes Substituted with Carbonyl Groups

The first report on what turned out to be a reduction of a carbonyl substituted cyclopropane was published in 1949 [41]. Reaction of methyl cyclopropyl ketone *33* with sodium in liquid ammonia in the presence of ammonium sulfate yielded instead of the expected methyl cyclopropylcarbinol *34* a mixture of 2-pentanone *35* and 2-pentanol *36*.

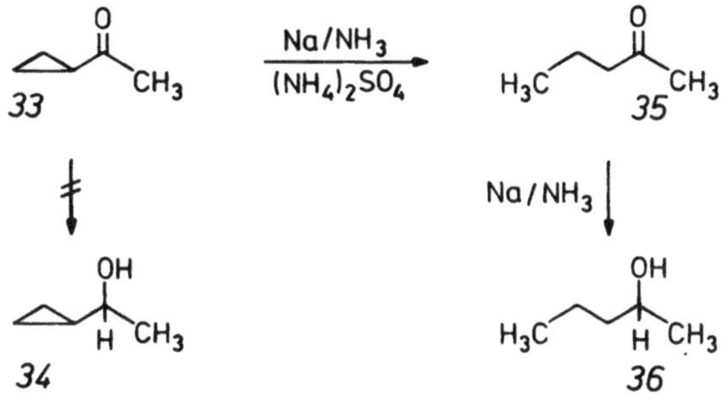

From an investigation of several conjugated cyclopropyl ketones, as e.g. *37*, Norin discovered [42] that the steric course of the reaction depends on the configuration of the cyclopropyl ketone in such a way that the cyclopropyl bond which is cleaved is the one possessing maximum overlap of the Walsh orbitals with the π-orbitals of the carbonyl group. *37* thus leads to *38*.

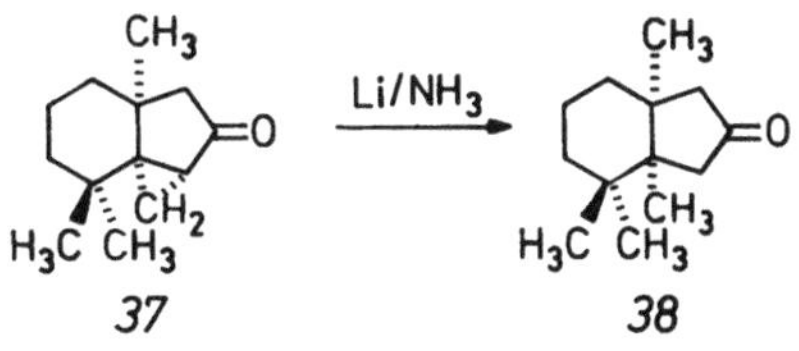

By means of the cyclopropyl ketones *39*, *cis-40* and *trans-41*, in which the two bonds of the cyclopropane ring, C-1–C-2 and C-1–C-3, are free to overlap with the carbonyl π-system, Dauben evaluated the importance of stereoelectronic versus steric factors [43].

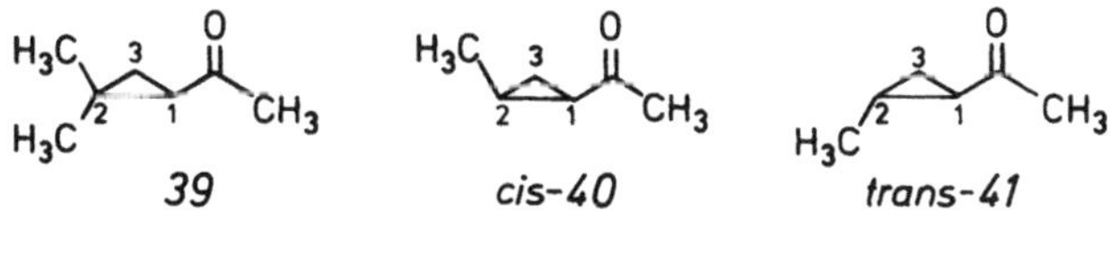

The main products in the reaction mixtures of the 2,2-dimethylcyclopropyl ketone *39* and the *cis*-2-methylcyclopropyl ketone *cis-40*, respectively, result from C-1–C-2 bond breaking. In contrast, the *trans*-2-methylcyclopropyl ketone *trans-41* breaks the C-1–C-3 bond. This strongly suggests that steric factors control the direction of cleavage, presumably through asymmetric overlap of the carbonyl π-system with one of the cyclopropane bonds [43]. In the absence of these steric effects, as in *trans-41*, the bond that cleaves is the one to give the more thermodynamically stable (= less substituted) "carbanion" intermediate.

2.3.2 Cyclopropanes Substituted with Phenyl(Aryl) Groups

The reduction of 1-methyl-2,2-diphenylcyclopropane *42* and one of its enantiomers, (+)-(R)-*42*, with Na/NH_3 to give 1,1-diphenylbutane *43* and 1,1-diphenyl-2-methylpropane *44* in a ~5.5:1 ratio over a wide concentration range has been studied by Walborsky and Pierce [46].

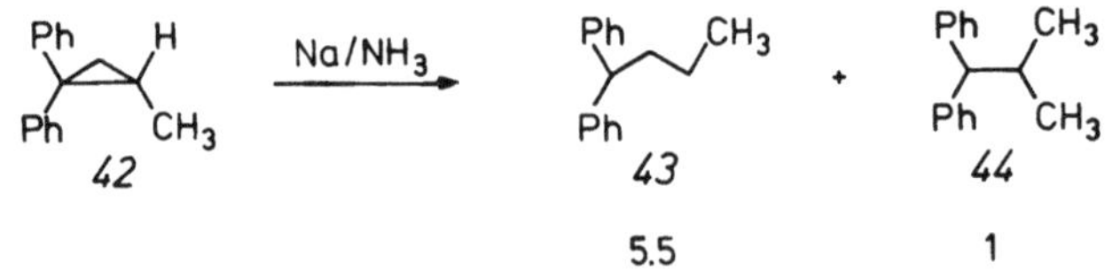

From the well-known ability of the phenyl group to accept electrons from sodium in NH_3 [47] the following mechanism (Scheme 3) [46] was proposed for the opening of the cyclopropane ring in *42* (and other phenyl(aryl)-substituted cyclopropanes).

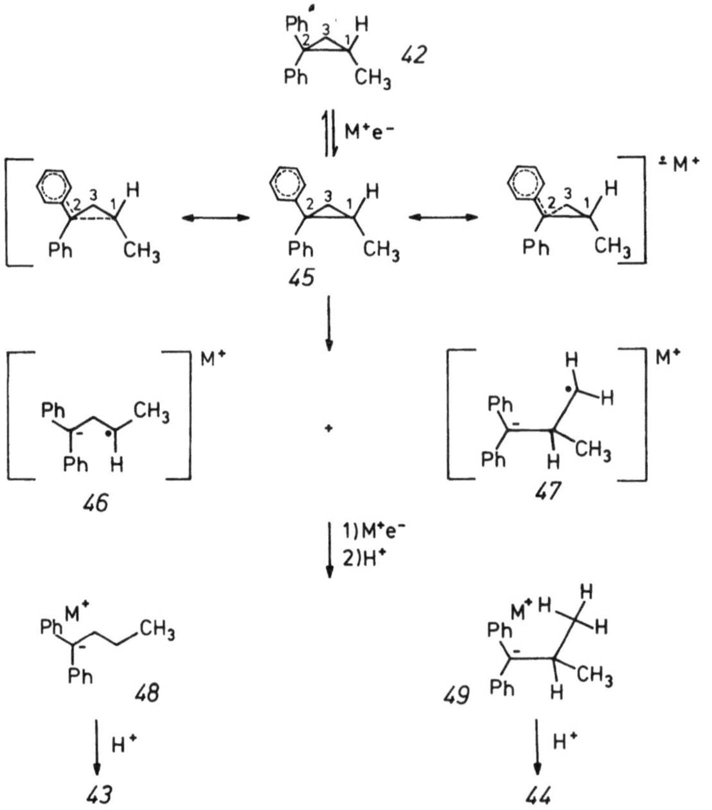

Scheme 3

The role of the phenyl groups (and similarly of the carbonyl groups in *33*, *37*, *39*, *cis-40* and *trans-41*) is to accept an electron to give the short-lived radical "anion" *45*. ESR experiments have so far failed to demonstrate the existence of *45* [48], and of the products of rearrangement of *45*, the trimethylene radical "anions" *46* and *47*. This means that *45* must *readily* open to *46* and *47* which themselves must *readily* add another electron to form a "dianion" which is protonated by the solvent to give the "anions" *48* and *49*, respectively. Further protonation gives *43* and *44*, respectively. Moreover, under the reaction conditions (Na in NH_3) the ring-opening of *45* to *46* and *47* is *irreversible* since optically active (+)-(R)-*42* is recovered without loss of optical activity [49].

The predominant formation of *43* from *42* via 1,2-bond cleavage of *45* to give *46* is expected on the basis that the trimethylene radical "anion" *46* would be predicted to be more stable than the trimethylene radical "anion" *47*. This argument is based on the reasonable assumption that in *46* there is more negative charge on the carbon atom bearing the two phenyl groups; otherwise the isomer *47* should be more stable.

It seemed unlikely that, instead of the radical "anions" *46* and *47*, the corresponding 1,3-"dianions" had been formed *directly* from their common cyclopropane "dianion" precursor. This, however, has not been rigorously excluded by means of these experiments.

Electronic *and* steric factors are important in the electron transfer induced ring-opening reactions of the methyl-substituted phenylcyclopropanes *cis-50* and *trans-51* (as is the case with *cis-40* and *trans-41*).

cis-50 *trans-51*

Staley and Rocchio [50] observed in Li/NH_3 reductions of *cis-50* that the C-1–C-2 bond is cleaved 70 times faster than the C-2–C-3 bond, while in *trans-51* the C-2–C-3 bond is cleaved with a 320-fold preference.

The high regioselectivity in the case of *trans-51* (in which there is no steric bias for either pathway) shows that in this very example a methyl group exerts a destabilizing effect relative to hydrogen for the cleavage. This is consistent with a description of the activated complex in which a substantial negative charge is also present on the cyclopropyl β-carbon of the bond undergoing cleavage. This explanation agrees nicely with a simple MNDO calculation for the phenyltrimethylene radical anion *52*: not only C-1 which bears the phenyl group, but also C-3 has an appreciable amount of negative charge [51]. It should be emphasized that *52* — like *trans-51* but opposite to *42* — has *one* phenyl group.

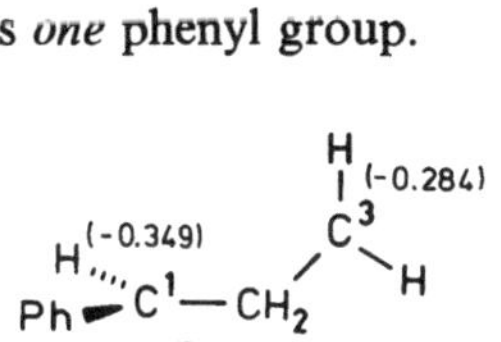

52

In the case of *cis-50* the conformation of maximum overlap for cleavage of the C-2 to C-3 bond possesses a substantial steric interaction between the methyl group and the ortho hydrogen atom of the phenyl ring. Therefore, the cleavage of C-1–C-2 is favored, Staley, too, points out that these results are strongly suggestive of radical "anion" instead of "dianion" intermediates. A distinction between these two mechanisms (or a combination thereof) on the basis of these data is, however, not possible [50].

2.3.3 Other Examples

The literature offers many more examples of electron transfer induced rearrangements of compounds containing three-membered rings. In general it is not known whether the rearrangement occurs at the radical "anion" or "dianion" stage, as is the case in the above-mentioned examples.

Reaction of dibenzonorcaradiene *53* with lithium, sodium or potassium naphthalenide followed by quenching with water led to 9-methyl-phenanthrene *54* (16–24%), 9-methyl-9,10-dihydrophenanthrene *55* (33–43%) and 6,7-dihydro-5-H-dibenzo-[a,c]cycloheptane *56* (23–34%) [52].

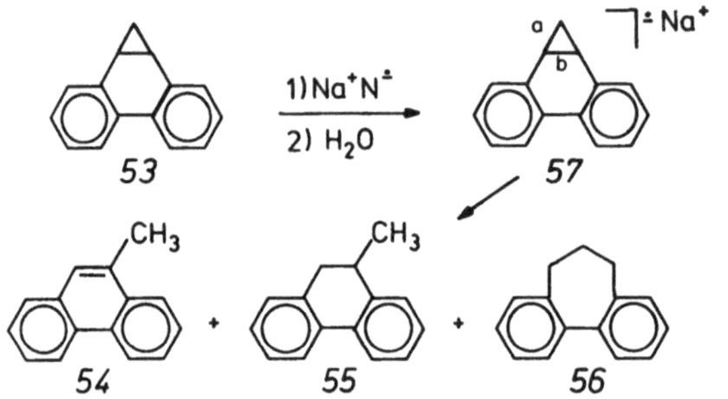

Preferential cleavage of bond a in the presumed radical "anion" *57* is caused by the familiar stereoelectronic effect (see Sect. 2.3). The formation of *56* via cleavage of bond *b*, however, is not established.

Electron transfer on *58* gives the rearranged *59*; *60*, in contrast, gives the 'closed' radical anion *61* [53].

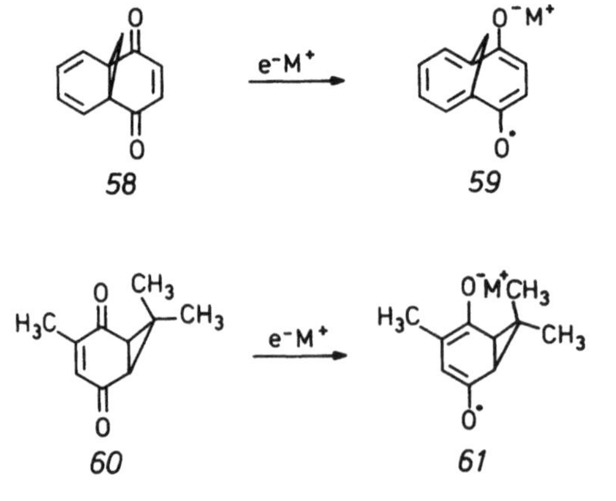

Reduction of *cis*-bicyclo[6.1.0]nona-2,4,6-triene *cis-62* proceeds through the nine electron homoaromatic radical "anion" *63* to give the monohomocyclooctatetraene "dianion" *64* [54)]; in contrast to this, the bicyclic radical "anion" *66* is produced exclusively upon reaction of *trans-65* with a potassium mirror in THF or dimethoxyethane (DME) solution at —90 °C [55)].

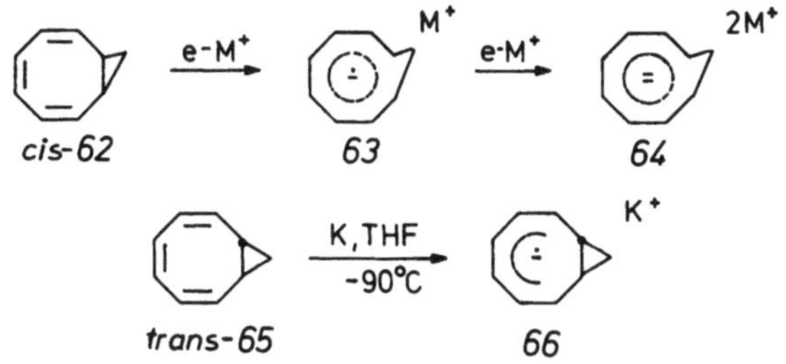

These observations agree fully with orbital-symmetry considerations if the highest occupied MOs of *cis-62* and *trans-65* are the levels which control reactivity thus requiring disrotatory bond cleavage of the central cyclopropane bond [54d)]. Disrotatory ring-opening of *cis-62* to give *cis,cis,cis,cis*-cyclononatetraene radical "anion" *63* is accordingly favorable while the analogous reaction channel starting from *trans-65* is prohibited by the high-strain energy and poor p-π overlap in the *trans,cis,cis,cis*-cyclononatetraene radical "anion". However, the ring-opening is not clear if the symmetries of all occupied orbitals are considered. Again there is no evidence as to whether the rearrangement occurs at the "dianion" stage.

Reaction of bullvalene *67* with sodium-potassium alloy in THF or DME led to the bicyclo[3.3.2]decatrienyl "dianion" *68*. The facile formation of *68* in contrast to the unsuccessful reduction of dihydrobullvalene *69* which does not give *70*, has been discussed along the lines of longicyclic stabilization of *68* as opposed to the bishomo-antiaromatic nature of the bicyclo[3.3.2]decadienyl "dianion" *70* [56a, b)].

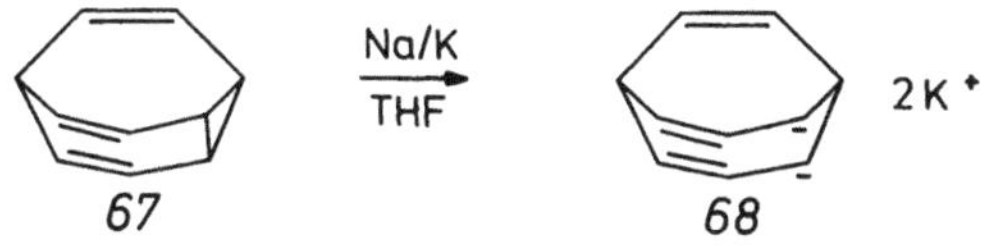

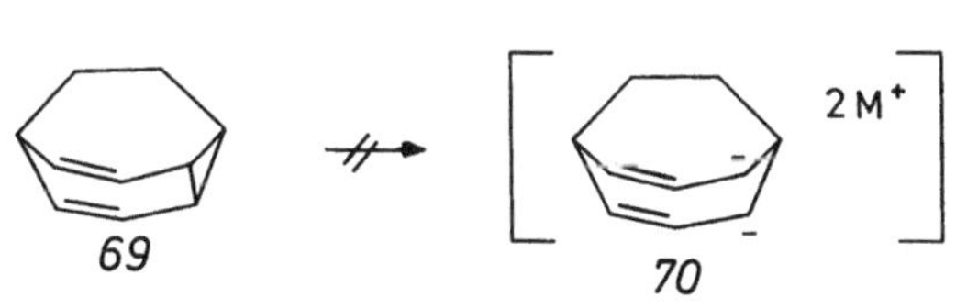

Reduction of semibullvalene *71* could lead to the destabilized [3.3.0]"dianion" *72*. A symmetry-allowed and thermodynamically attractive rearrangement to the cyclooctatetraenyl "dianion" *73*, however, would seem likely.

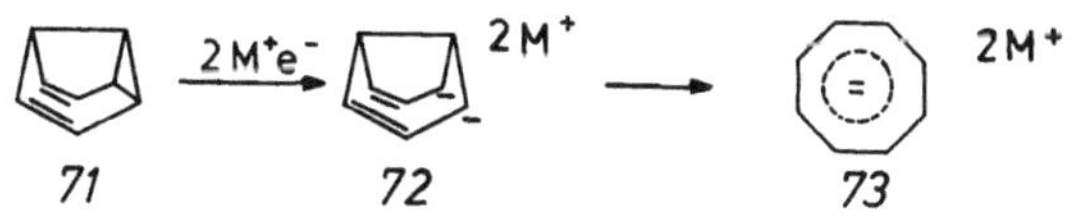

Reaction of *71* with lithium in THF or dimethyl ether at −78 °C led to "dilithium semibullvalenide" [56b−d, 57a] which according to Goldstein's and Wenzel's NMR investigations exists as the C_{2h} and D_2 diastereoisomers of bis(bicyclo[3.3.0]octa-3,7-diene-2,6-diyl)tetralithium C_{2h}-*74* and D_2-*74*, respectively [56d] (Scheme 4).

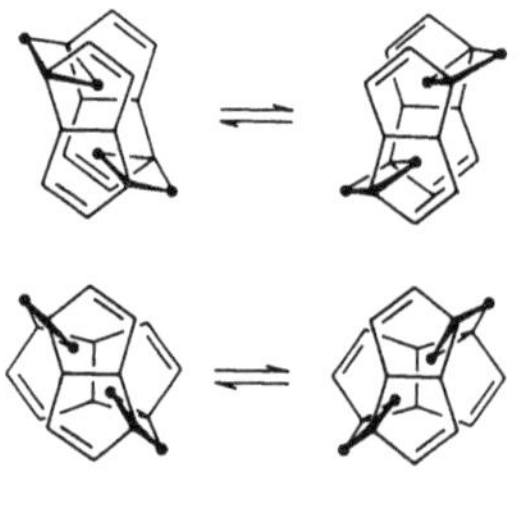

C_{2h}-74

D_2-74

• = lithium atoms

Scheme 4

Furthermore, the NMR investigations revealed two intramolecular rearrangements, the automerization of the achiral, meso-C_{2h}-*74* diasteroisomer and the racemization of the chiral, (±)-D_2-*74* diastereoisomer. The diastereoisomeric equilibration, an intermolecular process, occurs at higher temperatures than the other two processes. It will be interesting to find out whether the solution strctures C_{2h}-*74* and/or D_2-*74* will be confirmed by solid state investigations.

The intra- and intermolecular rearrangements of C_{2h}- and D_2-*74* clearly indicate the importance of rearrangements hitherto not normally covered in a chapter on "carbanion" rearrangements. Organometallic chemists, however, are learning more and more about the complex structure(s) of such compounds in solution and in the solid state, as well as their rearrangements, e.g. within aggregates. It is thus predictable that the near future will provide us with more examples of this sort.

As expected, "dilithium semibullvalenide" *74* rearranges at 0 °C with an apparent first-order rate constant $k = 9.0(1) \cdot 10^{-5}\ s^{-1}$ to the cyclooctatetraenyl "dianion" *73*.

The dipotassium compound of *72* (or its dimer *74*) could not be prepared by deprotonation of a mixture of tetrahydropentalenes (e.g. *75*) with a 1:1 n-butyllithium/potassium t-pentoxide mixture [58]. Instead, only the dipotassium salt of the cyclooctatetraenyl "dianion" *73* was detected.

75 —(n-BuLi / t-PentOK)→ *73* ←(K or Na/K)— *71*

This result was confirmed when semibullvalene *71* was reacted with potassium or Na/K alloy. Even at −78 °C this reaction resulted only in the rearranged *73* [56c]. A common intermediate should be the dipotassium compound *72* (and/or its dimers *74*). The different behavior of the lithium and potassium species *72* (and/or *74*) illustrates again the problem of "carbanion" rearrangements: changing the gegenion from lithium to potassium leads to species with rather different properties.

Yet another species which rearranges via electron transfer is *76*. Reaction with lithium leads to the "dianion" *77* [57b)].

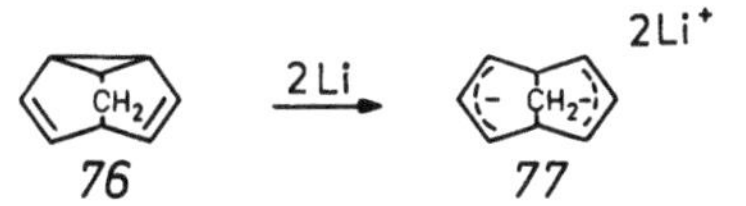

76 *77*

2.3.4 Trimethylene Radical "Anions" as Intermediates in Cyclopropane Isomerizations and Transformations into Isomeric Olefins

2.3.4.1 Electron Transfer Catalyzed Stereoisomerization

Although it seems appropriate to formulate rearrangements of cyclopropanes induced by electron transfer reactions via trimethylene radical "anions" (see Sect. 2.3.2) the existence of such species so far had not been proven unambigously. Especially noteworthy in this context is the observation of Walborsky [49)] that in the reaction of optically active (+)-(R)-*42* with sodium in ammonia the three-membered (+)-(R)-*42* is recovered *without loss of optical activity*. Thus, the formation of the trimethylene radical "anions" *46* and *47*, if it occurs, is irreversible (see page 12 and 13). Therefore, electron transfer reactions of various cyclopropanes with special attention to the question of the formation of radical "anion" intermediates have been studied by Boche and coworkers [59)].

A typical example is given by the reaction of *cis*-1,2-diphenylcyclopropane *cis-78a* with Na/K alloy at 0 °C to yield on protonation *trans*-1,2-diphenylcyclopropane *trans-79a*, *trans*-1,3-diphenylpropene *84a* (together with some *cis*-1,3-diphenylpropene), and 1,3-diphenylpropane *82a*. Of special importance is the time dependence of the yields of *cis-78a*, *trans-79a*, *84a* and *82a*. The results are given in Table 4.
Table 4 can be summarized as follows:

1. *cis-78a* disappears steadily.
2. *trans-79a* and the propane *82a* pass through a maximum. This demonstrates that the stereoisomer *trans-79a* is formed as an intermediate; likewise, the ring-opened

Table 4. Time-dependent relative yields [%] of *cis-78a*, *trans-79a*, *84a* and *82a* in the reaction of *cis-78a* with Na/K alloy in THF at 0 °C, followed by protonation with water; total yields 86%

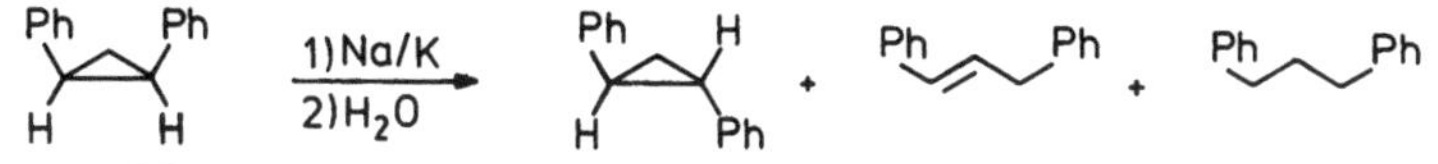

Time (min)	*cis-78a*	*trans-79a*	*84a*	*82a*
0	95	5	—	—
10	83	5	—	—
45	8	57	5	30
120	1	28	30	41
240	—	9	53	38

1,3-"dianion" *81a* is an intermediate since *81a* gives the propane *82a* on protonation, as confirmed by deuteration, see Scheme 5.
3. The concentration of the propene *84a* increases steadily.

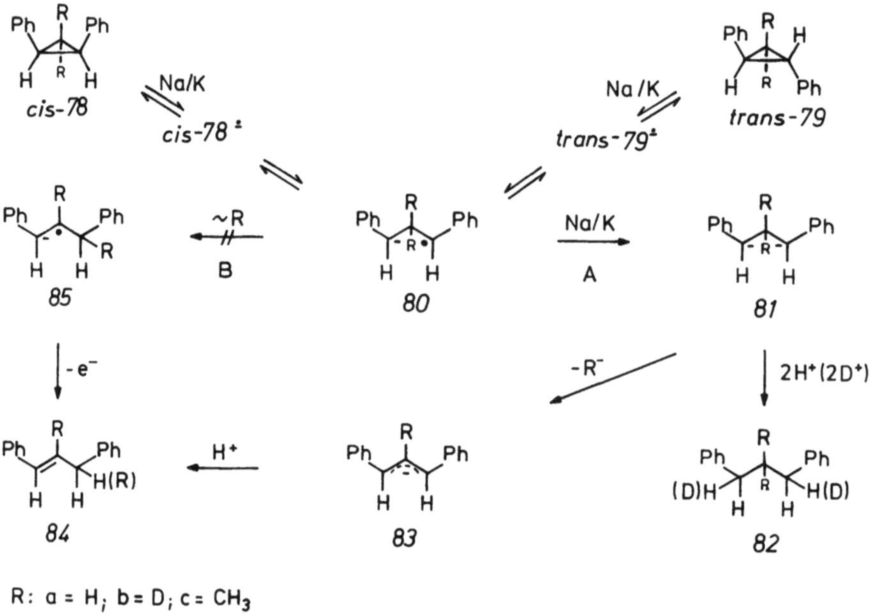

Scheme 5 (gegenions omitted)

The cyclopropane *stereoisomerization cis-78a* ⇌ *trans-79a* and the transformation of the cyclopropanes into the *ring-opened* products *82a* and *84a* are explained by the reversible formation of the trimethylene radical "anion" *80a* and pathway A in Scheme 5:

Stereoisomerization: *cis-78a* and *trans-79a* accept an electron to give the diphenylcyclopropane radical "anions" *cis-78a*$^{\dot{-}}$ and *trans-79a*$^{\dot{-}}$ which rearrange *reversibly* into the trimethylene radical "anion" *80a*. The stereoisomerization takes place at the trimethylene radical "anion" stage. The reversible formation of *80a* in *THF* is noteworthy because the trimethylene radical "anions" *46* and *47* have been formed irreversibly, however in this case in NH_3, see Scheme 3, page 12.

Ring-opened products: A second electron transfer transforms the trimethylene radical "anion" *80a* into the 1,3-"dianion" *81a*. In the absence of a proton (deuterium) source *81a* eliminates hydride as a function of time (Table 4) to give the allyl "anion" *83a*. Protonation of *83a* leads to *84a*. Protonation of *81a* gives *82a*, as mentioned earlier.

The following results support the *stereoisomerization* as an electron transfer catalyzed reaction.

1. The kinetics of the *thermal* cyclopropane isomerization *cis-78a* → *trans-79a* extrapolate to a half-life at 0 °C of ~10^9 *years*[60]. In the presence of Na/K *cis-78a* disappears almost completely at the same temperature after 120 min; clearly, this is not a thermal reaction.

2a. The *base-catalyzed* isomerization of a 72:28 mixture of *cis-78a* and *trans-79a* with t-BuOK to give somewhat more *trans-79a* is much slower (70 d at 25 °C)[61] than the Na/K catalyzed isomerization.

$$\frac{cis\text{-}78a}{trans\text{-}79a} = \frac{72}{28} \xrightarrow[\text{2. } H_2O]{\text{1. t-BuOK/DMSO or HMPT, 25 °C, 70 d}} \underset{27\%}{cis\text{-}78a} + \underset{50\%}{trans\text{-}79a} + \underset{23\%}{84a}$$

Thus, the Na/K catalyzed isomerization of *cis-78a* is not a base-catalysed reaction via the cyclopropyl "anions" *86* and *87*.

86 87

2b. The *base-catalyzed* t-BuOK, DMSO, *100* °C, *17* h isomerization of r-1-phenyl-c-2,c-3-dimethyl- and r-1-phenyl-t-2-,t-3-dimethylcyclopropanes (*cis,cis-88* and *trans,trans-89*, respectively)[62] and the *electron-transfer-catalyzed* (Na/K, THF, *20* °C, *16* min) reaction[59] are shown below.

Base catalysis

cis,cis-88 trans,trans-89

Electron transfer catalysis

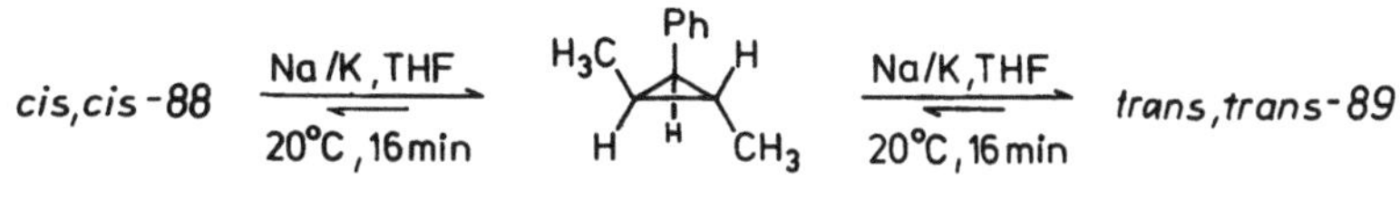

cis, trans-90

It is not only the conditions of temperature and time which are strongly different in these two reactions. In the *base-catalyzed* reaction only *cis,cis-88* and *trans, trans-89* are formed as expected if the formation of the cyclopropyl "anion" *91* is the rate and product determining step.

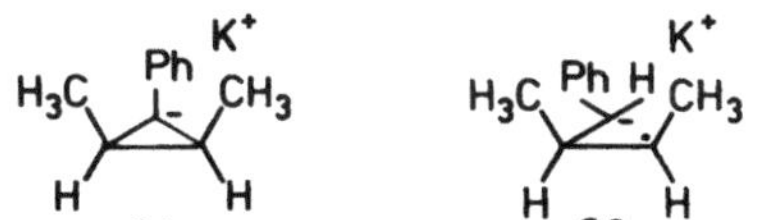

91 92

In the much faster *electron transfer catalyzed* rearrangement a new stereoisomer, r-1-phenyl-c-2,t-3-dimethylcyclopropane *cis,trans-90* shows up in the equilibrium with *cis,cis-88* and *trans,trans-89*. The formation of *cis,trans-90* requires that in the course of the equilibration a *cyclopropane bond* (instead of a C—H bond) is broken. As in the case of the 1,2-diphenylcyclopropanes *cis-78a* and *trans-79a* a trimethylene radical "anion", here *92*, is the most likely intermediate.

That the radical "anion" *92* is indeed the intermediate, is shown by a prolonged reaction of the equilibrium mixture of the three cyclopropanes *cis,cis-88*, *trans,trans-89* and *cis,trans-90* with Na/K alloy which leads, among other products, to a dimer of the radical "anion" *92* in 37% yield.

The observation that *cis,trans-90* is not formed in the base-catalyzed reaction provides important information. It excludes the possibility that under base-catalyzed conditions ET-catalyzed reactions generally play a significant role. This aspect is treated in more detail in the following Sections 2.3.4.2 and 2.3.5.

3. The cyclopropane stereoisomerization *cis-78a* ⇌ *trans-79a* via the *1,3-"dianion"* *81a* (a *two* electron pathway) is also excluded. A priori, this is not a totally unlikely pathway. There are at least two reports in the literature dealing with a C—C bond fission at the "dianion" stage [63,64]. The formation of the 1,3-"dianion" *81a* would exactly correspond to the observations reported in these publications (see Scheme 5).

   If 1,3-"dianions" of the typ *81* were intermediates in the ET-catalyzed reversible stereoisomerization of the cyclopropanes *78* and *79*, the *reverse reaction* — formation of the cyclopropanes from such "dianions" — should also take place. Preparation of the "dianion" *81a* from the cyclopropane *cis-78a* with Na/K or with Li at —78 °C shows clearly that *81a* is stable at this temperature — although *cis/trans* rearrangement of the cyclopropanes *cis-78* and *trans-79* takes place!

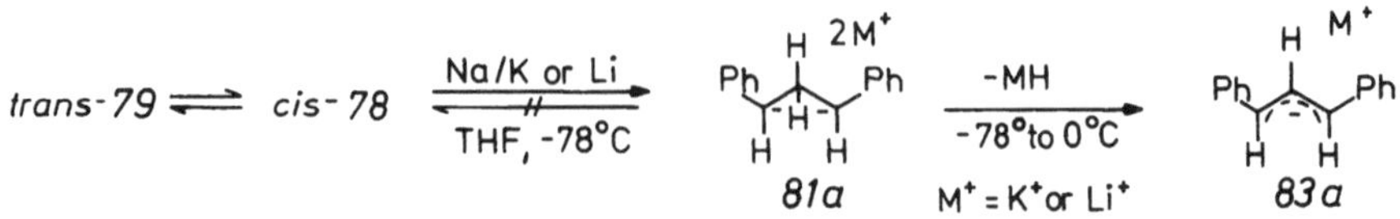

81a *83a*

Whether the "dianion" *81a* gives the cyclopropanes *cis-78* and *trans-79* at higher temperatures such as 0 °C cannot be decided because *81a* loses hydride between —78 and 0 °C to give the allyl "anion" *83a* [65].

If the methylene hydrogens in *81a* are replaced by two methyl groups as in *81c* (Scheme 5) then elimination cannot take place. *81c* is stable up to 0 °C; ring closure to give *cis-78c* or *trans-79c* does not occur. These cyclopropanes, however, rearrange in the presence of Na/K at that temperature.

In summary, there is no doubt that the *reversible stereoisomerizations* of the cyclopropanes *cis-78*, *trans-79*, *cis,cis-88*, *trans,trans-89*, and *cis,trans-90* in the presence of Na/K alloy occur via the trimethylene radical "anions" *80* (Scheme 5) and *92*. These *ET-catalyzed reactions* do not occur in a thermal or base-catalyzed or "dianion" fashion. They differ greatly from the *irreversible* ET reactions of the cyclopropanes *42*, *cis-50* and *trans-51* (pages 12 and 13) with Na in NH_3 [49,50]. Undoubtedly, in the latter reactions trimethylene radical "anions" are also intermediates which, however, are quickly protonated by the rather acidic NH_3. Spectroscopic evidence for trimethylene radical "anions" is not available to date.

2.3.4.2 Electron Transfer Initiated Formation of Isomeric Olefins

The mechanism of the formation of the ring-opened species 1,3-diphenylpropene *84a* and 1,3-diphenylpropane *82a* from the cyclopropanes *cis-78a* and *trans-79a* has been outlined in Scheme 5, route A, page 18. Common intermediate is the 1,3-"dianion" *81a* which in the case of the formation of *84a* first loses hydride and then is protonated. The 1,3-"dianion" *81a* is formed by electron transfer from the trimethylene radical "anion" *80a* which thus is the key intermediate 1) in the *ET-catalyzed-stereoisomerization* and 2) in the "structural isomerization" of the cyclopropanes *cis-78* and *trans-79*. *Importantly, the "structural isomerization" is not an ET-catalyzed reaction.*

The literature offers exactly this alternative for the formation of propenes from cyclopropanes in the presence of electron sources [66, 67]. Accordingly, in the example given in Scheme 5 the group R in the radical "anion" *80a* (R = H) should undergo a 1,2-rearrangement to provide a new radical "anion" *85a* which after expulsion of an electron (possibly to *cis-78* or *trans-79*) should give the propene *84a* (pathway B).

Although this latter pathway is unlikely on the basis of the time dependence of the formation of *trans-79a*, *82a* and *84a* from *cis-78a* and Na/K presented in Table 4, page 17, the migration of a group R has been checked additionally by means of the bis-deuterated cyclopropane *trans-79b* (R = D).

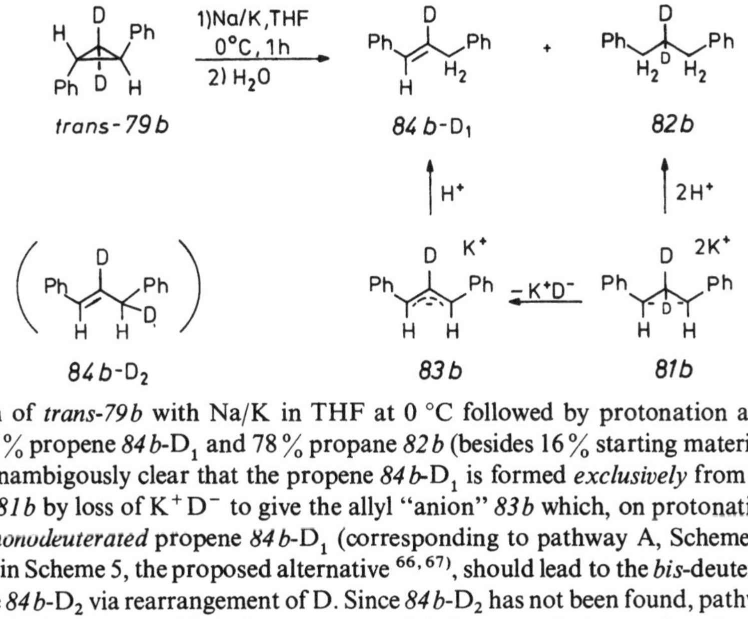

Reaction of *trans-79b* with Na/K in THF at 0 °C followed by protonation after 1 h led to 6% propene *84b*-D_1 and 78% propane *82b* (besides 16% starting material). It is thus unambigously clear that the propene *84b*-D_1 is formed *exclusively* from the "dianion" *81b* by loss of K^+D^- to give the allyl "anion" *83b* which, on protonation, gives the *monodeuterated* propene *84b*-D_1 (corresponding to pathway A, Scheme 5). Pathway B in Scheme 5, the proposed alternative [66, 67], should lead to the *bis*-deuterated propene *84b*-D_2 via rearrangement of D. Since *84b*-D_2 has not been found, pathway B is not a viable process. This is in agreement with the general observation that the *intra*-molecular 1,2-migration of hydrogen (deuterium) is not a facile reaction in either radicals [68] or "carbanions" [69].

2.3.5 Proton Transfer or Electron Transfer Initiated Rearrangements of Cyclopropanes in the Presence of a Base

Although it has been demonstrated that the base-catalyzed (t-BuOK, DMSO, 100 °C, 17 h) isomerization of the cyclopropane *cis,cis-88* to give exclusively the cyclopropane *trans,trans-89* (see page 19) does not involve an electron transfer reaction, the question of whether a reaction of a carbon acid with base is initiated by proton transfer or by electron transfer is of general interest. One has furthermore to consider the possibility that the "carbanions" primarily formed on deprotonation (or addition of the base) might be better electron donors for electron transfer reactions than the base itself [70, 71, 72].

The rearrangement of 2,3,4-triphenyl-*endo*-tricyclo[3.2.1.0^{2,4}]octane *93a* in the presence of base, e.g. t-BuOK in DMSO (70 °C, 20 h), to give 2,3,4-triphenylbicyclo-[3.2.1]oct-2-ene *96a* (Scheme 6) is an interesting example along these lines. Originally it was proposed that this rearrangement occurs via forbidden disrotatory ring opening of the cyclopropyl "anion" *94a* to give the allyl "anion" *95a* which is protonated to give *96a* (pathway A)[73].

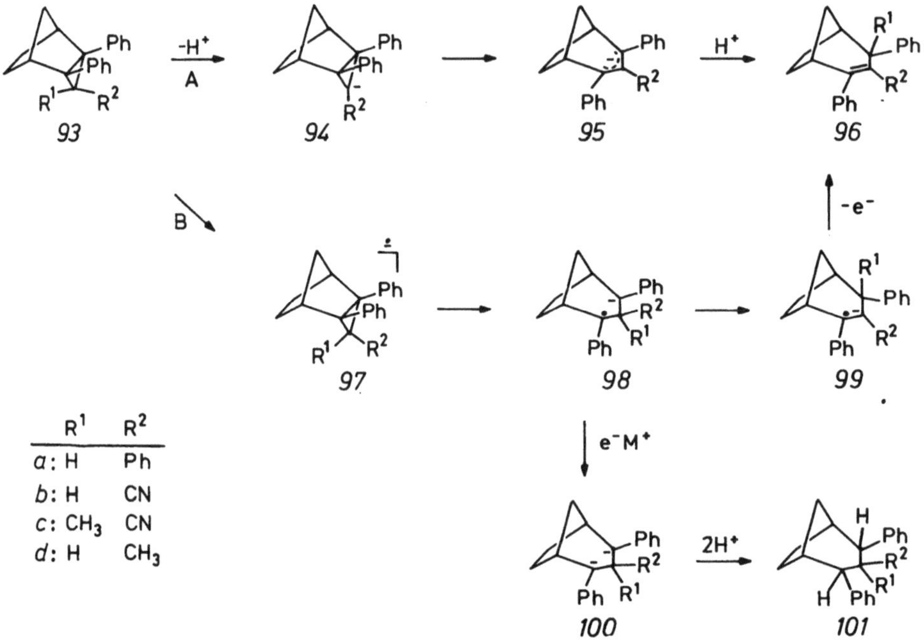

Scheme 6 (gegenions omitted)

Other authors concluded later [66] that the reaction of *93a* with t-BuOK in DMSO or HMPT (25 °C, 24 h), or with dimsylpotassium in DMSO (70 °C, 24 h) appears to proceed by a radical "anion" pathway. It was proposed that an initial electron transfer from base to *93a* affords the radical "anion" *97a* which opens to *98a*. The latter was envisaged to rearrange to *99a* which loses an electron, possibly to *93a*, to give *96a* (pathway B).

A reexamination of the reactions of *93a–d* with different bases and ET reagents led to the following results and conclusions [74]. The cyclopropyl "anion" *94b* transforms completely and in a disrotatory manner into the allyl "anion" *95b*, even at −75 °C within 1 h; *95b* gives *96b* on protonation. This strongly suggests a similar pathway in the reactions of *93a* with base. Indeed, *93a* is completely transformed into *96a* after

1 h at 0 °C by the "superbase" [75] potassium 3-aminopropylamide (KAPA) in 1,3-diaminopropane.

An entirely different result is observed with *93c* and *93d*. Reaction of *93c* with base (LDA, THF, −75 °C, 1 h; t-BuOK, DMSO, 25 °C, 5 h) led to complete recovery of starting material. *93d* is also unreactive towards base (t-BuOK, DMSO, 70 °C, 6 h; KAPA, 1,3-diaminopropane, 25 °C, 24 h). Thus, the impossibility of preparing a cyclopropyl "anion" of the type *94* from *93c* and the extremely low acidity of *93d* prevent a reaction with base *although electron transfer from base to initiate pathway B (Scheme 6) could still occur.*

With a real electron transfer reagent (Na/K alloy in THF), *93d* is transformed into *101d*, undoubtedly via *97d*, *98d* and *100d*. A similar sequence has been observed starting with *93a* and sodium naphthalenide [66]. Other reactions of this sort have been mentioned in Sect. 2.3.4.

One can therefore conclude that the base-catalyzed cyclopropane → propene rearrangement *93a(b)* → *96a(b)* is induced by an acid-base and not by an electron transfer reaction. The mechanism of the facile cyclopropyl → allyl "anion" rearrangement *94a(b)* → *95a(b)*, however, is not clear [74].

3 Rearrangements Involving Four-Membered Rings: Electron Transfer Induced Reductive Rearrangements of the Generalized Cyclobutene/Butadiene Radical "Anion" System

The *thermal* cyclobutene *102* ⇌ butadiene *103* transformation has been predicted by Woodward and Hoffmann to occur in a conrotatory fashion which is in agreement with experimental results [9].

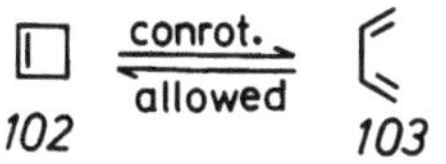

102 *103*

The same mode was also observed for the *thermal* ring-opening of the *cis*- and *trans*-3,4-diphenylbenzocyclobutenes *cis-104* and *trans-105* to give the (Z,E)- and (E,E)-5,6-diphenyl-o-xylylenes *106* and *107*, respectively [76].

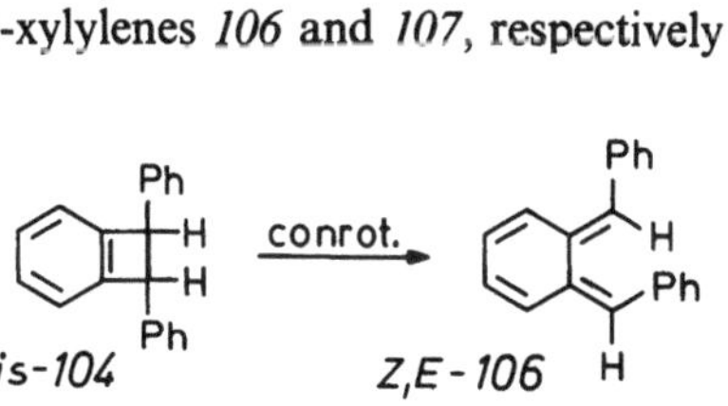

cis-104 *Z,E-106*

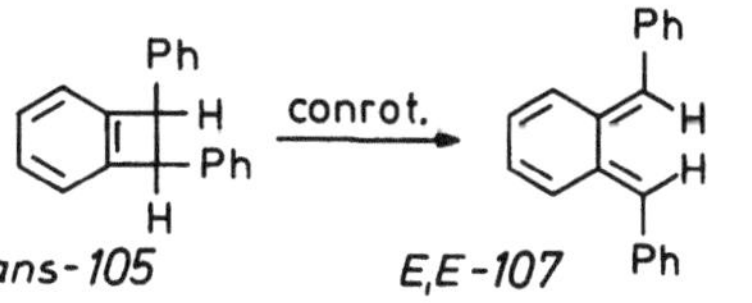

trans-105 *E,E-107*

What is the preferred mode of rearrangement after electron transfer to (derivatives of) *102* or to *104* and *105*? Does the ring-opening reaction occur on the radical "anion" or the "dianion" stage? These questions have been dealt with in a series of papers first by Bauld and coworkers [77]. They are of general interest because the theoretical treatment of electrocyclic reactions of open-shell systems, as e.g. the cyclopropyl ⇌ allyl *radical* rearrangement, led to ambiguous results [9, 78–81].

Bauld and coworkers reached the following conclusions [77g)]:

1a. The cyclobutene/butadiene radical "anion" rearrangement *108*⇌*109* is allowed neither in the disrotatory nor in the conrotatory mode.

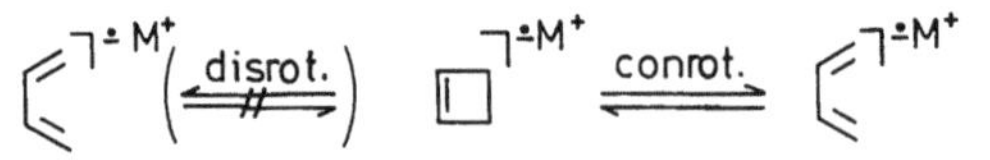

109 *108* *109*

1b. A variety of quantitative treatments, however, select the conrotatory mode as favored, the same mode as in the case of the neutral rearrangement *102*⇌*103*.

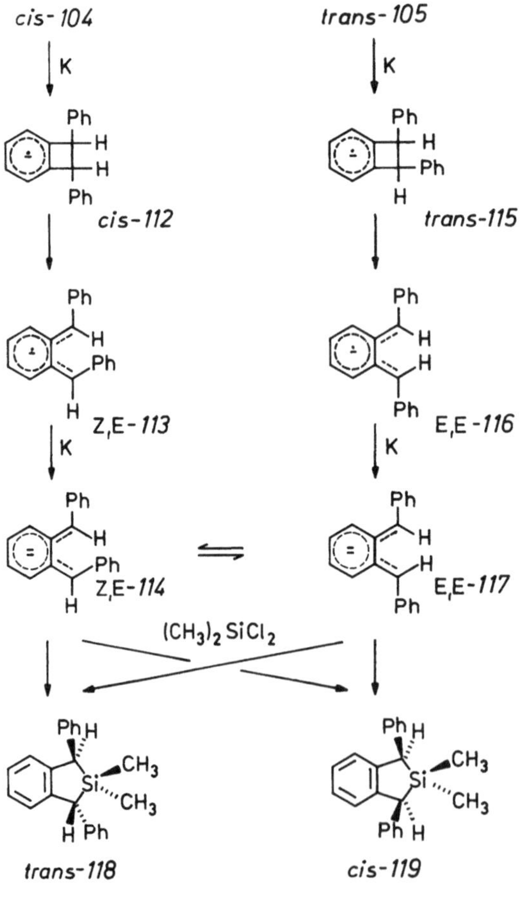

Scheme 7 (gegenions omitted)

2. The electrocyclic conversion of the benzocyclobutene into the o-quinodimethane radical "anion" *110⇌111* is *allowed* in the conrotatory mode [77c, g].

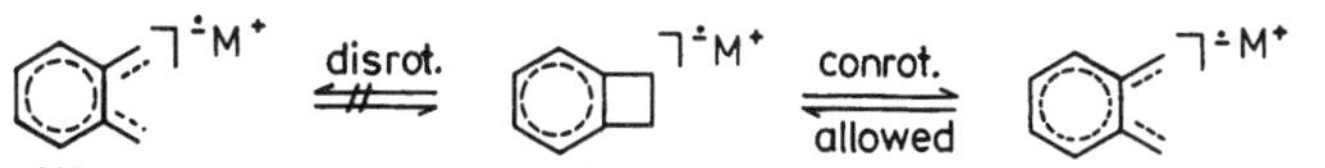

111 *110* *111*

The experimental tests have been performed by means of electron transfer reactions onto the diphenylbenzocyclobutenes *cis-104* and *trans-105* (Scheme 7) [77c, g].

Bauld and coworkers [77c, g] concluded on the basis of the following results that the ring-opening reactions of the diphenylbenzocyclobutene radical "anion" isomers *cis-112* and *trans-115* had occurred in a conrotatory fashion:

1. Reaction of *cis-104* with K in 2-methyltetrahydrofuran (2-MTHF) or THF should lead to *cis-112* which opens to Z,E-*113*. This radical "anion" is further reduced to give the Z,E-α,α'-diphenyl-o-xylylenediide Z,E-*114* which is trapped by dimethylsilyldichloride to give the adducts *trans-118*:*cis-119* in a 84:16 ratio.
2. Similarly, *trans-105* led to *cis-119*:*trans-118* in a 70:30 ratio.

One should mention that the two adducts *trans-118* and *cis-119* had not been separated nor isolated in crystalline form. The structural assignment is based on NMR evidence.

It is also of interest that the α,α'-diphenyl-o-xylylenediide Z,E-*114* is reported to be more stable than the isomer E,E-*117* both at low and at room temperature (although steric considerations clearly favor E,E-*117*). This conclusion was reached after deprotonation of dibenzylbenzene *120* with n-butyllithium in 2-MTHF followed by reaction with dimethylsilyldichloride giving *trans-118*:*cis-119* in a 83:17 ratio.

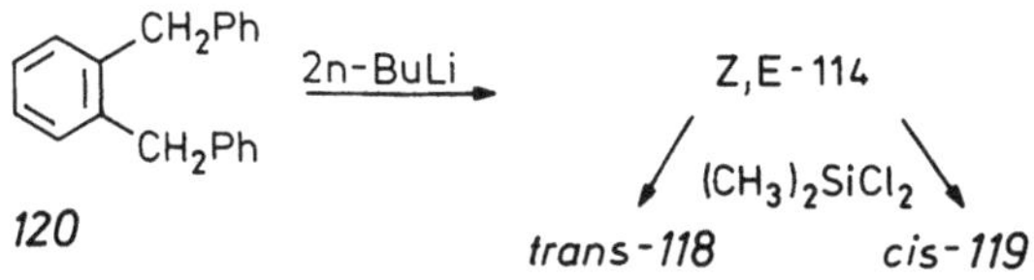

120 *trans-118* *cis-119*

Finally the ^{1}H-NMR spectra of Z,E-*114* and E,E-*117* prepared from *cis-104* and *trans-105*, respectively, with K in 2-MTHF at —78 °C have been reported. Both isomers should be configurationally stable at that temperature [77c, g].

Most of the results mentioned above, however, are at variance with later investigations [82–86].

1. ^{1}H- and ^{13}C-NMR spectra of THF-D_8 solutions of crystalline dilithio-α,α'-diphenyl-o-xylylenediide · 2 tetramethylethylenediamine (E,E-*117*-2(Li · TMEDA)), disodio-and dipotassio-diphenyl-o-xylylenediide gave no hint for the existence of the isomer Z,E-*114*; in all cases only (>97%) the isomer E,E-*117* was found [82, 84, 85].
2. An X-ray structure determination reveals the E,E-conformation of E,E-*117*-2(Li · TMEDA) also in the solid state [86].

3. Reactions of E,E-*117*-2(Li · TMEDA) in THF, 2-MTHF or hexane solutions at 20, 0 and —73 °C, respectively, with dimethylsilyldichloride led only to *cis-119* in up to 95% yields. *Cis-119* is isolated in crystalline form (mp. 107 °C) [82,84].
4. Similarly, in reactions of *cis-104* [84] and *trans-105* with K in THF or 2-MTHF at 20 and —73 °C, respectively, only *cis-119* was isolated in up to 81% yields; a *trans*-adduct *trans-118* was never detected.

Thus, on the basis of these results the experimental data reported earlier [77c,g] with *cis-104* and *trans-105* do not support the theoretical conclusion of an allowed conrotatory rearrangement of the benzocyclobutene radical "anion" *110*.

The earlier reported rearrangement of the benzocyclobutene radical "anion" *110*, prepared from benzocyclobutene *121* with Na/K alloy, to give the o-xylylene radical "anion" *111* ($M^+ = K^+$) [77a], was later retracted [77b].

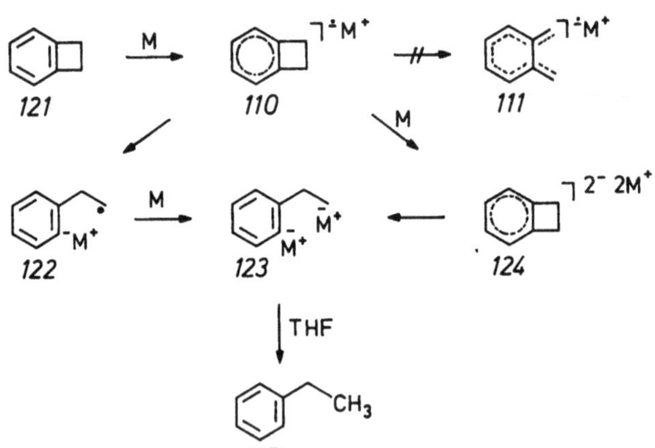

Rather, in the reaction of *121* with lithium metal in THF a C_{aryl}—CH_2 bond is broken to give finally, among others ethylbenzene *125*[87]. The bond fission is thought

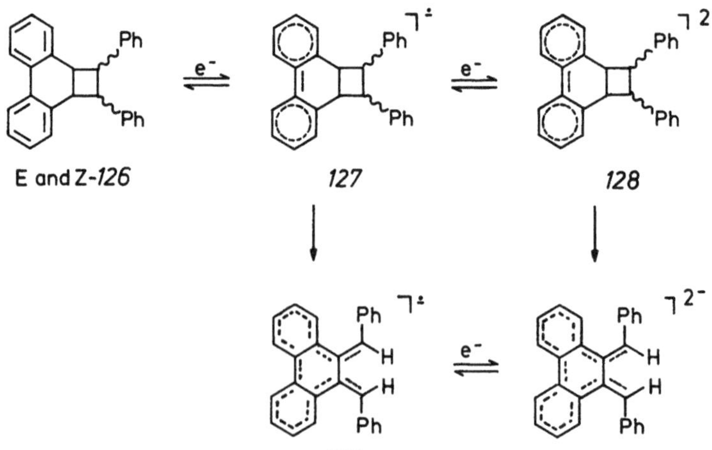

to occur at the "dianion" stage (*124*→*123*, 2 M^+ = 2 Li^+) and not via the radical "anion" pathway *110*→*122*, M^+ = Li^+.

Electrochemical and ESR investigations with E- and Z-1,2-dihydro-1,2-diphenyl-cyclobuta[l]phenanthrene, E- and Z-*126*, indicate that the ring-opening reaction to give the rearranged "dianion" *130* may occur at the radical "anion" (*127* → *129*) or the "dianion" (*128* → *130*) stage, depending on the ion pair conditions [88].

4 Cyclization of 5-Hexenyl "Anions"

The 5-hexenyl radical *131* rearranges in a highly regiospecific manner to give mainly the cyclopentylcarbinyl radical *132*. The cyclohexyl radical *133* is formed only as a side product [89].

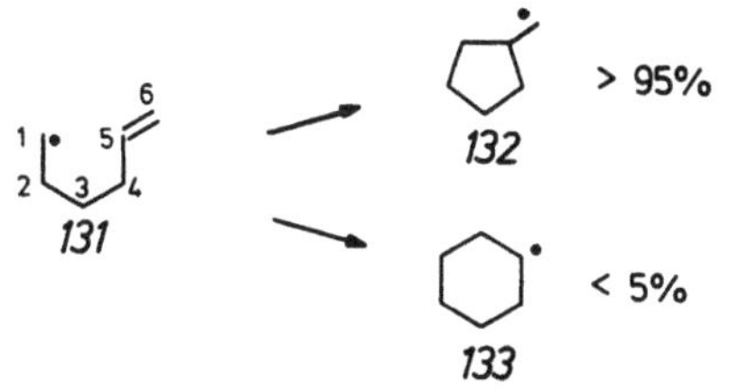

Because of the rather fast cyclization *131* → *132* ($k_1 = 1 \cdot 10^5\ s^{-1}$ at 25 °C) [90] this reaction or the cyclization of the related 1-methyl-5-hexenyl radical *134* [91] and the o-(3-butenyl)phenyl radical *135* [92] is often used as a "radical clock" in reactions possibly passing through radical intermediates [93].

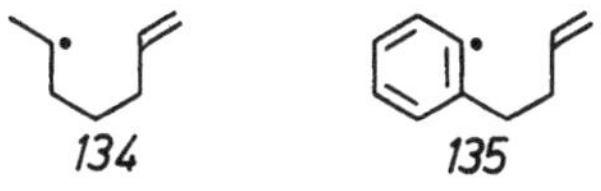

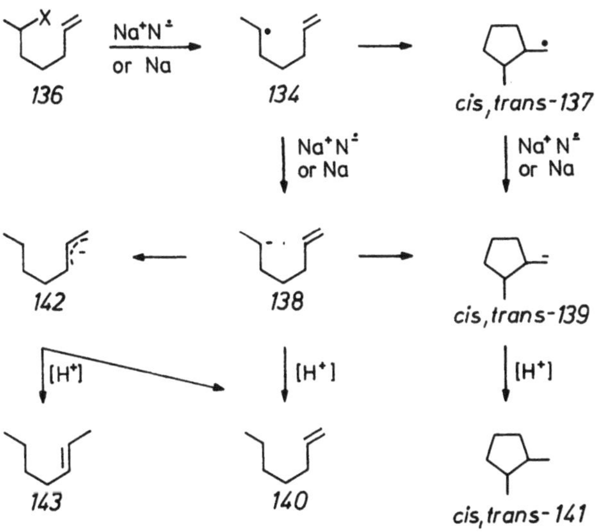

Scheme 8 (gegenions omitted)

131, *134*, or *135*, however, are useful radical probes only if "*anion*" cyclization is negligible. As outlined in Scheme 8 cyclization can also take place at the "anion" stage: Formation of the "anions" *cis*- and *trans-139* is not only possible via the radical route *134* → *cis,trans-137* → *cis,trans-139*; *cis,trans-139* is also formed via the "anionic" intermediate *138* [94].

Garst and coworkers have used the reactions of 1-methyl-5-hexenyl halides *136*, (X = Hal) with alkali metals or alkali metal naphthalenides such as sodium naphthalenide $Na^+N^{\dot{-}}$, as a mechanistic probe over the years [95]. A 1984 publication [96] discloses the possibility of differentiating between the radical *134* and "anion" *138* cyclization.

In *radical* cyclizations of *134* the ratio of the finally formed *cis*- and *trans*-cyclopentanes *cis*- and *trans-141* is normally 3.8 [91]. If 1-methyl-5-hexenyl halides *136* (X = Cl, Br) are reacted with $Na^+N^{\dot{-}}$ in dimethoxyethane (DME) or THF at room temperature the ratio varies between 0.62 and 2.8; for sodium mirror reactions the *cis/trans* ratio of *141* can be as low as 0.32 [96]. This suggests that under the reaction conditions, 1-methyl-5-hexenylsodium also cyclizes, with a *trans* preference, giving [(2-methylcyclopentyl)methyl]sodium *trans-139*, $M^+ = Na^+$. Most of the other "anion" cyclizations show a similar strong *trans* preference [94].

A correlation of the observed *cis/trans-141* ratios with the yields of 2-heptenes *143* supports the hypothesis that the "anion" *138* cyclizes. 2-Heptenes are formed from (1-methyl-5-hexenyl)sodium *138* through an intramolecular 1,4-proton transfer that competes with cyclization to give *142* (Scheme 8) [97]. Thus, cyclizations of 1-methyl-5-hexenyl "*anions*" and *radicals* can be distinguished, since they occur with different *cis/trans* ratios and since products of "anion" cyclizations are accompanied by those of the 1,4-proton transfer.

It is interesting to note that Ashby's investigations of the reactions of alkyl halides with (trimethyltin)sodium and dialkylcuprates by means of the 1-methyl-5-hexenyl probe which led to the conclusion that only radicals cyclize [98], have been questioned by Lee and San Fillipo in 1983 [99]. These authors correctly state that the cyclization alone does not prove a radical pathway. The *cis/trans* ratios from reactions of the halides *136* with (trimethyltin)sodium (revised: ~4.8 at 0 °C) [96], however, are consistent with 100% radical cyclization.

In his paper dealing with the separation of radical and "anion" cyclizations Garst also comments on the apparent halogen effects on product distributions in both sodium naphthalenide and sodium mirror reactions [96]: "While there are other possible explanations, the possibility that alkyl halide radical "anions" could be intermediates that undergo reactions other than fragmentation to alkyl radicals and halide ions should be kept in mind."

Very recently, Garst and coworkers [100] have provided a means to suppress the cyclization of (1-methyl-5-hexenyl)sodium with t-butyl amine [101]. t-Butyl amine is a sufficiently reactive proton donor to compete successfully with the "anion" cyclization *138* → *cis/trans-139* and the intramolecular 1,4-proton transfer *138* → *142*, leaving 1-heptene *140* as well as *cis*- and *trans*-1,2-dimethylcyclopentane *cis,trans-141* formation through *radical* cyclization only. For sodium metal reactions excess t-butyl amine nearly eliminates *cis,trans-141* suggesting that radical cyclization is negligible. For the $Na^+N^{\dot{-}}$ reaction, 2-heptenes *143* are eliminated by added t-butyl amine, but the yields of *cis,trans-141* are merely diminished. As expected, the *cis/trans* ratio of the residual *cis,trans-141* is near four, corresponding to pure radical cyclization.

Bailey and coworkers [102] prepared 5-hexenyllithium *145* from 6-iodo-1-hexene *144* with two equivalents of t-butyllithium in n-pentane/diethyl ether (3:2) at —78 °C and studied the kinetics of the "anion" rearrangement to give *146*.

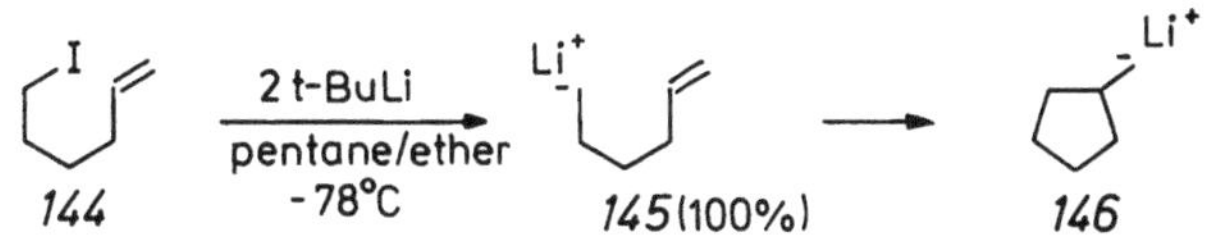

144 *145* (100%) *146*

The results are outlined in Table 5.

Table 5. First-order rate constants and activation parameters for the cyclization *145* → *146*

Temp. [°C]	$10^4 \cdot k$ [s^{-1}]	$\Delta H^{\neq}$ [kcal/mol]	$\Delta S^{\neq}$ [e.u.]
—11.1	1.75	11.8 ± 0.5	—30 ± 2
—0.5	4.18		
9.4	10.3		
20.0	20.6		

The data of Table 5 indicate that the conversion of the "anion" *145* into *146* is very much slower (by a factor of 10^8–10^{10}) than the cyclization of the 5-hexenyl radical *131*. However, nonnegligible quantities of product containing the cyclopentylmethyl group may still arise from cyclization of the "anion" *145* since the half-life for this process at temperatures above 0 °C ($t_{1/2}$ = 23 min at 0 °C; 5.5 min at 23 °C) is short relative to the time scale of many experiments that seek radical intermediates [103]. A differentiation between radical and "anion" cyclization, as in the case of *134* and *138*, is not available with *131* and *145*. The cyclization of the "anion" *145* is, however, very slow at lower temperatures like —78 °C.

Bailey and coworkers utilized this scenario to check whether or not radicals are involved in the halogen-metal exchange reaction with t-butyllithium [104]. Their results indicate that alkyl bromides and iodides may behave quite differently. There is no evidence for radical intermediates in the reaction with primary alkyl iodides at low temperatures. In the case of a primary alkyl bromide at least 15% of the reaction involves radical intermediates. This result, which is consistent with the recent reports of two other groups [105], implies that at least a portion of the exchange proceeds via ET from the alkyllithium to the primary bromide [106].

The rearrangement of such alkenyllithium compounds is also used for synthetic purposes [107].

Treatment of the alkenyl iodide *147* with t-butyllithium in pentane/ether at —78 °C led to the "anion" *148* which rearranged in a regiospecific manner to give the *cis*-hydrindane skeleton *149* via generation of a quarternary center.

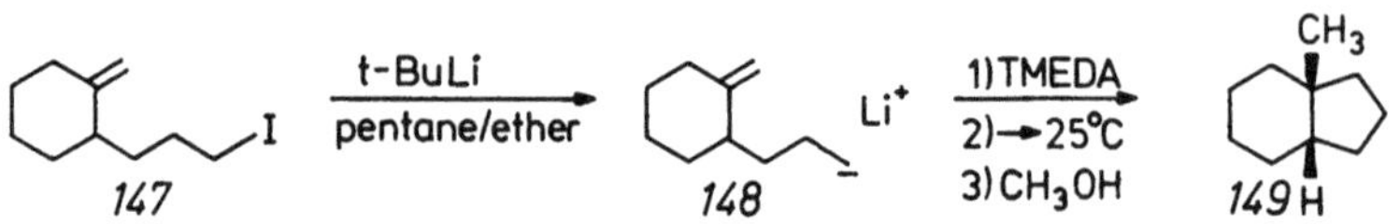

147 148 149

The rearrangement of the aryllithium species *150* was investigated by Woolsey and coworkers [108] in different solvents at 23 °C.

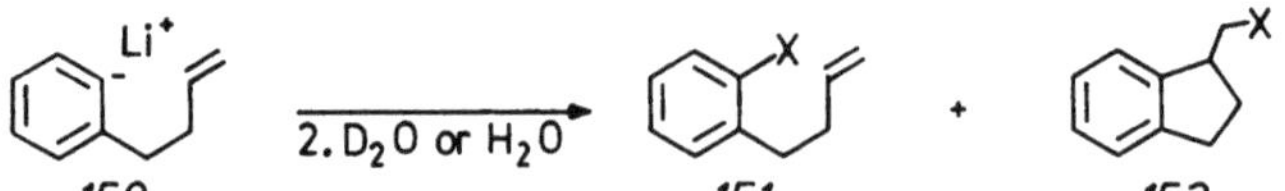

150 151 X = D; H 152

In THF after 60 min 10% *151* and 90% *152* were found. In diethylether the rearrangement is somewhat slower (60% *152*). Addition of TMEDA to the ether solution accelerates the ring closure: after 30 min 94% of *152* was detected. These results lead to the conclusion that radical tests by cyclization of what is thought to be the phenyl radical *135* must be carried out at low (—78 °C) temperature at which the "anion" *150* cyclizes very slowly: after 120 min in THF at —78 °C only the non-cyclized *151* is detected.

A mechanistic study concerning the variation of the product ratio in reactions of different o-(3-butenyl)halobenzenes (Hal = F, Cl, Br and I), and of 6-bromo-1-hexene with different alkali metals (Li, Na, K) in ammonia/tert-butyl alcohol solution [110] is related to the problem of "anion" versus radical cyclization as outlined in this chapter.

In connection with the question of whether the reduction of benzophenone by a lithium dialkyl amide containing β-hydrogen atoms proceeds via electron transfer or not, Newcomb and coworkers prepared N-lithio-N-butyl-5-methyl-1-hex-4-enamine *153* as a mechanistic probe [111].

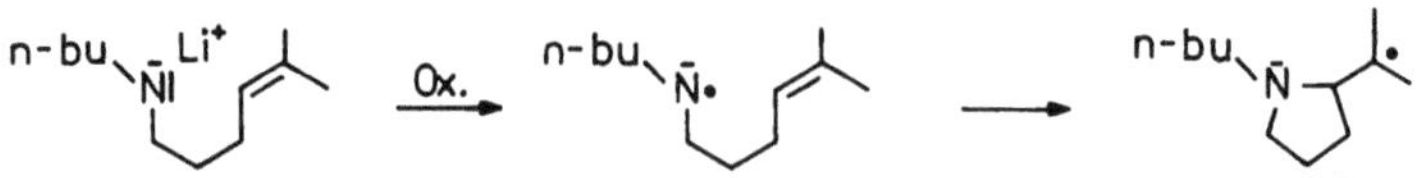

153 154 155

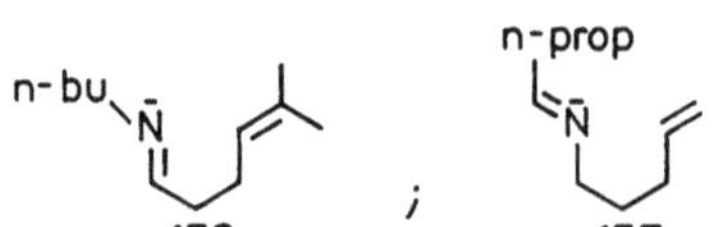

156 157

Treatment of benzophenone with *153* in THF gave good yields of benzhydrol and imines *156* and *157* but no trace of cyclic products derived from *155* (formed via oxidation of *153* to give *154*). The "anion" *153* does not cyclize under the reaction conditions. Cyclic products would be expected if *153* had been oxidized to radical *154* since *154* cyclizes [111]. This and other experiments led to the conclusion that the reduction of benzophenone by dialkyl amides containing β-hydrogen atoms occurs via hydride and not via electron transfer.

In summary, the rearrangement of radicals and "anions" of the 5-hexenyl type is a useful mechanistic probe if the two reaction types can be clearly separated.

5 Configurational Isomerization of α-Substituted Vinyl "Anions"

5.1 α-Aryl-Vinyllithium Compounds

The configurational lability *158* ⇌ *159* of α-aryl-vinyllithium species in etheral solvents was first observed by Curtin [112] and more recently studied kinetically by Knorr and coworkers [113]. The investigations show nicely the influence of different donor solvents on the rate of the rearrangement of the following organolithium compounds.

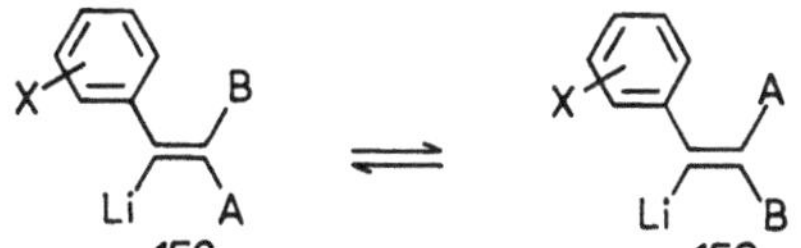

158 *159*

	A	B	X
a:	H	D	H
b:	H	D	2-$CH(CH_3)_2$
c:	H	D	2,6-$(CH_3)_2$
d:	CH_3	CH_3	H
e:	2-$C_6H_4CH_3$	2-$C_6H_4CH_3$	H

From the dependence of pseudo first-order rate constants k_ψ on the formal concentrations (0.3–0.7 M) in THF an order of reaction of 0.5 was found for *158a* and *b* [113]. Therefore a dissociation step must precede the configurational isomerization. Dissociation to give anions and lithium cations, however, is very unfavorable in THF [26], and is not rate controlling in the case of *158a* since addition of lithium bromide causes no common ion rate depression. It is hence concluded that the 0.5 order of reaction is due to disaggregation into two subunits. Although not compelling it has been assumed that *158a–c* occurs as dimers *160* with one THF per lithium atom. Partial dissociation produces the monomer or contact ion pair *161* which is expected to bind about two solvent molecules.

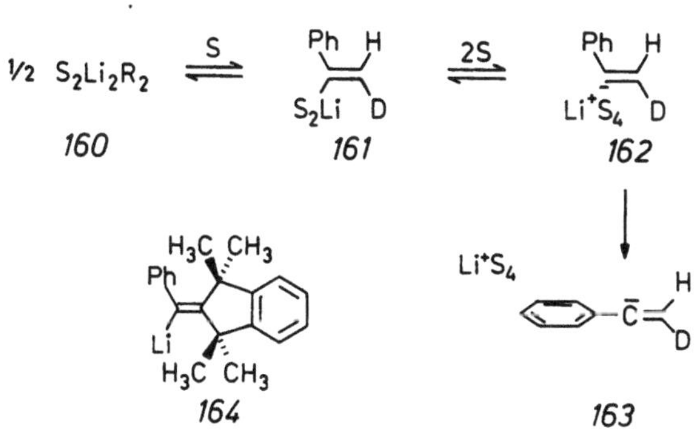

Table 6. Pseudo first-order rate constants k_ψ and activation parameters for the rearrangement *158 a–c* ⇌ *159 a–c*

158	Formality	Solvent	$\Delta H^{\neq}$ [kcal/mol]	$\Delta S^{\neq}$ [e.u.]	k_ψ [s^{-1}] at 27 °C
a	0.5 M	ether	—		ca. 10^{-4}
a	0.7 M	1.1 M TMEDA/ether	—		ca. 0.05
a	0.9 M	TMEDA	9.9 ± 0.7	−31 ± 3	0.062
a	0.9 M	2.5 M TMEDA/THF	7.3 ± 0.3	−38.4 ± 1	0.13
a	0.6 M	THF	7.8 ± 0.6	−36 ± 2	0.17
a	0.28 M	1.9 M HMPA/THF	5.4 ± 1.5	−24 ± 6	4040
b	1.7 M	THF-D_8	11.1 ± 0.3	−22.4 ± 1	0.63
c	0.7 M	THF-D_8	8.9 ± 0.2	−22.5 ± 0.6	24.1

The activation parameters and the pseudo first-order rate constants k_ψ of the configurational inversion *158 a–c* ⇌ *159 a–c* are listed in Table 6.

The results of Table 6 have been tentatively explained as follows: The balance for the transformation of *160* into the solvent-separated ion pair *162* should amount to $\Delta H^\circ \sim -10$ kcal/mol and $\Delta S^\circ \sim -37$ e.u.; an ion-paired transition state like *163* can now be reached with an additional enthalpy of about 18 kcal/mol, in close agreement with the inversion of N-phenyl-imines as a model for *162* → *163*.

The donor dependence of k_ψ in Table 6 agrees with current notions except for TMEDA. Moderate amounts of HMPA cause an impressive rate enhancement; unfortunately, the decomposition temperature of *158 a* in the presence of HMPA drops from around 80 to below 0 °C, precluding a precise study.

Whereas *158 d* with β-CH_3 groups does not invert faster than *158 a*, E/Z topomerization of *158 e* is rapid in THF even at −25 °C ($k_\psi = 7.4\ s^{-1}$ for CH_3 coalescence). In the case of *164* with even bulkier t-alkyl β-substituents nonequivalences were detected only in the ^{13}C NMR spectra with coalescences at ∼ −45 °C ($k_\psi = 50\ s^{-1}$ in THF). A comparison of the relative rate constants with $k_{158a} = 1$ leads to $k_{158e} = 820$ and

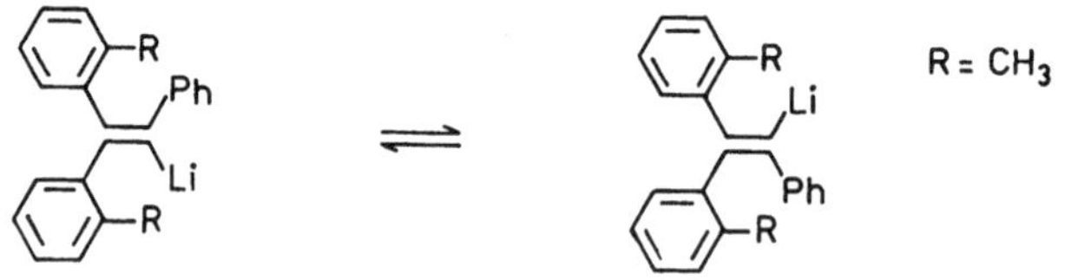

158 e *159 e*

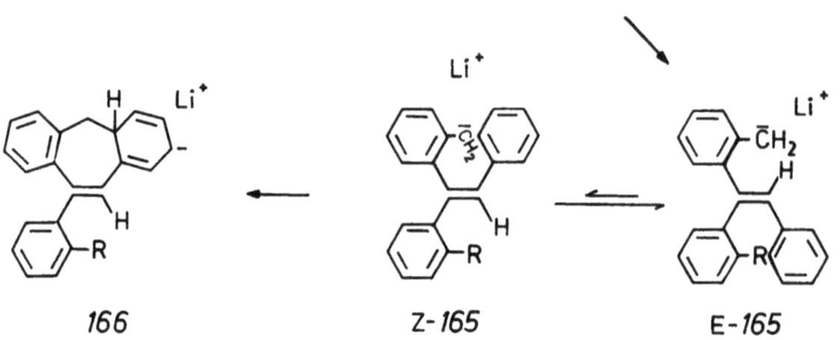

166 Z-*165* E-*165*

k_{164} = 24200. The tremendous steric acceleration is possibly due to the prevalence of disaggregation such that the analog of *161* becomes the ground state. Interestingly, the chemical reactivity of *164* is also greatly enhanced.

As far as *158e* is concerned, in addition to the isomerization *158e* ⇌ *159e*, at 0 °C in THF yet another isomerization takes place, the intramolecular vinyl-benzyl isomerization to give E-*165* predominantly ($\tau_{1/2}$ = 17 min) [113b)].

E-*165*, however, is also unstable. At room temperature, especially in the presence of some of the corresponding olefin, it rearranges in an electrocyclic reaction to give the seven-membered ring "anion" *166*. Since this rearrangement cannot occur from E-*165* and since this reaction is catalyzed by the olefin, Z-*165* should be the reactive intermediate.

5.2 α-Trimethylsilyl- and α-Alkoxy-Vinyllithium Compounds

In a more recent study Knorr and von Roman [114)] investigated the configurational stability of the synthetically interesting [115)] α-trimethylsilyl- and α-alkoxy-vinyllithium species *167* and *168*.

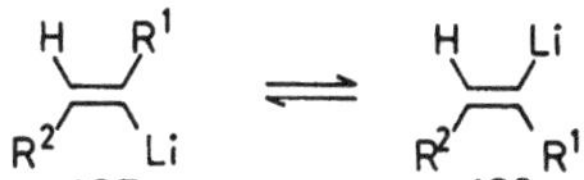

167 *168*

	R^1	R^2
a:	$Si(CH_3)_3$	H
b:	OC_2H_5	H
c:	OCH_3	H
d:	OC_2H_5	CH_3

1-Trimethylsilyl-vinyllithium *167a* in THF has a half-life for the Z/E rearrangement of 0.11 s at 48 °C. Extrapolation ($\Delta H^{\neq}$ = 7.4(±1.5 kcal/mol) and $\Delta S^{\neq}$ = −32(±4) e.u.) to −70 °C leads to a half-life of 0.5–15 min. This is in agreement with qualitative observations: in THF 1-silyl-vinyllithium species are configurationally stable only at −70 °C. The same holds for diethylether solutions. In alkane solutions in the presence of TMEDA, on the other hand, configurational lability is only observed at 25 °C.

1-Alkoxy-1-alkenyllithium species are prepared and react in THF between −78 and −20 °C with retention of configuration [116)]. 1-Ethoxy-vinyllithium *167b* [117)] as well as 1-methoxy-vinyllithium *167c* [118)] are configurationally stable in TMEDA up to 106 °C. At this temperature decomposition starts. In the NMR spectra in THF or in diethyl ether at 25 °C no line broadening is observed which indicates configurational stability. This is supported by different E/Z mixtures of *167d* and *168d* which are chemically and configurationally stable in THF at 25 °C for more than 13 days.

The importance of kinetic studies of this sort for synthetic investigations is self-evident.

6 The Reversible Rearrangement of Acyl-Substituted [9]Annulene "Anions" into Nonafulvenolates as a Function of the Ion Pair Character

In previous chapters the influence of the gegenion and the solvent on rearrangements of "carbanions" mostly concerned the *rate* of a certain reaction (and thus *kinetic* aspects), as, e.g., in the case of the cyclopropyl-allyl "anion" rearrangement (Sect. 2.1), the allyl "anion" isomerization (Sect. 2.2), the cyclization of 5-hexenyl alkali metal compounds (Sect. 4) or the configurational isomerization of α-substituted vinyllithium compounds (Sect. 5).

In this section ion pair effects will be discussed which influence the structures of the "carbanionic" parts of two ion pairs (and thus the *ground states* of "carbanions") as a function of the gegenion M^+, solvent S and temperature T. The influences of M^+, S and T on "carbanions" have, of course, been noted before as, e.g., in the pioneering work of Szwarc [119] and Smid [26, 120]. The structural changes in the "carbanionic" parts of the compounds studied by these authors, however, are most probably limited to minor perturbations.

One can visualize an entirely different situation if one deprotonates the cyclononatetraenyl ketones *170* (prepared from the [9]annulene "anion" *169* and acyl chlorides in THF at —20 °C) with a base M^+B^- [121].

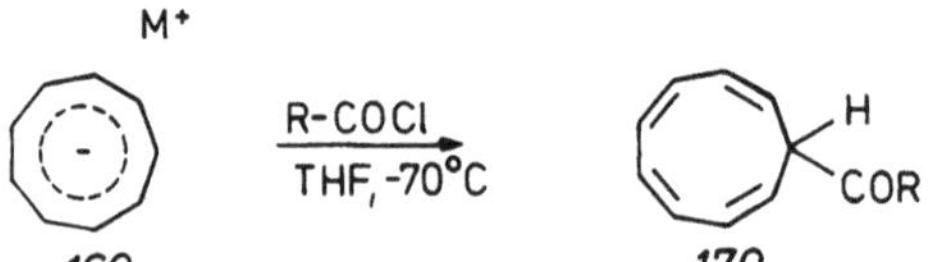

169 *170* a: R = CH_3 b: R = C_6H_5

Does the deprotonation lead to a "nonafulvenolate"-type structure 171^-M^+ (an enolate contact ion pair [122, 123, 124]) or to an acyl-substituted [9]annulene "anion" 171^-⌇M^+, possibly a solvent separated ion pair? [127]

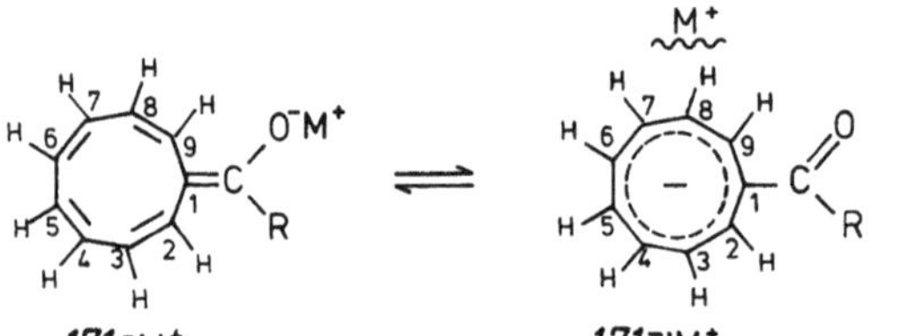

171^-M^+ 171^-⌇M^+

Reaction of the phenyl ketone *170b* with *lithium* bis(trimethylsilyl)amide in THF at —78 °C led to the nonafulvenolate $171b^-Li^+$ which is strongly olefinic in nature.

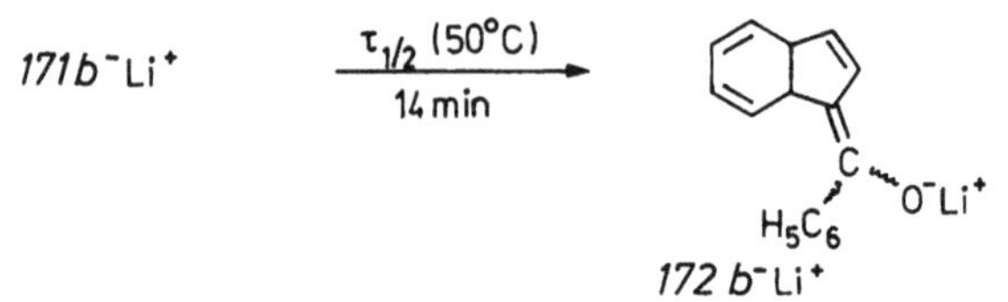

$172b^-Li^+$

This is shown by the ^{1}H nmr spectrum ($\delta_{H^2-H^9}$ 5.0–6.4) and by the facile valence isomerization ($\tau_{1/2(50\,°C)}$ = 14 min) to give *172b*$^-$Li$^+$.

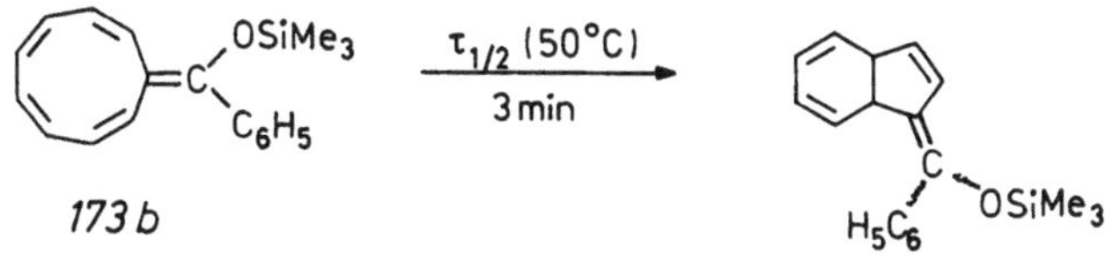

173b *174b*

That the gegenion Li$^+$ in *171b*$^-$Li$^+$ is indeed similarly attached to the enolate oxygen atom as the (covalently bonded) trimethylsilyl group in the trimethylsilylenol ether *173b* is indicated by the very similar isomerization rates of *171b*$^-$Li$^+$ and *173b* (*173b* → *174b*, $\tau_{1/2\,(50\,°C)}$ = 3 min).

The ^{1}H nmr spectra of *171b*$^-$Li$^+$ and *173b* deserve a closer look. In the enol ether *173b* all eight hydrogen atoms at the nine-membered ring are magnetically different — as expected — leading to a complex spectrum [121 c)]. In *171b*$^-$Li$^+$, however, an AA'BB'CC'DD'-spectrum results from pairs of identical hydrogen atoms (H^2/H^9, H^3/H^8, H^4/H^7, H^5/H^6). The coupling constants $J_{2,3} = J_{8,9} = J_{4,5} = J_{6,7}$ = 12.5 Hz and $J_{3,4} = J_{7,8}$ = 3.5 Hz are in agreement with single and double bonds, respectively, and a non-planar shape of the nine-membered ring [128)].

What leads to the pairs of identical hydrogen atoms in the ^{1}H-nmr spectrum of *171b*$^-$Li$^+$? A fast rotation around the exocyclic CC double bond in this enolate should be excluded especially at room temperature since enolates, as mentioned in Ref. 124, are configurationally stable even at much higher temperatures [125 a, 126)]. It is therefore concluded, and this is supported by the results of the experiments described below, that the contact ion pair *171b*$^-$Li$^+$ is in fast equilibration with some small amount of an ion pair *171b*$^-$ ⦚ Li$^+$ in which the acyl group is either quickly rotating or orthogonal to the [9]annulene "anion" ring.

$$171b^-\,Li^+ \overset{fast}{\rightleftharpoons} 171b^- \wr Li^+$$

That the contact ion pair *171b*$^-$Li$^+$ is indeed easily transformed into an ion pair *171b*$^-$ ⦚ Li$^+$ *with totally different properties* is nicely demonstrated by the addition of either HMPT or DMSO to the THF-solution of *171b*$^-$Li$^+$: 1) The hydrogen atoms at the nine-membered ring are shifted to lower field ($\delta_{H^3-H^8}$ ~ 6.7; H^2 and H^9 are buried underneath the phenyl hydrogens) which is characteristic of [9]annulene "anions" [129)], and 2) The newly formed species *171b*$^-$ ⦚ Li$^+$ is thermally stable at 60 °C for more than 12 hours(!) — in strong contrast to *171b*$^-$Li$^+$ which has a half-life of 14 min at 50 °C! Unsubstituted [9]annulene "anions" are also thermally stable and do not show an isomerization similar to that of *171b*$^-$Li$^+$ or *173* [129)].

Removal of Li$^+$ from the oxygen atom by HMPA or DMSO thus leads to a charge redistribution (the negative charge essentially moves from the oxygen atom into the nine-membered ring) and consequently to the formation of an aromatic substituted [9]annulene "anion" from an olefinic nonafulvenolate.

If the methyl ketone *170a* is reacted with *potassium* bis(trimethylsilyl)amide in THF, only the ion pair *171a*$^-$ ⦚ K$^+$ is observed in the ^{1}H-nmr spectrum. Again, the compound is thermally stable even on warming to 60 °C for 4 days! It is not surprising that in *171a*$^-$ ⦚ K^+ the gegenion is removed from the enolate oxygen atom even by the solvent THF, while HMPA and DMSO are required in the case of the Li^+-species *171b*$^-Li^+$.

If the methyl ketone *170a* is deprotonated by *sodium* bis(trimethylsilyl)amide in *THF* the following observations are made: At 30 °C a ^1H nmr spectrum results which is similar to that of *171b*$^-Li^+$ in THF, indicating the formation of the contact ion pair *171a*$^-Na^+$. At *—45 °C*, however, only *171a*$^-$ ⦚ Na^+ is observed – the spectrum is very similar to that of *171a*$^-$ ⦚ K^+. At *intermediate temperatures* weighted averages of the chemical shifts and the coupling constants of the spectra of *171a*$^-Na^+$ and *171a*$^-$ ⦚ Na^+ reveal a fast equilibration between the two ion pairs [130].

The temperature dependence of the equilibrium between *171a*$^-Na^+$ and *171a*$^-$ ⦚ Na^+ allows the evaluation of the enthalpy and the entropy of reaction:

171a$^-Na^+$(THF) ⇌ *171a*$^-$ ⦚ Na^+(THF)
$\Delta H^\circ = -6.9$ kcal/mol
$\Delta S^\circ = -30$ e.u.

In the case of fluorenyl*sodium* in *THF*, the following values for the equilibrium between the contact and the solvent separated ion pair are found:

$\Delta H^\circ = -7.6$ [26] (-6.7) [131] kcal/mol
$\Delta S^\circ = -33$ [26] (-27) [131] e.u.

It is the similarity of these data with those of the equilibrium *171a*$^-$ Na^+ (THF) ⇌ *171a*$^-$ ⦚ Na^+ (THF) which led to the suggestion that there exists a contact ion pair – solvent-separated ion pair relationship also between *171a*$^-Na^+$ and *171a*$^-$ ⦚ Na^+ (and the other species mentioned above).

The intermediate nature of the Na^+ between the Li^+ and K^+ species is not surprising. It is, however, surprising that the olefinic nonafulvenolate *171a*$^-Na^+$ is reversibly transformed into the aromatic [9]annulene "anion" derivative *171a*$^-$ ⦚ Na^+ simply by lowering the temperature. This emphasizes the importance of ion pair effects (and their consequences) not only as a function of the gegenion M^+ and the solvent S but also of the temperature T.

^{13}C-nmr investigations support the [9]annulene "anion" character of the ring carbon atoms in *171b*$^-$ ⦚ Li^+, *171a*$^-$ ⦚ Na^+ and *171a*$^-$ ⦚ K^+ [121 c)]. Unexpectedly, however, one finds the following chemical shifts [δ] for the "carbonyl" C atoms in these compounds: *171a*$^-$ ⦚ Li^+ 206.7, *171a*$^-$ ⦚ Na^+ 203.1 and *171b*$^-$ ⦚ K^+ 205.9. The observed chemical shifts are very close to those of "normal" carbonyl C atoms (ca. 205), but they are far away from those of normal enolate C atoms ($\delta = 160$–170) [123]. This led to the conclusion that in the solvent-separated ion pairs of the [9]annulene "anion" type mentioned above, the carbonyl group is oriented *orthogonally* to the plane of the 9-membered ring. In a coplanar conformation the carbonyl group should adopt at least some enolate character. Interestingly, the orthogonal conformation is in agreement with MNDO calculations which favor the orthogonal over the sterically very crowded planar conformation by 7.9 kcal/mol [121 c)]. Although this result has to be substantiated by an X-ray structure determination it illustrates nicely the structural changes which may be induced by ion pair variations.

7 Electron Transfer Valence Tautomerism

Electron removal from, and electron transfer to a compound is related to conformational and configurational changes as well as to an alteration of the reactivity. This was shown, e.g., in Sect. 2.3 entitled "ET-induced rearrangements of cyclopropanes and consecutive reactions" [132]. The effect of charge redistribution *within* a "carbanion", resulting from different ion pair situations, on the conformations of the "carbanion" was also illustrated by the examples given in the last section. In this section "electron transfer valence tautomerism", as proposed by Staley [133], is reviewed, and the reader will recognize the similarity of the phenomena discussed here to those outlined in Sect. 6.

The charge in the ground states of bicyclooctatetraenyl dipotassium *175* and 1,2-dicyclooctatetraenylene dipotassium *176* is localized in one planar eight-membered ring, while the other ring adopts a distorted tub conformation, as confirmed by ^{1}H nmr studies in liquid ammonia [133].

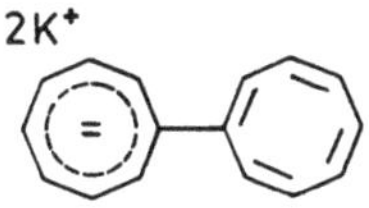

175

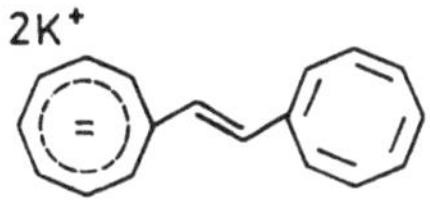

176

The simplicity of the spectrum of *175* at room temperature (20 °C), as well as the exchange broadening evident at lower temperatures, can be rationalized only if an inter-ring electron exchange process occurs at a rapid enough rate on the nmr time scale. The ^{1}H nmr spectrum furthermore discloses that the "neutral" ring in *175* is being flattened somewhat at one end relative to cyclooctatetraene, presumably to permit a greater delocalization of charge.

The rate of "electron transfer" of $>75\ s^{-1}$ at -65 °C requires the eight-membered rings of *176* to complete a planar-to-tub (or vice versa) conformational change at the same rate. However, the rate of inversion of neutral cyclooctatetraene is only $26\ s^{-1}$ at -70 °C. There are several ways of reconciling the greater rate in *176* compared with that in cyclooctatetraene. First, the latter inversion forces the ring into a planar conformation having an 8π electron antiaromatic perimeter. In the case of *176*, the cyclooctatetraene ring undergoing flattening is simultaneously gaining charge density and eventually attaining a 10π electron aromatic configuration. Second, the "neutral" ring in *176* may not fold up completely as observed in the case of *175*.

It is considered that the conformations *177* and *178* might be transition states or intermediates for the electron redistribution process.

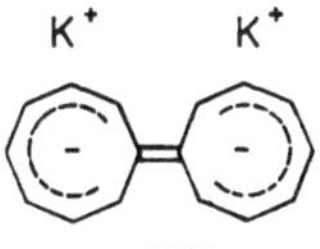

177

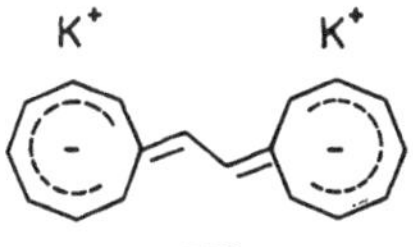

178

Finally, it should be mentioned that an analogous process appears to occur in a related fused-ring system, viz., the octalene dianion *179* [134].

179

8 Polymorphism of Organolithium Compounds

In most of the examples given so far in this article the free enthalpy difference $\Delta\Delta G^\circ$ between two "isomers" of an organometallic compound is so high, and the free activation energy $\Delta G^{\#}$ for a transformation so low, that only one of the two isomers is observed. An example to the contrary *in solution* was given in Sect. 6 by the two ion pairs *171a*$^-$Na$^+$ and *171a*$^-$ ‖ Na$^+$ which are similar in energy although structurally rather different.

In recent years increasing effort has been devoted to the elucidation of solid state structures of organolithium species [135]. This led at least in two cases to the discovery of polymorphism and thus, in a certain respect, to two "isomers" of a "carbanion" in the *solid state*.

The first report to deal with polymorphs concerned the structure of the "dianion" *180* [136].

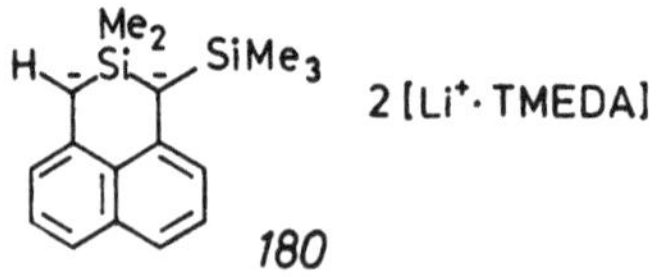

180

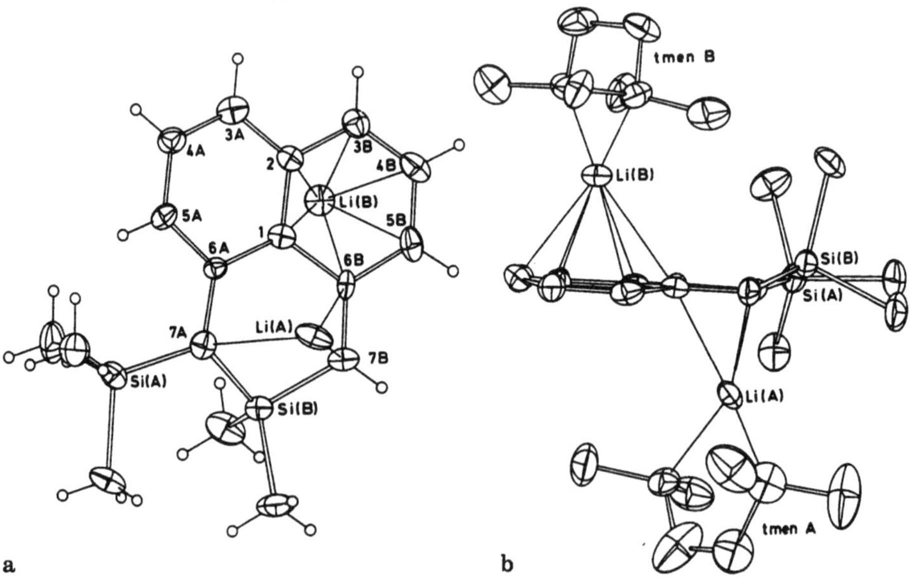

Fig. 1a. Projection of the "dianion" *180* of the α-phase and its associated lithium atoms normal to the aromatic plane. Hydrogen atoms are shown as spheres of radius 0.1 Å; **b** The full "molecule" *180* of the α-phase projected through the "dianion" plane

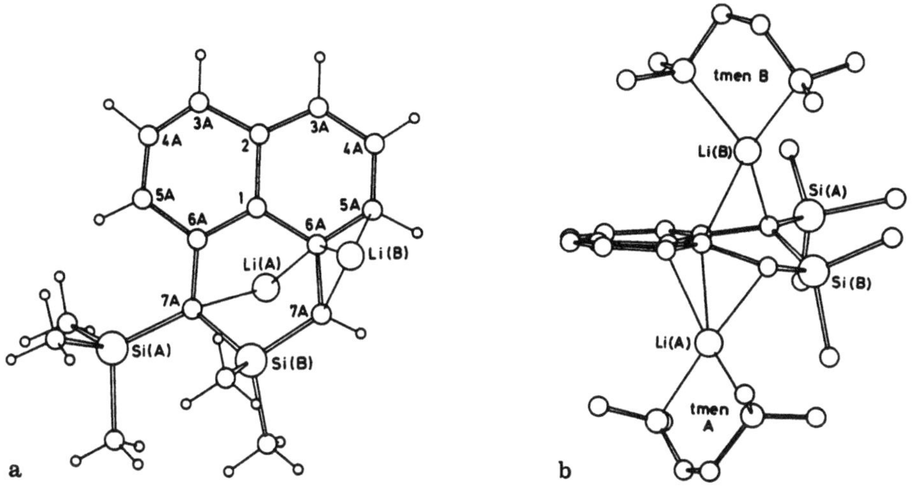

Fig. 2a. Projection of the "dianion" *180* of the β-phase and its associated lithium atoms normal to the aromatic plane; **b** Projection of the full "molecule" *180* of the β-phase through the "dianion" plane

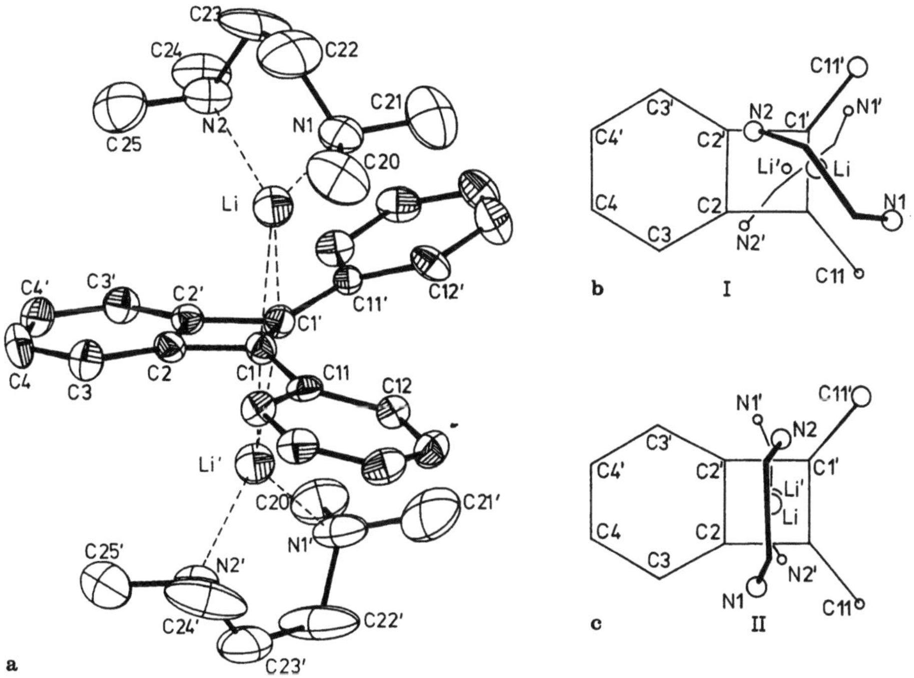

Fig. 3a. ORTEP diagram of the crystal structure of *181*-I; **b** Projection of *181*-I normal to the benzocyclobutadiene-diide-rings; **c** Projection of *181*-II normal to the benzocyclobutadiene-diide-rings

The structure of the α-phase of *180* is given in Fig. 1.
As shown in Fig. 1, Li(B) sits upon one of the two six-membered rings of the naphthalene part of *180*, while Li(A) is located on the other side of the "dianion" below the Si-heterocyclus.

The β form of *180*, although the precision of its determination is lower than that of the α form, is shown in Fig. 2. In the β-phase Li(A) is also located below the six-membered heterocyclus containing Si(B). Li(B), however, sits outside the six-membered carbocyclus leading to a bonding situation which is similar to that in benzyllithium [137].

Different positions of the Li^+ ions and the TMEDA molecules are also observed in the two phases of dilithium 1,2-diphenylbenzocyclobutadiene-diide · TMEDA (*181*) [138]. The X-ray structure of phase I is shown in Fig. 3a. A projection of *181*-I normal to the benzocyclobutadiene-diide rings is given in Fig. 3b. One Li^+ ion is below the four-membered ring, the other one sits on the other side of the "dianion" slightly outside the four-membered ring. The two TMEDA molecules attached to the Li^+ ions are oriented more or less orthogonally to one another.

In *181*-II, as shown in Fig. 3c, the Li^+ ions are positioned above and below the center of the four-membered ring, and the TMEDA molecules are essentially parallel to each other.

The polymorphism observed in the cases of *180* and *181* clearly shows that it is not at all justified to consider a certain X-ray structure of an organometallic compound as being *the* solid state structure of this species! Rather, a facile — as it seems — rearrangement of the gegenions Li^+ (and of the TMEDA molecules) relative to the dianion backbone leads into a second energy minimum; the two minima are probably not very different in energy. Whether these observations are due to "molecular" properties and/or to packing forces in the solid state remains an unanswered question at the moment. Interestingly, even simple calculations of the positions of the gegenion relative to the carbon skeleton in delocalized "carbanions" led to different minima of comparable energy [139]. This was similarly the case with more sophisticated calculations [140]. The polymorphism observed in the delocalized "carbanions" mentioned above thus illustrates once again the complex bonding situation in such compounds and the facile relocation at least of the cation.

9 Rearrangements of and within Alkyllithium Aggregates

Although the aggregation of alkyllithium compounds in etheral and other solvents has been known for quite a few years [125b, 141–144], detailed information about the kinds of aggregates and the rearrangements within a certain aggregate or between different aggregates became available only in recent years [145–147]. Because of their importance in mechanistic and synthetic "carbanion" chemistry [143, 148], it therefore seems appropriate to include some of the pertinent results in this review article.

A systematic study of the structures of thirteen lithiated hydrocarbons and of eighteen α-halolithium carbenoids in donor solvent (R_2O, R_3N) mixtures down to −150 °C has been performed by Seebach and coworkers [149, 150, 151]. At room temperature no ^{13}C—^{6}Li coupling was observed in the ^{13}C nmr spectrum because of fast *inter*- and *intra*-aggregate exchange processes. At lower temperatures (< −70 °C) these reactions freeze out thus allowing a correlation of the multiplicities of the signals with the

Table 7. Expected multiplicities in the ^{13}C nmr spectra of dynamic and static aggregates of $^{6}Li(I = 1)$-organometallic compounds; • ≙ ^{6}Li atoms; multiplicity $m = 2n + 1$, n = number of ^{6}Li atoms; d = dynamic; s = static

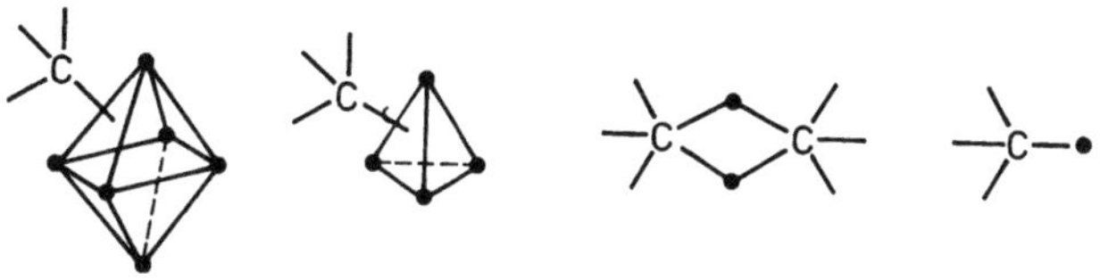

	Hexamer		Tetramer		Dimer		Monomer	
	d	s	d	s	d	s	d	s
equivalent next partners of ^{13}C	6	3	4	3	2	2	1	1
multiplicity m	13	7	9	7	5	5	3	3

Table 8.

Type of R—Li	Δδ(H, Li)	Coupling constants $^{1}J(^{13}C-^{6}Li)$ and multiplicities comp. Table 7	Remarks
1. Derivatives of saturated hydrocarbons	small		hexamer or tetramer in hydrocarbons ref. 150c, 151
a. Alkyl-Li	up to 15 ppm upfield	3–4 Hz (up to 13 lines) 4–6 Hz (9 lines)	dimer or tetramer in R_2O- and R_3N-containing solvents
b. Cyclopropyl-Li	up to 12 ppm up- or downfield	8–11 Hz (quintuplet)	investigated only in R_2O- or R_3N-containing solvents
2. σ-Derivatives of unsaturated hydrocarbons			
a. Vinyl-, phenyl-Li	50–65 ppm downfield	8–12 Hz (quintuplet)	dimer
b. Alkinyl-Li		no coupling observed down to −150 °C in R_2O solvents	probably monomer
3. Compounds with an α-hetero substituent			monomer or heteroatom bridged dimer
a. Halogen-lithium-carbenoids	40–280 ppm downfield	16.5 ± 0.5 Hz (triplet)	
b. α-S- and α-Se-substituted lithium organyls	5–20 ppm	7–11 Hz (triplet)	
4. π-Derivatives of unsaturated hydrocarbons (allyl-, benzyl-type Li-compounds)	up to 40 ppm downfield	no coupling observed in R_2O-containing solvents down to −140 °C	monomer [152]

degree of aggregation in solution: in the ^{13}C nmr spectra of $^{6}Li(I = 1)$ compounds the *triplets* must arise from monomers or heteroatom bridged oligomers, the *quintuplets* from dimers with planar arrangement of two ^{6}Li and two ^{13}C atoms, and so on. Table 7 gives a summary of the most important aggregation states and the multiplicities expected in the ^{13}C nmr spectra. Some important results are given in Table 8.
The following conclusions can be drawn from Table 8:

1. C—Li coupling constants increase with increasing s-character of the C—Li bond (alkyl- < cyclopropyl- < vinyl-Li, phenyl-Li).
2. The averaged C—Li coupling constants increase with decreasing aggregation state of the organolithium species. Thus, in a 'fluctuating' tetramer (four equivalent ^{6}Li atoms) a smaller coupling (4–6 Hz) is observed than in a dimer (8 Hz). For a monomeric species one therefore extrapolates a coupling constant $^{1}J(^{13}C—^{6}Li)$ of ca. 16 Hz.

Some selected organolithium compounds are discussed in more detail below.

9.1 n-Butyllithium

It is well established that n-butyllithium is always aggregated, in hydrocarbons as a hexamer, and in ethers and amines as a tetramer [153]. Seebach, Hässig and Gabriel [149] have shown for the first time that n-butyllithium in THF and dimethylether (DME) also gives a dimer. The signal of the dimer is observed beside the signal of the tetramer at −70 °C in THF; further cooling to −90 °C reveals the quintet-nature of the dimer signal. Thus, at lower temperatures the equilibrium

$$(\text{n-butyllithium} \cdot \text{THF})_4 + 4\,\text{THF} \rightleftharpoons 2\,(\text{n-butyllithium} \cdot 2\,\text{THF})_2$$

is shifted to the right side. The formation of the tetramer (that is the shift to the left-hand side) should be favored by entropy.

This was nicely demonstrated by a high-field ^{1}H nmr study of the aggregation and complexation of n-butyllithium in THF by McGarrity and Ogle [148 a)]. For the above-mentioned equilibrium they determined the following values:

$$K = 0.021 \pm 0.004\ \text{M} \quad \text{at} \quad -85\ °\text{C}$$
$$\Delta H° = -6.3 \pm 0.4\ \text{kJ/mol}$$
$$\Delta S° = -58 \pm 2\ \text{J/mol} \cdot \text{K}$$

The entropy change observed for the dissociation of the tetramer to the dimer is indeed reasonably explained in form of restriction of four extra solvent molecules in the solvated dimer, relative to the tetramer.

McGarrity and Ogle also determined the rate constants k_1 [s^{-1}] for the conversion of the tetramer to the dimer at various temperatures and the activation parameters of this reaction:

T [K]	k_1 [s^{-1}]
245	171
224	23.3
216	8.30
211	3.93

$$\Delta H^{\neq} = 41 \pm 2 \text{ kJ/mol}$$
$$\Delta S^{\neq} = -30 \pm 10 \text{ J/mol} \cdot \text{K}$$

Undoubtedly, this *inter*aggregate rearrangement reaction is very fast even at low temperatures.

Earlier, Brown [154)] determined the activation energy for the dissociation of tetrameric to dimeric methyllithium to be $E_A = 47 \pm 4$ kJ/mol. The two values are in good agreement.

The predominance of different aggregates at different temperatures is also of synthetic utility: both the carbon and the lithium atoms in the *dimer* should be more accessible than in the tetramer. Hence, the dimer should be more reactive.

This assumption was also verified by measurements of McGarrity and Ogle and their coworkers [148 b)]. Introducing a rapid-injection nmr method to study fast reactions like the reaction of n-butyllithium with benzaldehyde they reached the following conclusions if n-butyllithium in toluene-D_8 is mixed with THF and benzaldehyde:

1. The rate of rearrangement of hexameric n-butyllithium in THF into tetrameric and dimeric is twice that of the dissociation of tetrameric into dimeric.
2. Tetrameric n-butyllithium reacts more readily with benzaldehyde than it dissociates into dimeric.
3. Dimeric n-butyllithium is more reactive towards benzaldehyde than tetrameric n-butyllithium by a factor of 10.

These results are an elegant demonstration of the power of the new 'rapid-injection nmr' technique. One can expect an increasing amount of new information especially about rearrangements and reactions of organometallic compounds by this method because reactions of this type are normally "very fast".

Interestingly, solvents like TMEDA also tend to shift the equilibrium of tetrameric and dimeric n-butyllithium to the side of the dimeric species. It is therefore not surprising that many reactions with compounds like n-butyllithium are performed at low temperatures and in the presence of TMEDA.

Alkyllithium compounds normally crystallize in the form of amine-complexed tetramers $(\text{alkyllithium} \cdot NR_3)_4$ [155)].

9.2 Phenyllithium

Wittig [156)] and Waack [157)] showed ebullioscopically and osmometrically, respectively, that phenyllithium is a dimer in etheral solvents. Thönnes and Weiss [158)] found a TMEDA complexed phenyllithium dimer in the solid state, and calculations performed by Schleyer et al. [159)] similarly showed the dimer to be the most stable species. The ^{13}C nmr spectrum of phenyl-^6Li in THF shows a quintuplet at −118 °C which also reveals a dimeric aggregate [149, 160)]. Thus experimental investigations of the structure in *solution* and in the *solid state* as well as a theoretical study (corresponding to the situation in the *gas-phase*) lead remarkably to the same result: a phenyllithium dimer structure seems to be the most stable one.

9.3 tert-Butyllithium and tert-Pentyllithium

In another ^{13}C—^{6}Li nmr study Thomas and coworkers investigated the *intra*aggregate ("fluxonial") exchange of tert-butyllithium tetramers in cyclopentane [161]. The nonet signal (peak separation 4.1 Hz) at 10.7 ppm in the room temperature nmr spectrum is the multiplet expected for a ^{13}C nucleus coupled to four equivalent ^{6}Li nuclei. Upon cooling the nonet peaks broaden, coalesce, and then reform below —10 °C into a seven-line multiplet with $J(^{13}C—^{6}Li) = 5.4$ Hz. The process is completely reversible and is indicative of slowing of *fluxonial exchange* of the tetramer at lower temperatures. The seven-line multiplet represents coupling to the three nearest-neighbor lithium nuclei. The coupling of 4.1 Hz in the fast fluxonial limit is the weighted average of three adjacent coupling interactions of 5.4 Hz and one remote coupling interaction of zero.

Since coupling is retained even in the fast exchange limit, the exchange process observed must be *intra*molecular. Correspondingly, there is no concentration dependence of the exchange rates.

The k-values between —5 °C (85–100 s^{-1}) and —22 °C (3–4 s^{-1}) yielded the following values of the overall activation enthalpy $\Delta H^{\neq}$ and activation entropy $\Delta S^{\neq}$ for the fluxonial exchange rates of tetrameric tert-butyllithium in cyclopentane:

$$\Delta H^{\neq} = 25.0 \pm 0.1 \text{ kcal/mol}$$
$$\Delta S^{\neq} = 44 \pm 1 \text{ e.u.}$$

The same authors [161] also investigated tert-pentyllithium (2-methyl-2-lithio-^{6}Li-butane) in cyclopentane, which is in the fast fluxonial exchange limit even at —80 °C. This corresponds to a $\Delta G^{\neq}$ (188 K) of less than or equal to 7.9 kcal/mol. The corresponding $\Delta G^{\neq}$ (188 K) for tert-butyllithium is 16.7 kcal/mol. The more sterically demanding tert-pentyl group causes a marked increase in the rate of the exchange process even though the aggregation state and the geometry of the aggregate does not change relative to tert-butyllithium.

The concentration independence of the exchange rates, as mentioned earlier, indicates that the exchange is first order in tert-butyllithium tetramers. The rather large positive $\Delta S^{\neq}$-value and the rate acceleration upon going from tert-butyl- to tert-pentyllithium both suggest a transition state which is less sterically crowded than the ground state. This is consistent with several mechanisms. However, the precise nature of the *intra*aggregate rearrangement remains speculative at the moment.

Finally, one should mention that the *inter*aggregate exchange reaction of tert-butyllithium in cyclohexane is nearly six orders of magnitude slower than the *intra*-aggregate rearrangement. This was already noticed by Hartwell and Brown more than twenty years ago [162]. They also noticed that, e.g., in toluene the difference between the rates of *intra*aggregate rearrangement and *inter*aggregate exchange is not as big as in cyclohexane.

10 "Carbanion"-Accelerated Rearrangements

π-Electron donor groups X at the 2-position of allyl vinyl ethers are known to accelerate the Claisen rearrangement markedly [163-166].

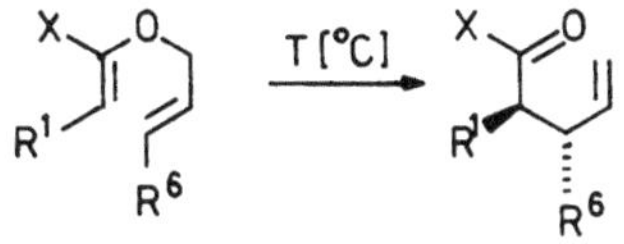

X	T[°]
H, alkyl	160–200
RO—	135
R_2N—	110–140
$(CH_3)_3SiO$—, $Li^+\bar{O}$—	25–67

Based on this observation Denmark and coworkers developed a "carbanion"-accelerated Claisen rearrangement with X = $\bar{C}HSO_2Aryl\ K^+$ [164]. While the $C_6H_5SO_2CH_2$-substituted vinyl-allyl-ether *182* does not rearrange to give *185* when kept at 50 °C in HMPA [167] for 3.5 h, rearrangement occurs under the same conditions in the presence of potassium hydride. Undoubtedly, with potassium hydride the

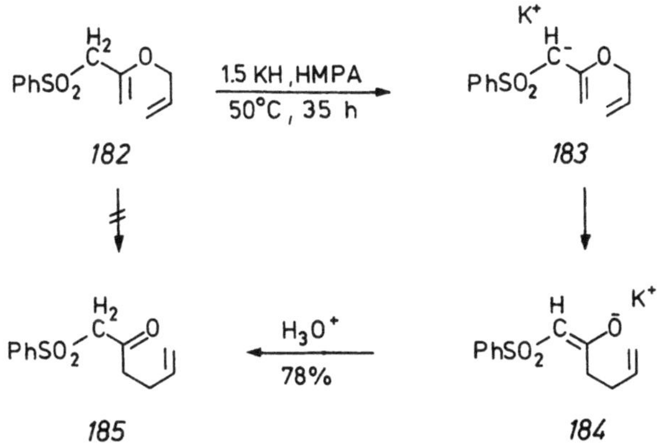

α-sulfonyl "carbanion" *183* is formed which rearranges to give regioselectively the β-keto sulfone "anion" *184*; protonation of *184* leads to *185* in 78% yield. The regioselective alternative — involvement of the anionic α-C atom of *183* in the Claisen rearrangement — is not observed. The product of this rearrangement would be a less favorable ketone enolate instead of the β-keto sulfone "anion" *184*.

In agreement with the mechanism outlined above, reaction of *186* — an isomer of *182* — with KH also leads to *185* in 78% yield. Alkyl substitution at the double bonds of *182* does not affect the general pattern of this rearrangement. A different reaction, however, is observed in the case of the phenyl substituted *187*. In this case the inter-

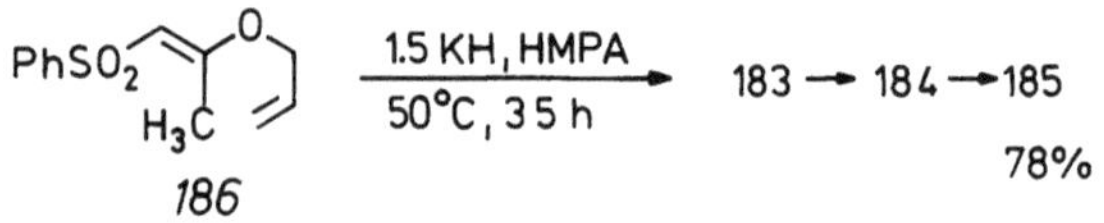

186

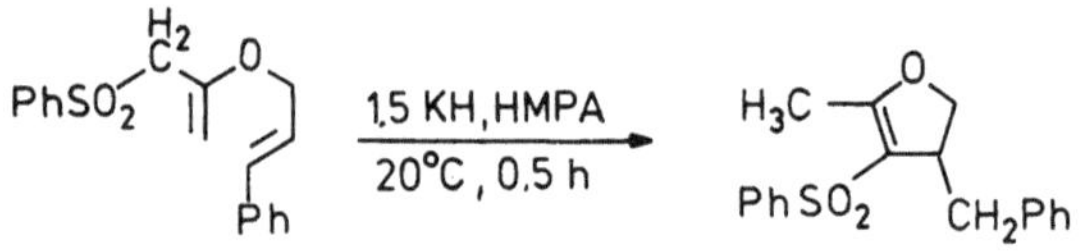

187 188

mediately formed α-sulfonyl "carbanion" adds to the styryl double bond to give *188* (on protonation).

The high stereoselectivity of the rearrangement was shown by means of the compounds *189* and *190* [164b]. Reaction of *189*, e.g., with 2.2 KH and 4.4 LiCl in DMSO

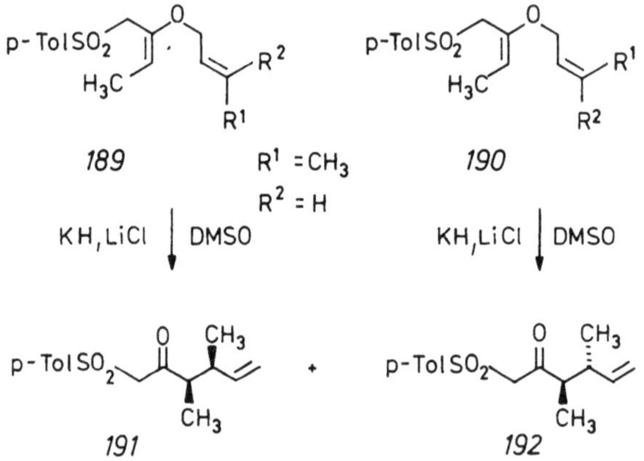

at 20 °C led after 4 h to *191* and *192* in a 97:3 ratio (73% yield). Similarly, reaction of *190* with 2.6 KH and 15 LiCl in DMSO at 50 °C led after 1.5 h to *191* and *192* in a 6:94 ratio (85% yield).

The "carbanion"-accelerated Claisen rearrangement is also a viable reaction for the creation of vicinal quaternary centers, as shown, e.g., by the rearrangement of *193* to *194* [164c]. A similar "carbanion"-accelerated Claisen rearrangement of *197* could

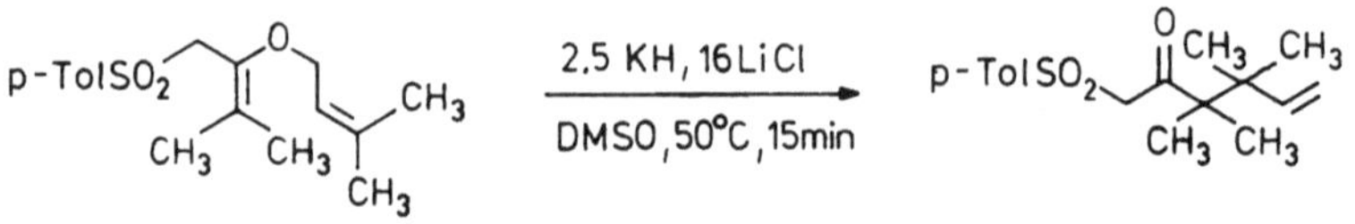

193 194, 98%

have occurred when the allene *195* was reacted with the allyl alcoholate *196* [168]. The formation of *both* structural isomers *200* and *201*, however, indicates that it was not the first formed "carbanion" *197* which underwent the rearrangement as was the

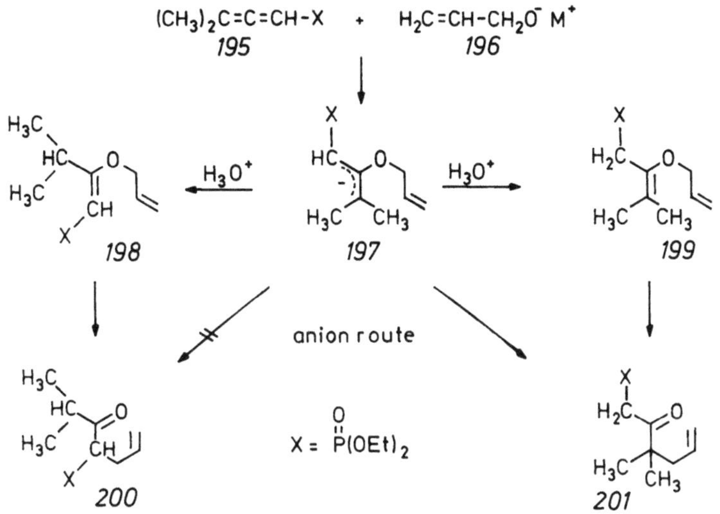

case with the $C_6H_5SO_2$-substituted *183*. Rather, protonation of the ambident *197* led to *198* and *199* which reacted via normal Claisen rearrangement to give *200* and *201*, respectively [169].

A "carbanion"-accelerated hetero-Cope rearrangement was reported by Blechert [170]. Reaction of the sodium salt of the hydroxamic acid (*202*) with the allene sulfone *203* led at 0 °C within 10 min to *206*. Undoubtedly, the "carbanions" *204* and *205* are the important intermediates in this transformation. A "carbanion"-accelerated

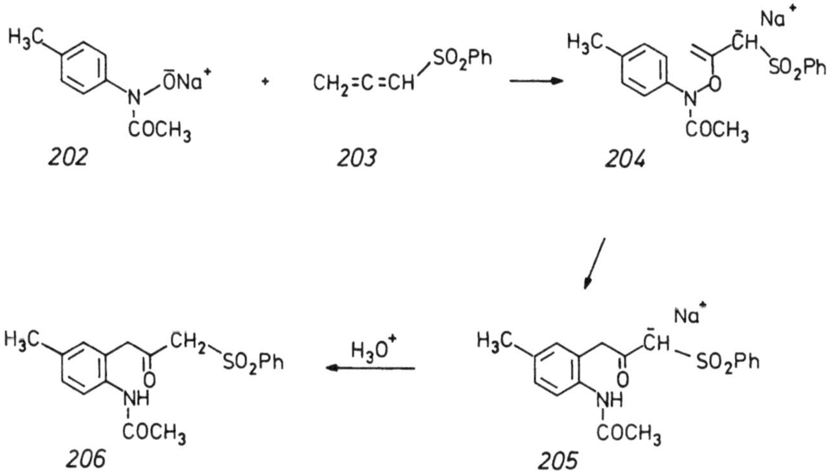

vinylcyclopropane → cyclopentene rearrangement was reported by Danheiser et al. [171]. The strategy is illustrated by means of the vinylcyclopropylsulfone *207* which, on reaction with n-butyllithium between −78 and −30 °C, smoothly rearranged to the cyclopentene *210*, undoubtedly via the "carbanions" *208* and *209*. The facility of this "carbanion"-accelerated process at low temperatures is in dramatic contrast

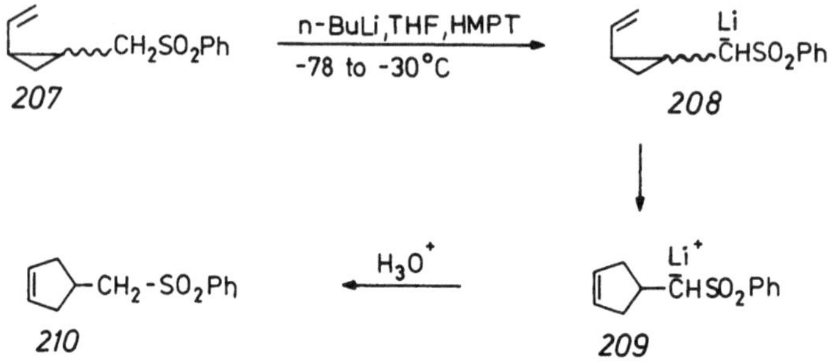

to the 258–600 °C normally required to effect the vinylcyclopropane rearrangement [172].

Two other examples of "carbanion"-accelerated rearrangements have been described in the literature. The first one leads from the "anion" of the ^{6}O-allylic guanine *211* in two consecutive [3.3] sigmatropic shifts (a combined Claisen-Cope rearrangement) via *212* and *213* to the 8-allylic guanine *214* [173].

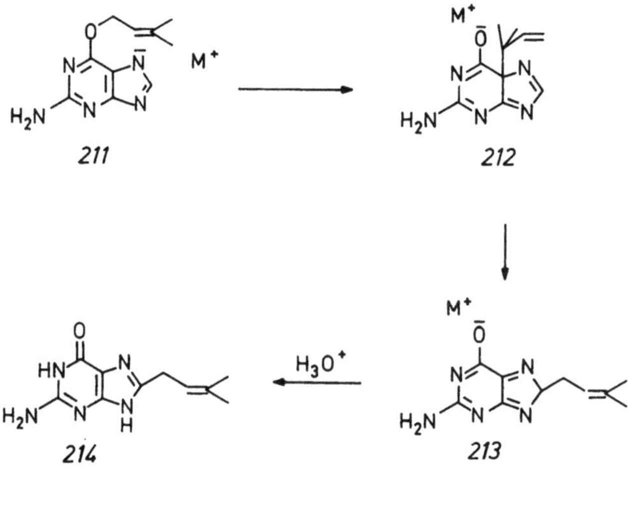

The second example refers to α-sulfonyl and α-sulfinyl "carbanion" substituted cyclobutenes and their facile ring-opening reaction [174]. Treatment, e.g., of the cyclobutene *sulfone 215* with n-butyllithium in THF/hexane between —78 and —30 °C led after 10 min (protonation with an aqueous solution of ammonium chloride) to the diene sulfone *218* in 90% yield. If the reaction mixture is treated at —78 °C with methyl iodide, the "carbanion" *216* is trapped before it rearranges to the "carbanion" *217*. If the methylation is performed at —30 °C, *217* alone reacts.

In the case of the cyclobutene *sulfoxide 219*, deprotonation with lithiumdiisopropyl amide (LDA) followed by protonation with $NH_4^+Cl^-/H_2O$, however, did not lead to the diene sulfoxide *220*. Instead, the sulfoxide *222* was isolated in 75% yield. The formation of *222* is easily explained by a double [2.3]sigmatropic rearrangement (*220* → *221* → *222*) of the first formed *220*.

Normally, cyclobutene → butadiene transformations occur at higher temperatures

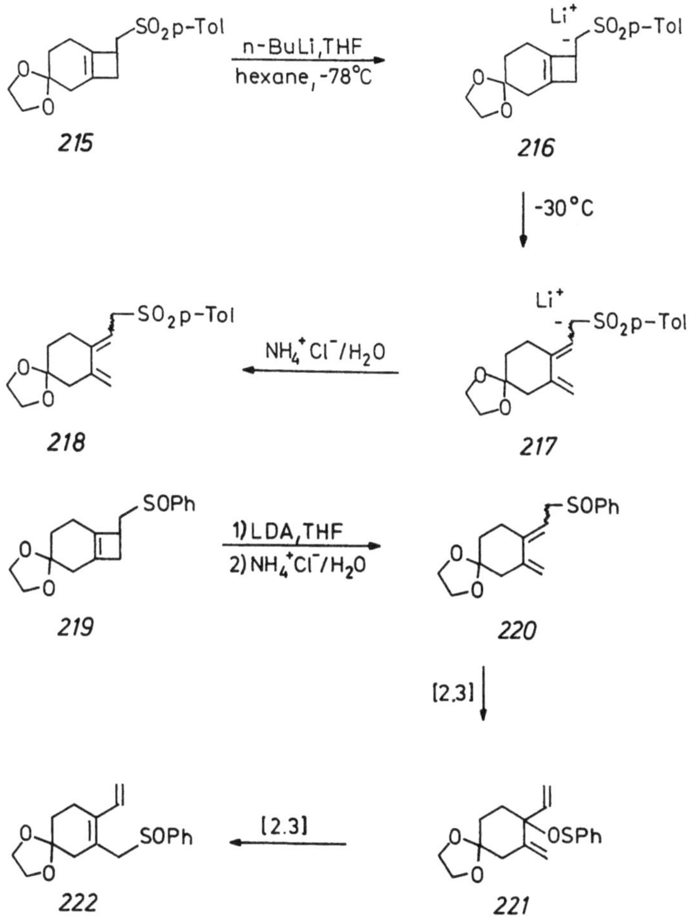

[9] than reported here for the "carbanion"-accelerated variation of this reaction. The importance of the cyclobutene → butadiene rearrangement for synthetic purposes is amply documented [175].

11 Conclusions and Outlook

Our understanding of reactivity and selectivity of "carbanions" is still at an early stage despite the widespread use and enormous success of these compounds in synthetic chemistry in recent years. Little is really known about the influence of the gegenion and the solvent. Rearrangements of "carbanions" are essentially no exception as shown in this review. Undoubtedly, we need to know much more about gegenion and solvent effects in "carbanion" chemistry. Similarly, it is necessary to study the inter- and intra-"molecular" dynamics of aggregates in more detail. We have included some of the first results along these lines in this article together with the more "classical" skeletal rearrangements of "carbanions".

The reader will recognize the heterogeneity of the subject — "carbanion" rearrangements are not "limited by rules" especially if one includes "carbanion" radicals. This, in turn, is responsible for the richness of the chemistry associated with the topic. Because of the intensive world-wide use and investigation of "carbanions" and their rearrangements, it is foreseeable that this field will remain an active one in the future.

12 References

1. Zimmerman HE (1963) Base-Catalyzed Rearrangements, in: Molecular Rearrangements de Mayo P (ed) Vol. 1, p. 345, New York, Wiley
2. Cram DJ (1965) Fundamentals of Carbanion Chemistry, New York, Academic Press
3. Buncel E (1975) Carbanions: Mechanistic and Isotopic Aspects, Amsterdam, Elsevier
4. Hunter DH Isotopes in Carbanion Rearrangements, in: Isotopes in Organic Chemistry Buncel E, Lee CC (ed) (1975) Vol. 1, p. 135, Amsterdam, Elsevier
5. Staley SW Pericyclic Reactions of Carbanions, in: Pericyclic Reactions Marchand AP, Lehr RE (ed) (1977) Vol. 1, p. 199, New York, San Francisco, London, Academic Press
6. a. Grovenstein Jr E (1977) Adv. Organomet. Chem. *16*: 167;
   b. Grovenstein Jr E (1978) Angew. Chem. *90*: 317; (1978) Angew. Chem. Int. Ed. Engl. *17*: 313
7. a. Hill EA (1975) J. Organomet. Chem. *91*: 123;
   b. Hill EA (1977) Adv. Organomet. Chem. *16*: 131
8. Hunter DH, Stothers JB, Warnhoff EW Rearrangements in Carbanions, in: Rearrangements in Ground and Excited States de Mayo P (ed) 1980 Vol. 1, p. 391, New York, London, Toronto, Sydney, San Francisco, Academic Press
9. Woodward RB, Hoffmann R (1965) J. Am. Chem. Soc. *87*: 395;
   b. Woodward RB, Hoffmann R (1969) Angew. Chem. *81*: 797; (1969) Angew. Chem. Int. Ed. Engl. *8*: 781
10. see also: a. ref. 5.;
    b. Boche G, Walborsky HM Cyclopropyl Radicals, Anion Radicals and Anions, in: The Chemistry of the Cyclopropyl Group Rappoport Z (ed) 1987 New York, John Wiley and Sons. This article contains a comprehensive review of the cyclopropyl-allyl "anion" rearrangement including the historical aspects and the photochemical reactions
11. Huisgen R, Scheer W, Huber W: (1967) J. Am. Chem. Soc. *89*: 1753
12. Kauffmann T, Habersaat K, Köppelmann E (1972) Angew. Chem. *84*: 262; (1972) Angew. Chem. Int. Ed. Engl. *11*: 291; see also Kauffmann T, Köppelmann E (1972) Angew. Chem. *84*: 261; (1972) Angew. Chem. Int. Ed. Engl. *11*: 290; Kauffmann T (1974) Angew. Chem. *86*: 715; (1974) Angew. Chem. Int. Ed. Engl. *13*: 627
13. Boche G, Martens D, Schneider DR, Buckl K, Wagner H-U (1979) Chem. Ber. *112*: 2961. As far as the calculations are concerned see also ref. 30, 33, 34 and the literature reviews given in ref. 30 and 34
14. Seyferth D, Cohen HM (1963/64) J. Organomet. Chem. *1*: 15
15. a. Walborsky HM (1952) J. Am. Chem. Soc. *74*: 4962;
    b. Walborsky HM, Hornyak FM (1955) ibid. *77*: 6026
    c. Walborsky HM, Hornyak FM (1956) ibid. *78*: 872;
    d. Walborsky HM, Youssef AA, Motes JM (1962) ibid. *84*: 2465;
    e. Walborsky HM, Motes JM (1970) ibid. *92*: 2445;
    f. Motes JM, Walborsky HM (1970) ibid. *92*: 3697;
    g. Levin J-O, Rappe C (1971) Chem. Scripta *1*: 233
16. Mulvaney JE, Savage D (1971) J. Org. Chem. *36*: 2592
17. Boche G, Martens D (1972) Angew. Chem. *84*: 768; (1972) Angew. Chem. Int. Ed. Engl. *11*: 724
18. Boche G, Buckl K, Martens D, Schneider DR Liebigs Ann. Chem. *1980*: 1135
19. Wittig G, Rautenstrauch V, Wingler F (1965) Tetrahedron Suppl. *1*: 189
20. a. Ford WT, Newcomb M (1973) J. Am. Chem. Soc. *95*: 6277;
    b. Newcomb M, Ford WT (1973) ibid. *95*: 7186;
    c. Newcomb M, Ford WT (1974) ibid. *96*: 2968

21. Coates RM, Last LA (1983) ibid. *105*: 7322
22. a. Fox MA (1979) ibid. *101*: 4008;
b. Fox MA (1979) Chem. Rev. *79*: 253;
c. Fox MA, Chen C-C, Campbell KA (1983) J. Org. Chem. *48*: 321
23. Thompson TB, Ford WT (1979) J. Am. Chem. Soc. *101*: 5459
24. Brownstein S, Bywater S, Worsfold DJ (1980) J. Organomet. Chem. *199*: 1
25. West P, Purmont JI, Mc Kinley SV (1968) J. Am. Chem. Soc. *90*: 797
26. Smid J (1972) Angew. Chem. *84*: 127; (1972) Angew. Chem. Int. Ed. Engl. *11*: 112
27. a. West P, Waack R (1967) J. Am. Chem. Soc. *89*: 4395;
b. West P, Waack R, Purmort JI (1970) ibid. *92*: 840
28. Brubaker GR, Beak P (1977) J. Organomet. Chem. *136*: 147
29. Winchester WR, Bauer W, Schleyer PvR J. Chem. Soc., Chem. Commun. *1987*: 177
30. Clark T, Rhode C, Schleyer PvR (1983) Organometallics *2*: 1344 [31)]
31. Calculations of organometallic compounds are usually in good agreement with experimental results, see Schleyer PvR (1984) Pure Appl. Chem. *56*: 151; (1983) *55*: 355
32. Schlosser M, Stähle M (1980) Angew. Chem. *92*: 497; (1980) Angew. Chem. Int. Ed. Engl. *19*: 487
33. Chandrasekhar J, Andrade JG, Schleyer PvR (1981) J. Am. Chem. Soc. *103*: 5609. For further calculations of the allyl anion and allylalkalimetal compounds, and a literature review on this subject, see ref. 30 and 34
34. Decher G, Boche G (1983) J. Organomet. Chem. *259*: 31
35. a. Charton M (1976) J. Org. Chem. *41*: 2217;
b. Charton M (1975) J. Am. Chem. Soc. *97*: 1552
36. a. Burley JW, Ife R, Young RN Chem. Comm. *1970*: 1256;
b. Burley JW, Young RN J. Chem. Soc. Perkin Trans. 2, *1972*: 835;
c. Burley JW, Young RN ibid. 2, *1972*: 1006;
d. Greenacre GC, Young RN ibid. 2, *1975*: 1661
37. This subject has been treated in more detail in ref. 10b
38. Bowers KW, Greene FD (1963) J. Am. Chem. Soc. *85*: 2331
39. Bowers KW, Nolfi Jr, GJ, Lowry TH, Greene FD (1966) Tetrahedron Lett. 4063
40. Gerson F, Heilbronner E, Heinzer J (1966) ibid. 2095
41. Van Volkenburgh R, Greenlee KW, Derfer JM, Boord CE (1949) J. Am. Chem. Soc. *71*: 3595
42. Norin T (1965) Acta Chem. Scand. *19*: 1289
43. Dauben WG, Wolf RE (1970) J. Org. Chem. *35*: 374. This result has been confirmed by other groups Refs. 44, 45
44. Fraisse-Jullien R, Frejaville C Bull. Soc. Chim. Fr. *1968*: 4449
45. House HO, Blankley CJ (1968) J. Org. Chem. *33*: 47
46. Walborsky HM, Pierce JB (1968) ibid. *33*: 4102
47. a. Birch JA (1950) Quart. Rev. (London) *4*: 69;
b. Krapcho AP, Bothner-By AA (1959) J. Am. Chem. Soc. *81*: 3658
48. It is thus quite understandable that the parent cyclopropyl radical "anion" *29* also has not been detected by ESR Refs. 39, 40
49. Walborsky HM, Aronoff MS, Schulman MF (1971) J. Org. Chem. *36*: 1036
50. Staley SW, Rocchio JJ (1969) J. Am. Chem. Soc. *91*: 1565
51. Boche G, Wintermayr H (1981) Angew. Chem. *93*: 923; (1981) Angew. Chem. Int. Ed. Engl. *20*: 874
52. Miller LL, Jacoby LJ (1969) J. Am. Chem. Soc. *91*: 1130
53. Russell GA, Ku T, Lokensgard J (1970) ibid. *92*: 3833
54. a. Rieke R, Ogliaruso M, McClung R, Winstein S (1966) ibid. *88*: 4729;
b. Katz TJ, Talcott CC (1966) ibid. *88*: 4732
c. Smentowski FJ, Owens RM, Faubion BD (1968) ibid. *90*: 1537;
d. Winstein S, Moshuk G, Rieke R, Ogliaruso M (1973) ibid. *95*: 2624;
e. Ley SV, Paquette LA (1974) ibid. *96*: 6670;
f. Okamura WH, Ito TI, Kellet PM J. Chem. Soc. Chem. Commun. *1971*: 1317;
g. Ito TI, Baldwin FC, Okamura WH ibid. *1971*: 1440
55. Moshuk G, Petrowski G, Winstein S (1968) J. Am. Chem. Soc. *90*: 2179
56. a. Goldstein MJ, Tomoda S, Whittacker G (1974) ibid. *96*: 3676;

b. Goldstein MJ, Wenzel TT, Whittacker G, Yates SF (1982) ibid. *104*: 2669;
c. Goldstein MJ, Wenzel TT J. Chem. Soc. Chem. Commun. *1984*: 1654;
d. Goldstein MJ, Wenzel TT ibid. *1984*: 1655

57. a. Schnieders C, Altenbach H-J, Müllen K (1982) Angew. Chem. *94*: 638; (1982) Angew. Chem. Int. Ed. Engl. *21*: 637;
b. Kohnz H, Schnieders C, Trinks R, Müllen K private communication to Boche G, August 1985
58. Wilhelm D, Clark T, Schleyer PvR, Davies AG J. Chem. Soc. Chem. Commun. *1984*: 558
59. Boche G, Schneider DR, Wernicke K (1984) Tetrahedron Lett. *25*: 2961; see also Schneider DR 1977 Dissertation, Universität München; Wernicke K 1982 Diplomarbeit, Universität Marburg; Wernicke K 1987 Dissertation, Universität Marburg; and Ref. 51
60. Rodewald LB, DePuy CH (1964) Tetrahedron Lett. 2951; see also Crawford RJ, Lynch TR (1968) Can. J. Chem. *46*: 1457
61. Leonova TV, Shapiro IO, Ranneva Yu I, Shatenshtein AI, Shabarov Yu S (1977) Zh. Org. Khim. *13*: 538; (1977) J. Org. Chem. USSR *13*: 491
62. Closs GL, Moss RA (1964) J. Am. Chem. Soc. *86*: 4042
63. Lagendijk A, Szwarc M (1971) ibid. *93*: 5359
64. Grovenstein Jr. E, Bhatti AM, Quest DE, Sengupta D, VanDerveer D (1983) ibid *105*: 6290. In this publication a thorough literature review of C—C bond cleavages by means of electrons is given
65. This reaction has also been observed by Hoell D, Schnieders C, Müllen K (1983) Angew. Chem. *95*: 240; (1983) Angew. Chem. Int. Ed. Engl. *22*: 243
66. Newcomb M, Seidel T, McPherson MB (1979) J. Am. Chem. Soc. *101*: 777
67. Dodd JR, Pagni RM, Watson Jr. CR (1981) J. Org. Chem. *46*: 1688
68. Wilt JW in: Free Radicals Kochi JK (ed) 1973 Vol. 1, p. 378, New York, Wiley
69. March J 1978 Advanced Organic Chemistry: Reactions, Mechanisms and Structure, p. 793, New York, McGraw Hill
70. Buncel E, Menon BC (1980) J. Am. Chem. Soc. *102*: 3499; see also ref. 71 and 72
71. Guthrie RD, Nutter DE (1982) ibid. *104*: 7478
72. a. Newcomb M, Burchill MT (1984) ibid. *106*: 2450;
b. Newcomb M, Burchill MT (1984) ibid. *106*: 8276
73. Londrigan ME, Mulvaney JE (1972) J. Org. Chem. *37*: 2823; see also Mulvaney JE, Londrigan ME, Savage DJ (1981) ibid. *46*: 4592
74. Boche G, Marsch M (1983) Tetrahedron Lett. *24*: 3225
75. Brown CA (1975) J. Chem. Soc. Chem. Commun. 222
76. Huisgen R, Seidl H (1964) Tetrahedron Lett. 3381
77. a. Bauld NL, Farr F (1969) J. Am. Chem. Soc. *91*: 2788;
b. Bauld NL, Farr F, Stevenson GR Tetrahedron Lett. *1970*: 625;
c. Bauld NL, Chang C-S, Farr FR (1972) J. Am. Chem. Soc. *94*: 7164;
d. Bauld NL, Farr FR, Hudson CE (1974) ibid. *96*: 5634;
e. Bauld NL, Cessac J Tetrahedron Lett. *1975*: 3677;
f. Bauld NL, Cessac J (1975) J. Am. Chem. Soc. *97*: 2284;
g. Bauld NL, Cessac J, Chang C-S, Farr FR, Holloway R (1976) ibid. *98*: 4561
78. Longuet-Higgins HC, Abrahamson EW (1965) ibid. *87*: 2045
79. a. Boche G, Szeimies G (1971) Angew. Chem. *83*: 978;
b. Szeimies G, Boche G (1971) ibid. *83*: 979;
c. Sustmann S, Rüchardt C, Bieberbach A, Boche G Tetrahedron Lett. *1972*: 4759
80. Dewar MJS, Kirschner S (1971) J. Am. Chem. Soc. *93*: 4290
81. Merlet P, Peyerimhoff S. D, Buenker RJ, Shih S (1974) ibid. *96*: 959
82. Etzrodt H 1981 Diplomarbeit, Universität Marburg;
Etzrodt H 1984 Dissertation, Universität Marburg
83. Boche G, Etzrodt H, Marsch M, Thiel W (1982) Angew. Chem. *94*: 141; Angew. Chem. Suppl. *1982*: 345; (1982) Angew. Chem. Int. Ed. Engl. *21*: 132
84. Decher G 1983 Diplomarbeit, Universität Marburg
85. Boche G, Etzrodt H Tetrahedron Lett. *1983*: 5477
86. Boche G, Decher G, Etzrodt H, Dietrich H, Mahdi W, Koss AJ, Schleyer PvR J. Chem. Soc. Chem. Commun. *1984*: 1493

87. Maercker A, Berkulin W, Schiess P (1983) Angew. Chem. *95*: 248; (1983) Angew. Chem. Int. Ed. Engl. *22*: 246
88. a. Kiesele H in: Abstracts of the Chemiedozententagung, Dortmund, March 14–18, 1983, p. 18, Weinheim, Verlag Chemie 1983;
b. Macheroux P 1982 Diplomarbeit, Universität Konstanz; see also ref. 77g
89. Beckwith ALJ, Ingold KU Free-Radical Rearrangements in: Rearrangements in Ground and Excited States de Mayo P (ed) 1980 Vol. 1, p. 185, New York, Academic Press
90. Schmid P, Griller D, Ingold KU (1979) Int. J. Chem. Kin. *11*: 333
91. a. Brace NO (1967) J. Org. Chem. *32*: 2711;
b. Walling C, Cioffari A (1972) J. Am. Chem. Soc. *94*: 6059;
c. Beckwith ALJ, Blair I, Phillipou G (1974) ibid. *96*: 1613
92. Beckwith ALJ (1981) Tetrahedron *37*: 3073
93. Some recent examples: a. ref. 89; b. Griller D, Ingold KU (1980) Acc. Chem. Res. *13*: 317;
b. Suzur J-M in Reactive Intermediates Abramovitch RA (ed) 1982 Vol. 2, chapter 3, p. 253, New York, Plenum Press
94. Cyclization of 1-methyl-5-hexenyl metallics: a. Richey HG Jr, Rees TC Tetrahedron Lett. *1966*: 4297;
b. Kossa WC, Rees TC, Richey HG Jr ibid. *1971*: 3455;
c. Richey HG Jr, Veale HS ibid. 1975: 615;
d. Drozd VN, Ustynyuk YuA, Tsel'eva MA, Dmitriev LB (1968) Zh. Obshch. Khim. *38*: 2114; (1968) J. Gen. Chem. USSR (Engl. Transl.) *38*: 2047;
e. Drozd VN, Ustynyuk YuA, Tsel'eva MA, Dmitriev LB (1969) Zh. Obshch. Khim *49*: 1991; (1969) J. Gen. Chem. USSR (Engl. Transl.) *39*: 1951;
f. Stefani A (1974) Helv. Chim. Acta *57*: 1346;
g. St. Denis J, Dolzine T, Oliver JPJ (1972) J. Am. Chem. Soc. *94*: 8260;
h. Whitesides GM, Bergbreiter DE, Kendall PE (1974) ibid. *96*: 2806;
i. Bahl JJ, Bates RB, Beavers WA, Mills NS (1976) J. Org. Chem. *41*: 1620
95. a. Garst JF, Ayers PW, Lamb RC (1966) J. Am. Chem. Soc. *88*: 4260;
b. Garst JF, Barbas JT, Barton FE II (1968) ibid. *90*: 7159;
c. Garst JF, Cox RH, Barbas JT, Roberts RD, Morris JI, Morrison RC (1970) ibid. *92*: 5761;
d. Garst JF (1971) Acc. Chem. Res. *4*: 400;
e. Garst JF, Barton FE II, Morris JJ (1971) J. Am. Chem. Soc. *93*: 4310;
f. Garst JF in: Free Radicals Kochi JK (ed) 1973 Vol. I, chapter 9, pp. 520–529, New York, Wiley;
g. Garst JF, Barbas JT (1974) J. Am. Chem. Soc. *96*: 3239;
h. see also: Felkin H, Meunier B (1977) Noveau J. de Chimie *1*: 281
96. a. Garst JF, Hines Jr. JB (1984) J. Am. Chem. Soc. *106*: 6443;
b. see also ref. 97
97. Garst JF, Pacifici JA, Felix CC, Nigam A (1978) ibid. *100*: 5974
98. a. Ashby EC, DePriest R (1982) ibid. *104*: 6144;
b. Ashby EC, DePriest RN, Tuncay A, Srivastava S (1982) Tetrahedron Lett. *23*: 5251
99. Lee K-W, San Filippo J Jr (1983) Organometallics *2*: 906
100. Garst JF, Hines Jr. JB, Bruhnke JD (1986) Tetrahedron Lett. *27*: 1963
101. a. Smith GF, Kuivila HG, Simon R, Sultan L (1981) J. Am. Chem. Soc. *103*: 833;
b. Patel DJ, Hamilton CL, Roberts JD (1965) ibid. *87*: 5144
102. Bailey WF, Patricia JJ, Del Gobbo VC, Jarret RM, Okarma PJ (1985) J. Org. Chem. *50*: 1999
103. The cyclization of the Grignard species is much slower, see ref. 94a, b, and Hill EA (1975) J. Organomet. Chem. *91*: 123
104. Bailey WF, Patricia JJ, Nurmi TT (1986) Tetrahedron Lett. *27*: 1865
105. a. Newcomb M, Williams WG, Crumpacker EL (1985) ibid. *26*: 1183;
b. Ashby EC, Pham TN, Park B (1985) ibid. *26*: 4691
106. see however: Bailey WF, Patricia JJ, Nurmi TT, Wang W (1986) ibid. *27*: 1861
107. Bailey WF, Nurmi TT, Patricia JJ (1985) Abstracts of the 190th ACS National Meeting, Chicago, Sept. 8–13, 1985, Division of Organic Chemistry, Abstr. Nr. 121
108. Ross GA, Koppang MD, Bartak DE, Woolsey NF (1985) J. Am. Chem. Soc. *107*: 6742; see also ref. 109
109. Koppang MD, Ross GA, Woolsey NF, Bartak DE (1986) ibid. *108*: 1441

110. Meijs GF, Bunnett JF, Beckwith ALJ (1986) ibid. *108*: 4899
111. a. Newcomb M, Burchill MT (1983) ibid. *105*: 7759;
b. see also ref. 72
After finishing this manuscript the following papers dealing with the cyclization of 5-hexenyl "anions" and radicals appeared.
c. Ashby EC, Pham TN (1987) J. Org. Chem. *52*: 1291 investigated single electron transfer mechanisms in connection with metal-halogen exchange reactions of alkyllithium reagents and alkyl halides;
d. Bailey WF, Nurmi TT, Patricia JJ, Wang W (1987) J. Am. Chem. Soc. *109*: 2442 published on the preparation and regiospecific cyclization of alkenyllithiums;
e. aminyl and aminium radicals and their cyclizations for the synthesis of the pyrrolidine nucleus have been studied by Newcomb M, Deeb TM (1987) J. Am. Chem. Soc. *109*: 3163;
f. Newcomb M, Sanchez RM, Kaplan J (1987) ibid. *109*: 1195 found fast halogen abstractions from alkyl halides by alkyl radicals. This led to a quantitization of the processes occurring in, and a caveat for, studies employing alkyl halide mechanistic probes. This latter publication is recommended to all those who are interested in using the 5-hexenyl system as a mechanistic probe of whether or not an electron transfer process occurs in a reaction of a certain nucleophile with an alkyl halide.
112. a. Curtin DY, Crump JW (1958) J. Am. Chem. Soc. *80*: 1922;
b. Curtin DY, Koehl Jr WJ (1962) ibid. *84*: 1967
113. a. Knorr R, Lattke E Tetrahedron Lett. *1977*: 3969;
b. Lattke E, Knorr R (1981) Chem. Ber. *114*: 1600
114. Knorr R, Roman von T (1984) Angew. Chem. *96*: 349; (1984) Angew. Chem. Int. Ed. Engl. *23*: 366
115. A literature review on vinyllithium compounds and their use in synthetic applications is given in ref. 114
116. a. Schmidt RR, Betz R Synthesis *1982*: 748;
b. Hartmann J, Stähle M, Schlosser M ibid. *1974*: 888;
c. Everhardus RH, Gräfing R, Brandsma L (1978) Recl. Trav. Chim. Pays-Bas *97*: 69
117. ^{13}C-nmr of *167b*: Oakes FT, Sebastian JF (1980) J. Org. Chem. *45*: 4959
118. ^{1}H-nmr of *167c*: Soderquist JA, Hsu G J-H (1982) Organometallics *1*: 830
119. Szwarc M 1974 Ions and Ion Pairs in Organic Reactions, Vol. I, J. Wiley, New York; 1974 Vol. II
120. a. Hogen-Esch TE, Smid J (1966) J. Am. Chem. Soc. *88*: 307;
b. Hogen-Esch TE, Smid J (1966) ibid. *88*: 318
121. a. Boche G, Heidenhain F (1978) Angew. Chem. *90*: 290; (1978) Angew. Chem., Int. Ed. Engl. *17*: 283
b. Boche G, Heidenhain F (1979) J. Am. Chem. Soc. *101*: 738;
c. Boche G, Heidenhain F, Thiel W, Eiben R (1982) Chem. Ber. *115*: 3167
122. Heiszwolf GJ, Kloosterziel H (1967) Rec. Trav. Chim. Pays-Bas *86*: 807
123. House HO, Prabhu AV, Phillips WV (1976) J. Org. Chem. *41*: 1209
124. In "normal" enolates the gegenion M^+ and the solvent S do not influence strongly the conformation of the anionic part of the ion pair. In particular, the rotation around the enolate CC double bond is high (not measurable) under any ion pair conditions [123,125,126]
125. a. Jackman LM, Haddon RC (1973) J. Am. Chem. Soc. *95*: 3687;
b. Jackman LM, Lange BC (1977) Tetrahedron *33*: 2737;
c. Jackman LM, Lange BC (1981) J. Am. Chem. Soc. *103*: 4494
126. House, HO, Trost BM (1965) J. Org. Chem. *30*: 2502
127. It should be emphasized that *171*$^-M^+$ and *171*$^-\{M^+$ are *not* resonance structures!
128. Thorough nmr studies on a wide variety of (nona)fulvenes which are in agreement with the results of *171*$^-Li^+$ have been performed by Neuenschwander M et al. Summary: Neuenschwander M (1986) Pure and Appl. Chem. *58*: 55
129. a. Katz TJ, Garratt PJ (1963) J. Am. Chem. Soc. *85*: 2852; (1964) ibid. *86*: 5194;
b. LaLancette EA, Benson RE (1963) ibid. *85*: 2853; (1965) ibid. *87*: 1941;
c. Boche G, Martens D, Danzer W (1969) Angew. Chem. *81*: 1003; (1969) Angew. Chem. Int. Ed. Engl. *8*: 984;
d. Boche G, Weber H, Martens D, Bieberbach A (1978) Chem. Ber. *111*: 2480;

e. Boche G, Bieberbach A (1978) ibid. *111*: 2850
130. Equilibriations between ion pairs are generally fast [119]
131. Grutzner JB, Lawlor JM, Jackman LM (1972) J. Am. Chem. Soc. *94*: 2306
132. Some recent reviews concerning this topic: a. Huber W, Müllen K (1986) Acc. Chem. Res. *19*: 300;
b. Gerson F, Huber W (1987) ibid. *20*: 85;
c. Müllen K (1987) Angew. Chem. *99*: 192; (1987) Angew. Chem. Int. Ed. Engl. *26*: 204
133. Staley SW, Dustman CK, Facchine KL, Linkowski GE (1985) J. Am. Chem. Soc. *107*: 4003
134. Müllen K, Oth JFM, Engels H-W, Vogel E (1979) Angew. Chem. *91*: 251; (1979) Angew. Chem. Int. Ed. Engl. *18*: 229
135. A summary of the structures of Li^+-compounds: Setzer WN, Schleyer PvR (1985) Adv. Organomet. Chem. *24*: 353; see also: Seebach D, Crystal Structures and Stereoselective Reactions of Organic Lithium Derivatives, in: Proceedings of The Robert A. Welch Foundation Conferences on Chemical Research XXVII, 1984, Houston, Texas; Boche G Altes und Neues zur Struktur der "α"-Lithioverbindungen von Sulfonen, Sulfoximiden, Sulfoxiden, Nitrilen und Nitroverbindungen submitted to VCH Verlagsg., Weinheim
136. Engelhardt LM, Papasergio RI, Raston CL, White AH J. Chem. Soc. Dalton Trans. *1984*: 311
137. Patterman SP, Karle IL, Stucky GD (1970) J. Am. Chem. Chem. Soc. *92*: 1150
138. Boche G, Etzrodt H, Massa W, Baum G (1985) Angew. Chem. *97*: 858; (1985) Angew. Chem. Int. Ed. Engl. *24*: 863
139. Bushby RJ, Tytko MP (1984) J. Organomet. Chem. *270*: 265
140. see, e.g., a ref. 30 and 31
b. Schleyer PvR, Kos AJ, Wilhelm D, Clark T, Boche G, Decher G, Etzrodt H, Dietrich H, Mahdi W (1984) Chem. Commun. *22*: 1495
141. Bates RB, Ogle CA 1984 Carbanion Chemistry, Berlin, Springer Verlag
142. Wardell JL in: Comprehensive Organometallic Chemistry Wilkinson G, Stone FGA, Abel EW (eds) 1982 Vol. 1, p. 43, Oxford, Pergamon Press
143. Seebach D, Amstutz R, Dunitz JD (1981) Helv. Chim. Acta *64*: 2622
144. a. Bartlett PD, Tauber SJ, Weber WP (1969) J. Am. Chem. Soc. *91*: 6362;
b. Bartlett PD, Goebel CV, Weber WP (1969) ibid. *91*: 7425
145. Lindman B, Forsen S in: NMR and the Periodic Table Harris RK, Mann BE (ed) (1978) pp. 129 to 181, London, Academic Press
146. a. Brown TL (1968) Acc. Chem. Res. *1*: 23;
b. Brown TL (1970) Pure Appl. Chem. *23*: 447
147. Fraenkel G, Henrichs M, Hewitt M, Su BM (1984) J. Am. Chem. Soc. *106*: 255
148. a. McGarrity JF, Ogle CA (1985) ibid. *107*: 1805
b. McGarrity JF, Ogle CA, Brich Z, Loosli H-R (1985) ibid. *107*: 1810
149. Seebach D, Hässig R, Gabriel J (1983) Helv. Chim. Acta *66*: 308
150. See also the following publications:
a. McKeever LD, Waack R, Doran MA, Baker EB (1968) J. Am. Chem. Soc. *90*: 3244; (1969) ibid *91*: 1057;
b. McKeever LD, Waack R J. Chem. Soc. Chem. Commun. *1969*: 750;
c. Bywater S, Lachance D, Worsfold DJ (1975) J. Phys. Chem. *79*: 2148
151. Similar investigations with aggregates of propyllithium have been published by a. Fraenkel G, Fraenkel AM, Geckle MJ, Schloss F (1979) J. Am. Chem. Soc. *101*: 4745
b. Fraenkel G, Henrichs M, Hewitt JM, Su BM, Geckle MJ (1980) ibid. *102*: 3345
152. see, however, ref. 29 in which the aggregation of allylithium in THF has been determined to be $n = 2.1 \pm 0.1$ at 165 K
153. Schlosser M 1973 Struktur und Reaktivität polarer Organometalle, Berlin, Springer-Verlag
154. Williams KC, Brown TL (1966) J. Am. Chem. Soc. *88*: 4134
155. a. Köster H, Thönnes D, Weiss E (1978) J. Organomet. Chem. *160*: 1;
b. After finishing the manuscript we became aware of a recent publication (Geissler M, Kopf J, Schubert B, Weiss E, Neugebauer W, Schleyer PvR (1987) Angew. Chem. *99*: 569; (1987) Angew. Chem. Int. Ed. Engl. *26*: 588 demonstrating the crystallisation of a tetrameric and a dodecameric tert-butyllithium ($(t\text{-}BuC{\equiv}CLi)_4(THF)_4$) and ($(tBuC{\equiv}CLi)_{12}(THF)_4$) from THF solutions

156. Wittig G, Meyer FJ, Lange G (1951) Liebigs Ann. Chem. *571*: 167
157. a. West P, Waack R (1967) J. Am. Chem. Soc. *89*: 4395;
b. low temperature nmr investigations revealed the phenyllithium monomer in the presence of the dimer, Reich H (1987) lecture in Marburg, June 29
158. Thönnes D, Weiss E (1978) Chem. Ber. *111*: 3157; one should, however, mention, that $(PhLi \cdot OEt_2)_4$ and $(PhLi \cdot OEt_2)_3 \cdot LiBr$ (Hope H, Power PP (1983) J. Am. Chem. Soc. *105*: 5320) and the monomeric PhLi · pentamethylethylene triamine (Schümann U, Kopf J, Weiss E (1985) Angew. Chem. *97*: 222; (1985) Angew. Chem. Int. Ed. Engl. *24*: 215) have been crystallized, too
159. Chandrasekhar J, Schleyer PvR J. Chem. Soc. Chem. Commun. *1981*: 260
160. The same observation has been made by Jackman LM, Pennsylvania State University; see ref. 11 and 13 in lit. 149
161. Thomas RD, Clarke MT, Jensen RM, Young TC (1986) Organometallics *5*: 1851. Earlier investigations of this topic have been performed by Bywater S et al. [150c)] as well as by Hartwell GE and Brown TL [162)]
162. Hartwell GE, Brown TL (1966) J. Am. Chem. Soc. *88*: 4625
163. Two (out of many) reviews of the Claisen-Rearrangement: Ziegler FE (1977) Acc. Chem. Res. *10*: 227; Gajewsky JJ (1980) ibid. *13*: 142 [164)]
164. A complete list of references of the Claisen rearrangement and the RO-, R_2N-, Li^+O^-- and Me_3SiO-accelerated Claisen rearrangement is given in the following publications:
a. Denmark SE, Harmata MA (1982) J. Am. Chem. Soc. *104*: 4972;
b. Denmark SE, Harmata MA (1983) J. Org. Chem. *48*: 3369;
c. Denmark SE, Harmata MA Tetrahedron Lett. *1984*: 1543
165. A theoretical explanation of the substituent effect was given by Burrows CJ, Carpenter BK (1981) J. Am. Chem. Soc. *103*: 6983, 6984
166. Similarly, in the alkoxide variant of the Cope rearrangement, rate accelerations of 10^{10}–10^{17} have been observed:

a. Evans DA, Golob AM (1975) J. Am. Chem. Soc. *97*: 4765;
b. Evans DA, Baillargeon DJ, Nelson JV (1978) ibid. *100*: 2242;
c. Evans DA, Nelson JV (1980) ibid. *102*: 774
167. In ref. 164b and 164c DMSO was used instead of HMPA
168. Cooper D, Trippett S J. Chem. Soc. Perkin I, *1981*: 2127
169. Denmark SE 1987 lecture in Marburg, June 1
170. a. Blechert S (1984) Tetrahedron Lett. *25*: 1547;
b. Blechert S Liebigs Ann. Chem. *1985*: 673
171. Danheiser RL, Bronson JJ, Okano K (1985) J. Am. Chem. Soc. *107*: 4579
172. Hudlicky T, Kutchan TM, Nagni SM (1985) Org. React. *33*: 247
173. Leonhard NJ, Frihart CR (1974) J. Am. Chem. Soc. *96*: 5894
174. a. Kametani T, Tsubuki M, Nemoto H, Suzuki K (1981) ibid. *103*: 1256;
b. Nemoto H, Suzuki K, Tsubuki M, Minemura K, Fukumoto K, Kametani T, Furuyama H (1983) Tetrahedron *39*: 1123
175. Reviews: a. Oppolzer W. Synthesis *1978*: 793;
b. Funk RL, Vollhardt KPC (1980) Chem. Soc. Rev. *9*: 41

Complex Eliminations; Eliminations with Rearrangements

Gerd Kaupp

Fachbereich Chemie — Organische Chemie I — der Universität Oldenburg,
D-2900 Oldenburg, P. O. Box 2503, FRG

Table of Contents

Complex eliminations proceed with rearrangements. They occur in most of the different branches of preparative organic chemistry. By relative notations of positions, they are ordered and named in a classifying way, independent from the respective reaction mechanism. Structural peculiarities of particular types are also included and distinguished. If only one or two groups or bonds migrate, the classifications are distinct and unequivocal. Nevertheless, in case of two migrating residues it is occasionally necessary to evaluate by labelling, which type the reaction belongs to. In the case of multiple migrations, it is necessary to divide the reaction into partial steps via (several) postulated or proved intermediates. The presented material shows synthetical potential and points out that most types can be realized according to different reaction mechanisms and under most different reaction conditions.

1 Introduction

The chemical literature contains numerous complex elimination reactions, but mostly they appear isolated. In these reactions, in addition to the cleaving off of the leaving groups, migrations of substituents or bonds in the reacting chain can also occur. Thus, very profound structural changes result. Until now, a systematic record has been largely omitted, and nomenclature has not been unified. Certainly, in isolated cases there are more or less comprehensive mechanistic studies, but they have not led to general considerations. Therefore, the undoubtedly high synthetic potential of complex eliminations has been partly overlooked. Correspondingly, a systematic classification is required. To allow for an overview of all single examples and for a broad extrapolation basis, it must be simple, clear, and convenient for indexing and retrieval. Only then will a broader development and mechanistic penetration of these versatile reactions be achieved.

Older reaction examples as well as more recent ones have been compared. Where a selection is available, it has been attempted to give priority to the preparative potential or elegance. However, it must be appreciated that some relevant examples which might be indexed in the literature solely under the headings "rearrangement", "ring closure", or "ring opening" may have escaped the notice of the author.

2 Nomenclature System for Complex Eliminations

The development of a nomenclature for complex eliminations should start from the approved classification system for simple eliminations ([1,1]-, [1,2]-, [1,n]-elimination), where the relative position numbers are used especially profitably. Also in case of rearrangements relative position numbers have often been used (e.g., [1,2]-, [1,3]-,

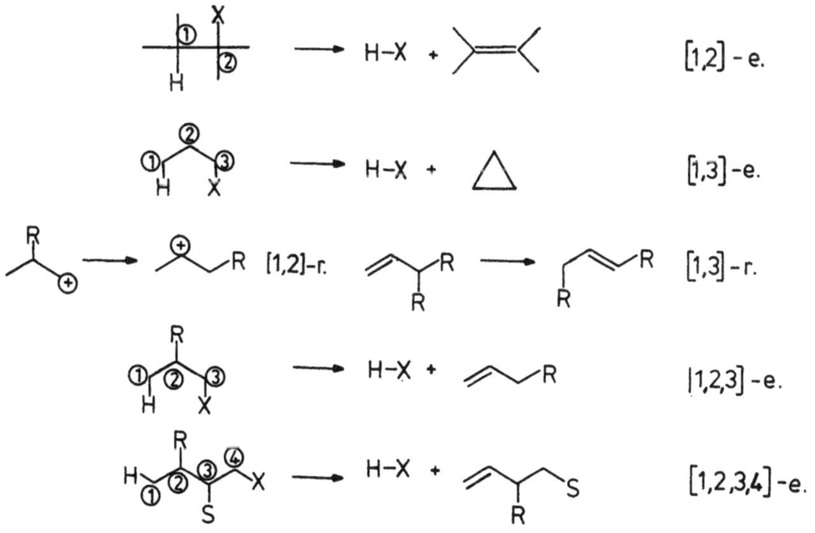

(1)

[1,n]-migrations, respectively-rearrangements). For complex eliminations it seems to be important to combine both systems. A [1,3]-elimination, where a [1,2]-shift (from position 2 to position 3) occurs additionally is simply called a [1,2,3]-elimination, with just three figures. The so-called "pinacol(ine)(one)-rearrangement" (see Sect. 3.3) appears to be the most wellknown example of this reaction type. If two substituents migrate during the elimination (e.g., formally a [1,4]-elimination and two [1,2]-shifts), four figures are necessary to unequivocally classify them (in this case a [1,2,3,4]-elimination).

Table 1 summarizes some of the most important basic types which are straightforwardly classified with three figures. The first and last figures always refer to the relative positions of the leaving groups. The middle figure indicates the starting point of the migration, which necessarily ends at the next (in this case the third) figure.

In the so called "Grob fragmentation" ([1,2,(3)4]-elimination), a bond of the reacting chain migrates, not a substituent. Thus, for a clear naming of the position of the secondly formed double-bond, the additional figure in parentheses is necessary. Also if a double bond, which is present in the chain, migrates, a clear classification requires a further figure (e.g., in the [1,2,(3)5]-elimination). However, if only an allylic double-bond shift takes place (principle of vinylogy), it is sufficient to insert a Δ between the corresponding positions to specify this unequivocally (e.g. in the [1,Δ,4,5]-elimination).

Although our system should remain free from mechanistic premises in order to avoid restrictive elements, for the sake of clarity we should fix that the more electropositive leaving group (e.g., Y) has the smaller position number and the more

Table 1. Important basic types of complex eliminations with one migrating group or bond

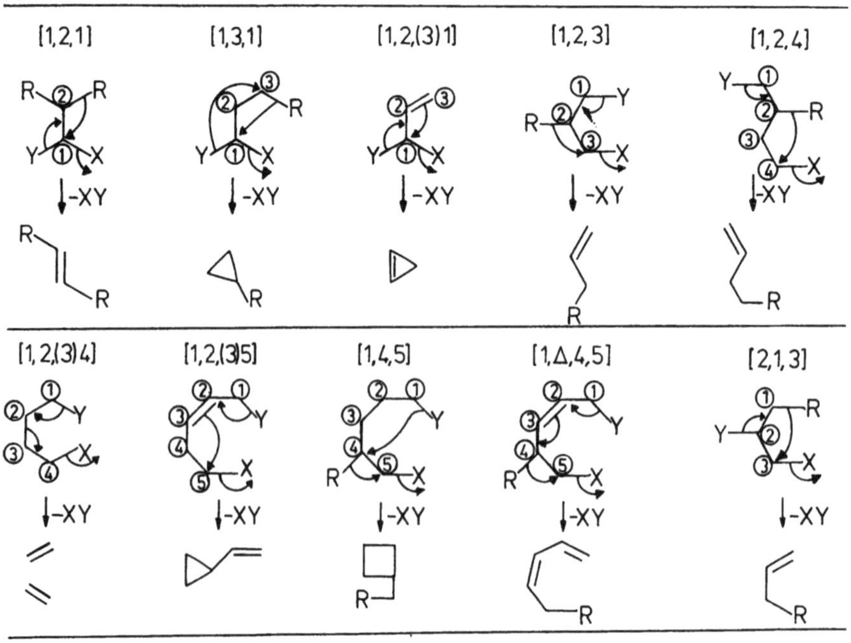

Table 2. Important basic types of complex eliminations with two migrating groups or bonds

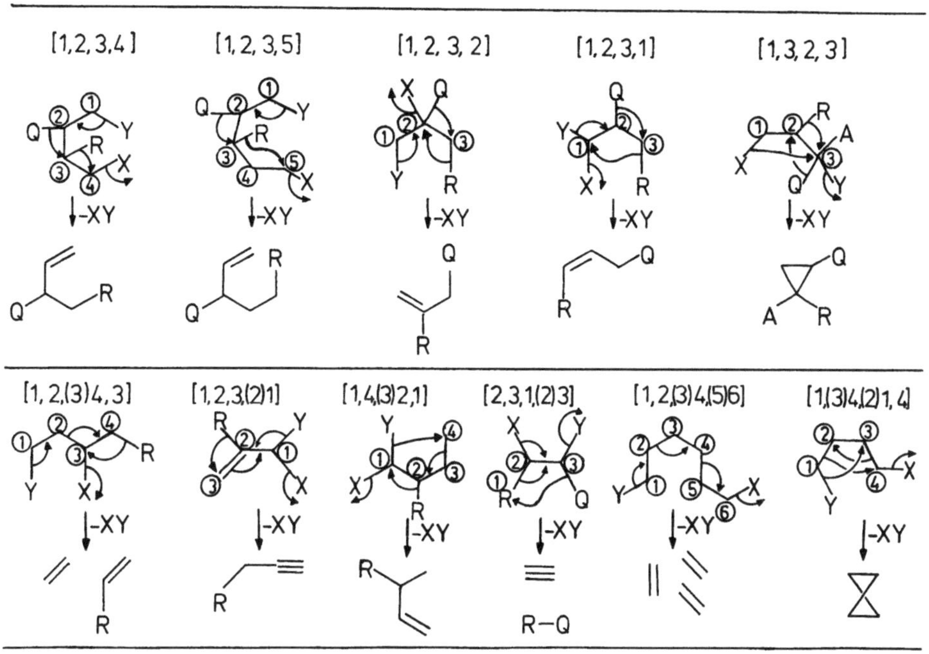

electronegative one (e.g., X) the larger position number and that those are consistently named first and last, respectively [1]. Furthermore, in cyclic compounds the best possible progression of figures should be selected in order to avoid regressive number sequences as far as possible (e.g., [1,2,5]-e., but not [2,1,4]-e. as in *159*).

Table 2 summarizes some complex eliminations with two migrating groups or bonds, for which there already exist reaction examples (see Sect. 4). Again, the more electropositive leaving group has the smaller first figure and the more electronegative one the larger last figure. Each of the two middle figures unequivocally specify the two migrations, and the additional figures indicate whether bonds of

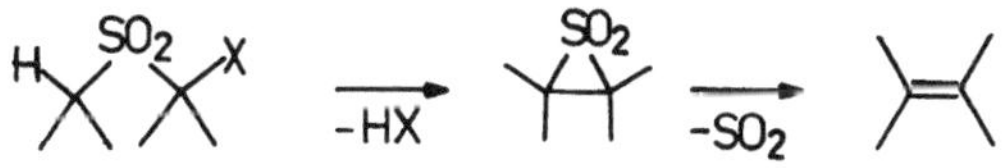

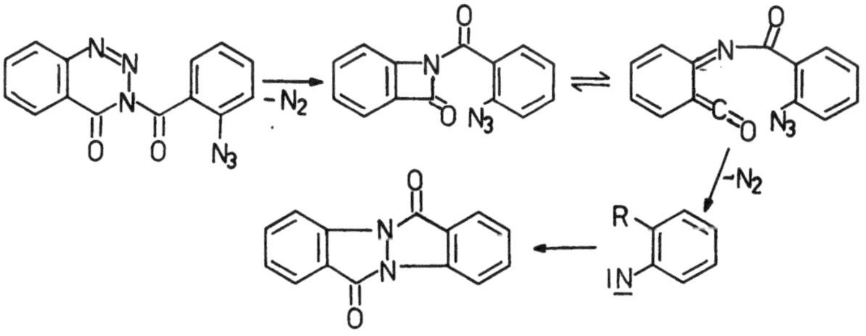

(2)

the chain or double-bonds are migrating, while, again, a "Δ" signalizes the allylic shift of a double-bond (principle of vinylogy) (see *287*, *307* below).

This review does not cover multiple eliminations at the same (cf. [156)]) or at different centres proceeding without rearrangement and which, [for e.g., the Ramberg-Bäcklund- [2)] or the benzotriazinone-arylazide-reaction [3)] (Scheme 2)], can be classified in simple stages ([1,3]-e. and [1,2]-e., or [1,4]-e., valence tautomerism, [1,1]-e., intramolecular nitrene cycloaddition, respectively).

With our purely topological system (Tables 1 and 2), reactions initially unclear can be easily comprehended and described in the following sections even if in most cases detailed mechanistic knowledge is lacking. This is important because the examples already available show that the complex types — as is also known from the simple types (e.g., [1,2]-eliminations) — can be realized under most different reaction conditions and (apparently) reaction mechanisms. The more precisely the numerous mechanistic alternatives will be explored, the more important the extrapolative potential will be.

3 Complex Eliminations with One Migrating Group (Bond)

3.1 [1,2,1]-Eliminations

Complex eliminations, where both leaving groups depart from the same centre and where a substituent migrates from the neighbouring centre (double-bond formation),

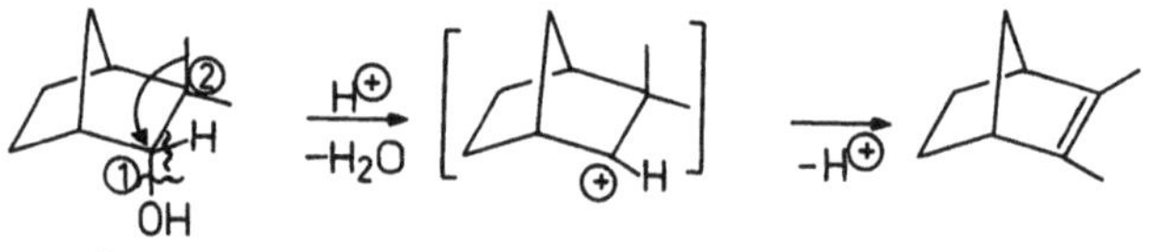

1 *2*

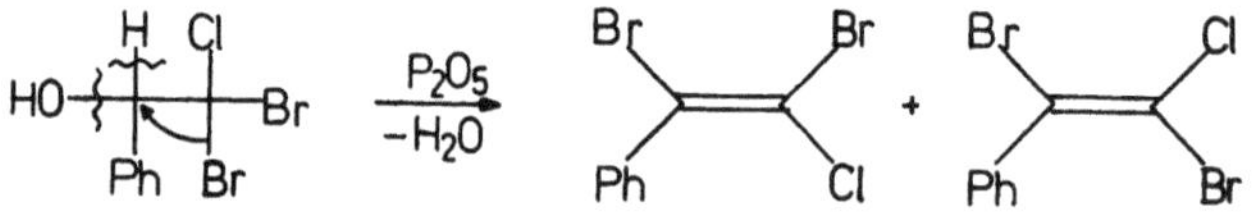

3 *4*

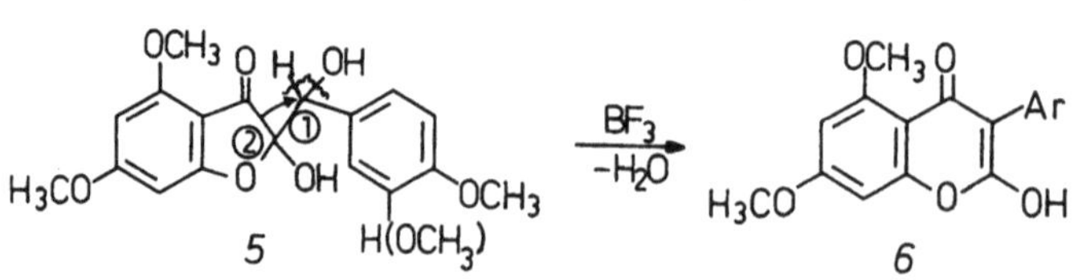

5 *6*

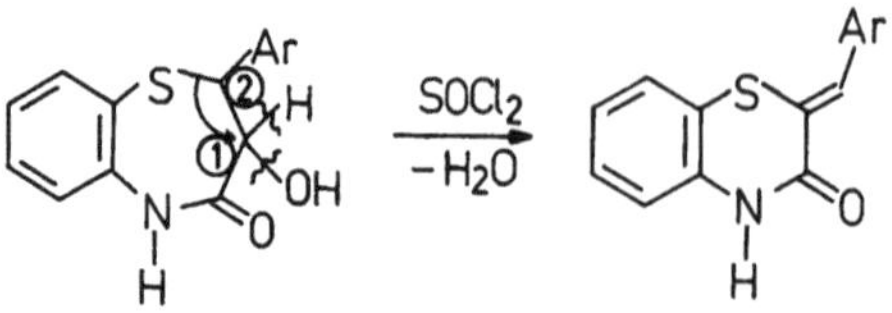

7 *8*

take place under many different reaction conditions. The synthesis of santene (*2*) from camphenilol (*1*) heated with $KHSO_4$ (190 °C) should take place via a more or less free carbocation [4]. Certainly, it would be interesting to know, whether the *exo*- or (and) the *endo*-methyl group migrates in the sense of a "Wagner-Meerwein-rearrangement". Further examples, catalyzed by acid, concern the conversions of *3* to (*E*)- and (*Z*)-*4* (59–75%) [5], as well as a biogenetically important coumarin synthesis (40; 45%) from easily accessible *5* [6], in which bromine- or aroyl-migrations take place. The conversion of *7* to *8* (14%) with sulphur-migration should start with the usual esterification by thionyl chloride [7].

Besides the acid-catalyzed examples, base-induced [1,2,1]-eliminations have been used under solvolysis conditions for the synthesis of heterocycles (*9* → *10*) [8], for the lactone synthesis *11* → *12* (85–91%) [9], and for steroid conversions (e.g. lupenyl tosylates *13* → *14* [10], 4 examples, $R^1 = R^2 = H$: 40%). Usually, it is not anticipated here that free carbenes (by [1,1]-elimination) occur as intermediates.

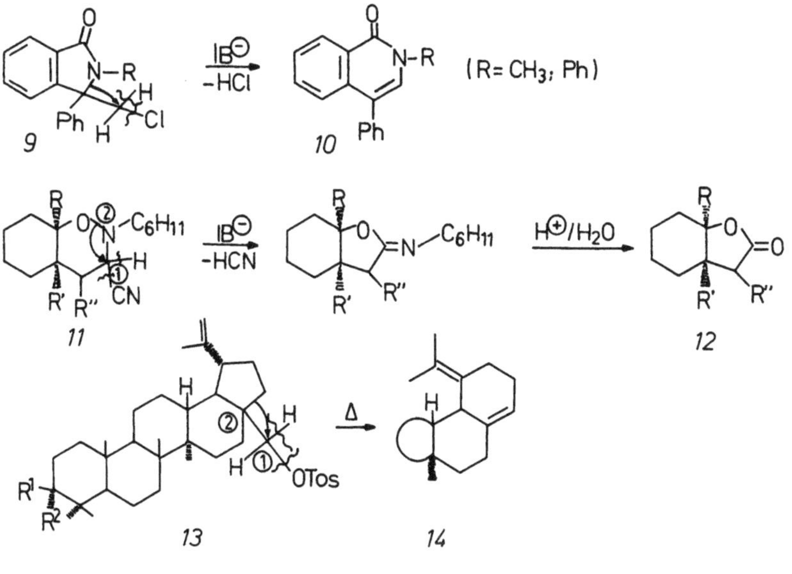

The [1,2,1]-eliminations as induced by bases or nucleophiles also succeed in the family of the polyhalogen compounds; generally under the influence of organolithiums at low temperatures. This suggests initial metalation. In this way, with a yield of 68–96% the educts *15* react stereoselectively to the (*Z*)-isomer *16* [11], and the alkene *17* opens up an interesting synthesis of the cyclopentyne (*18*) [12], which can be trapped, e.g., to *19* (36%). In the case of several alkenes *20*, the reductive [1,2,1]-halogen elimination can be achieved by electrolysis in DMF. Alkynes *21* form in good yields [13].

A standard method for the synthesis of allenes is the conversion of 1,1-dibromocyclopropanes with methyllithium according to Doering (30–90%) [14] (reductive [1,2,1]-dehalogenation). It can also be used for the synthesis of cyclic (*23*, which dimerizes) [15] and conjugated (*25*, isolated by vpc) [16] allenes. The unusual (trapable) bridgehead alkene *27* (possibly with the C-ylide-structure *27'*) arises from

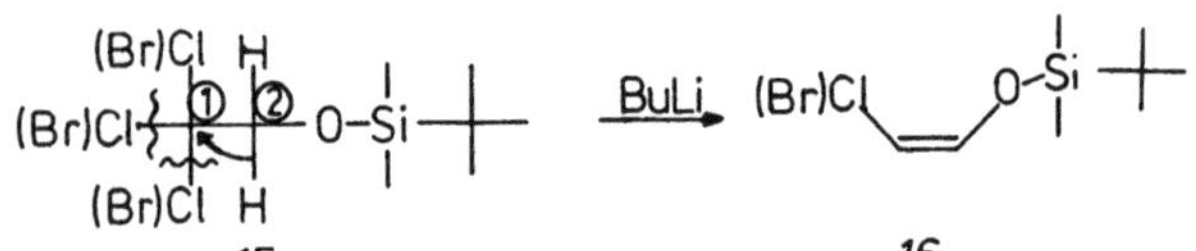

15 16

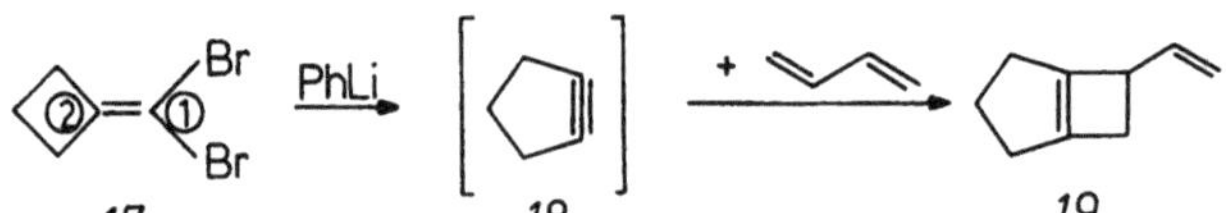

17 18 19

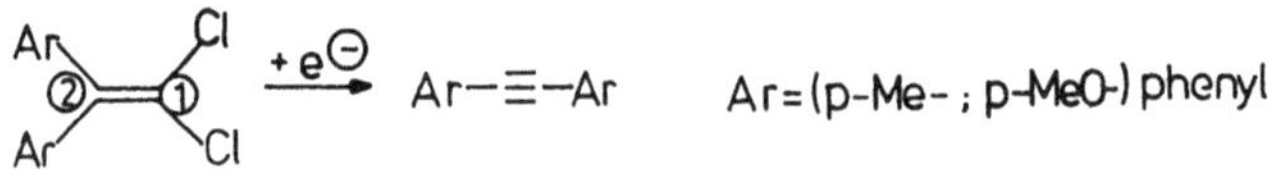

20 21

26 [17]. Mechanistic studies indicate that carbenoid intermediates occur in the reductive debrominations. If chiral educts are applied, the reactions usually run stereoselectively.

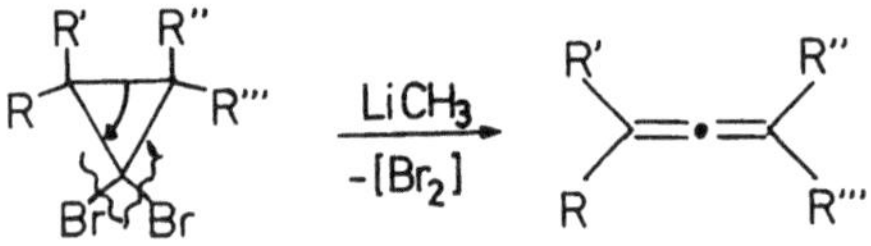

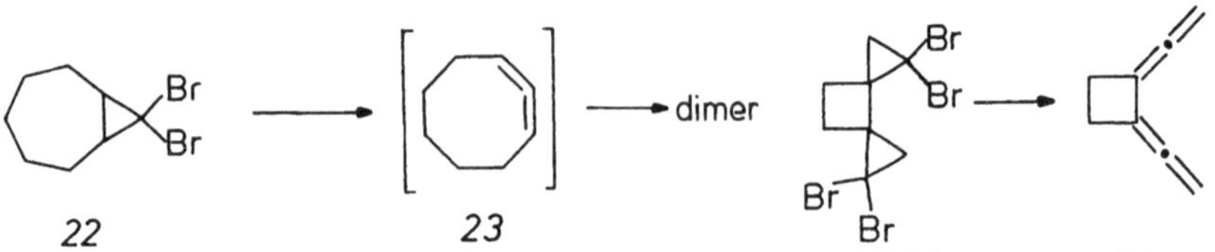

22 23 24 25

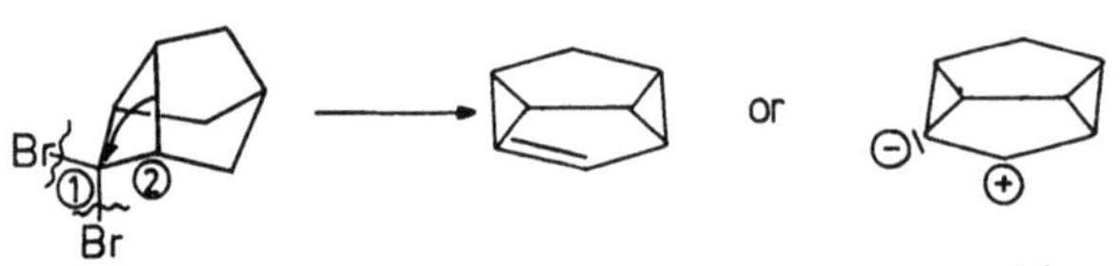

26 27 27'

Allenes can also be synthesized from diazocyclopropanes by chemical or photochemical [1,2,1]-elimination of nitrogen. In the thermolysis of *28* to give *30* the carbene-intermediate *29* could be trapped [18], and in the low-temperature photolysis of *31* the triplet carbene *32* could be detected by EPR-spectroscopy [17]. *32* is longlived in a polycrystalline matrix and rearranges to *33* (28%) at a temperature of −154 °C [19]. Numerous applications are included in Ref. [20]. Especially noteworthy are the syntheses of stable cyclobutadienes by Masamune (90%) [21] and Regitz (67%) [22].

There are numerous examples of [1,2,1]-eliminations of nitrogen from diazoketones

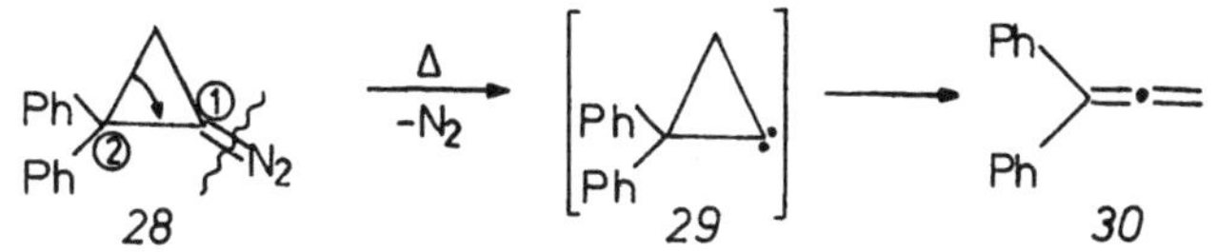

28 29 30

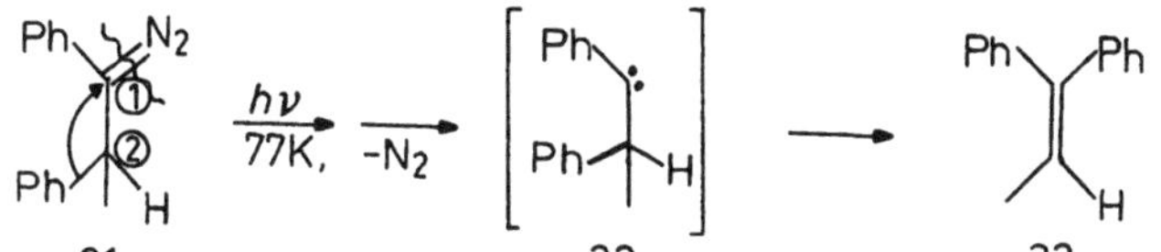

31 32 33

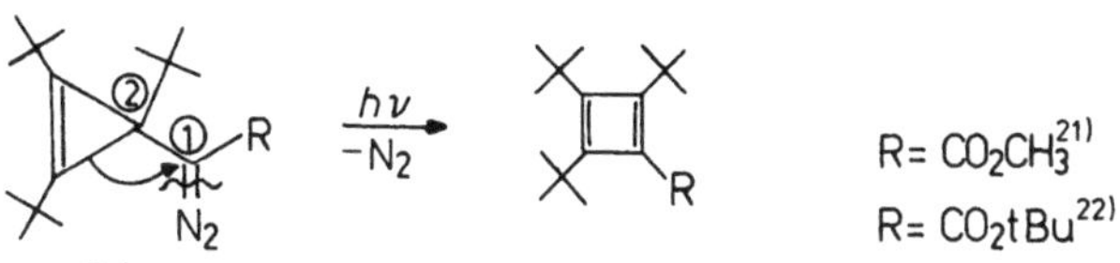

34 35

(socalled "Wolff-rearrangement"), which proceed thermally [23] and photochemically [20] and where solvolyzable ketenes are formed via ketocarbenes and, presumably, oxirenes [24]. The various synthetic possibilities are exemplified by the examples *36* → *37* (61 %, 2 epimeric methanol adducts isolated) [25] and *38* → *39* (65 %, methanol adducts isolated) [26]. Technical usage is exemplified by the corresponding "Süs-reaction" of o-quinonediazides such as *40* to give the alkali-soluble acids *42* in diazo-types [27]. Fully correspondingly, acyl azides yield isocyanates [28] by the "Curtius-degradation" (not always via free acylnitrenes).

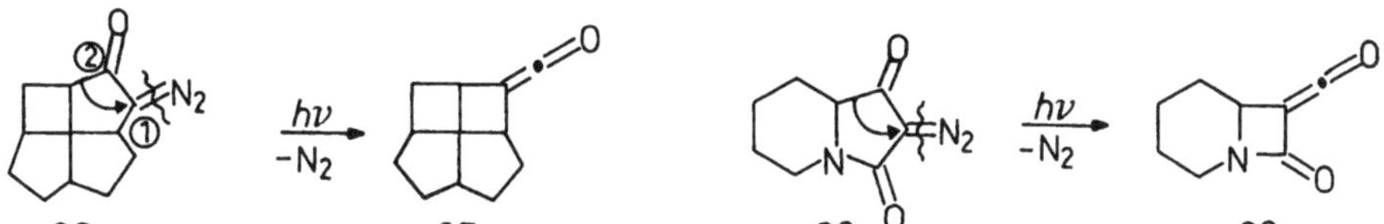

36 37 38 39

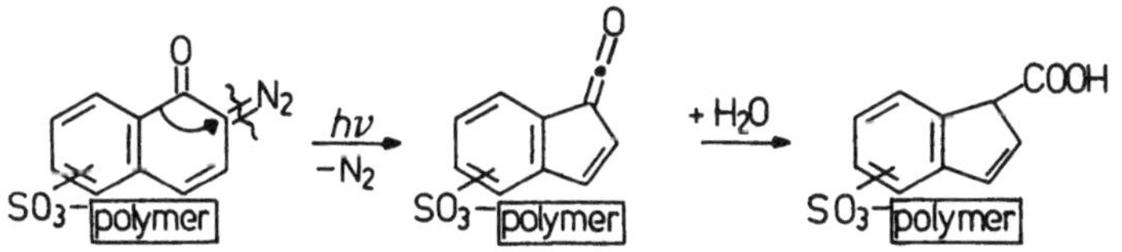

40 41 42

The versatility of the [1,2,1]-elimination can be shown by numerous modifications of the groups which are to be eliminated and in many cases a carbene or a carbenoid may be considered as a short-lived, rearranging intermediate (also further rearrangement possibilities — see Sect. 5 — have been discussed) [29].

As masked diazo compounds the salts of tosyl hydrazones can react thermally and photochemically ("Bamford-Stevens-reaction") to give alkenes (Scheme 3, variable yields)[30]. By the action of very strong bases and a subsequent protonation, the same result can be reached intermolecularly via dianion-intermediates ("Shapiro-reaction", frequently improved yields)[31]. In both the thermal and photochemical version, substituents can also migrate instead of hydrogen[30].

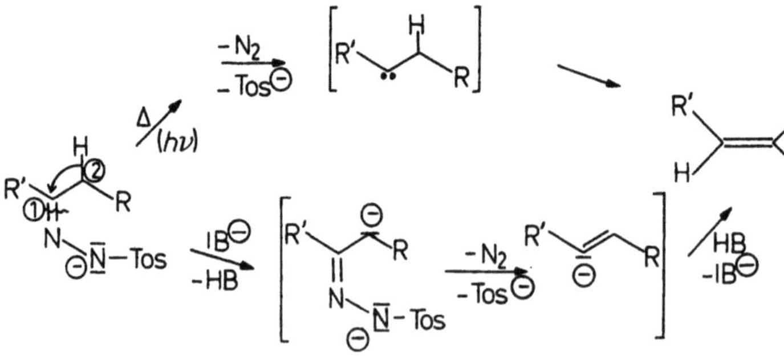

(3)

Similar [1,2,1]-eliminations are used in the photolyses of ketenes (e.g., *43* → *44*, quantitative)[32], thione acid derivatives (e.g., *45* → *46*)[33], sulphonium ylides (e.g., *47* → *48*, via the trapable benzoyl carbene, 26%, ethanol adduct isolated)[34] and thermolyses of phosphorus ylides such as *49* to give *50* (80; 79%)[35].

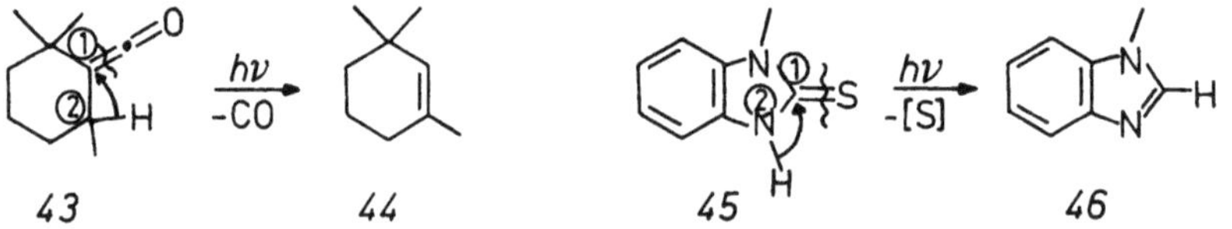

43 *44* *45* *46*

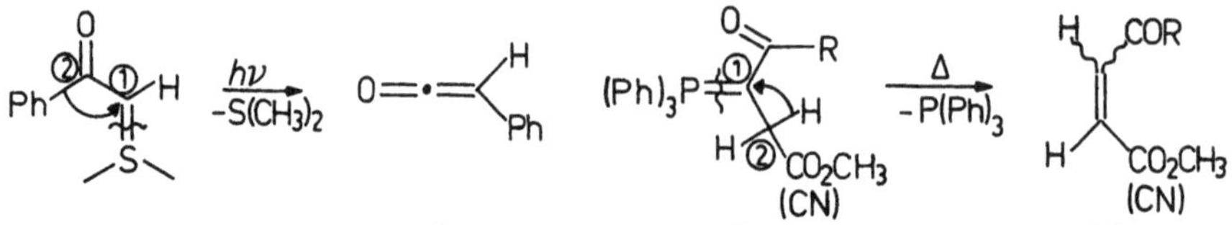

47 *48* *49* *50*

The "Hofmann-degradation" of acid amides (60–80%, via N-bromoamides) as well as the "Lossen-degradation" of hydroxamic acids by [1,2,1]-elimination of HBr, or H_2O, respectively, leads to isocyanates (Scheme 4), which usually yield primary amines RNH_2 by solvolysis[36].

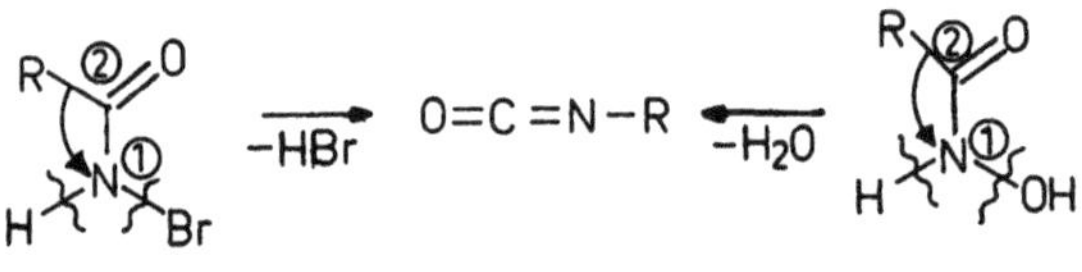

(4)

Numerous [1,2,1]-eliminations proceed under high-temperature pyrolyses in static or in flow systems. In this way, the dehydration of tetrahydrofurfuryl alcohol *51* contacted with aluminium-oxide at 340 °C forming the dihydropyrane (*52*, 70%) is

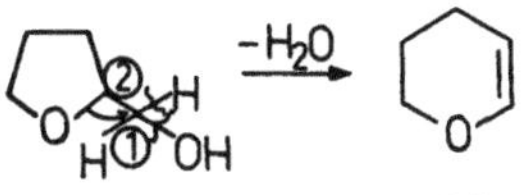

51 52

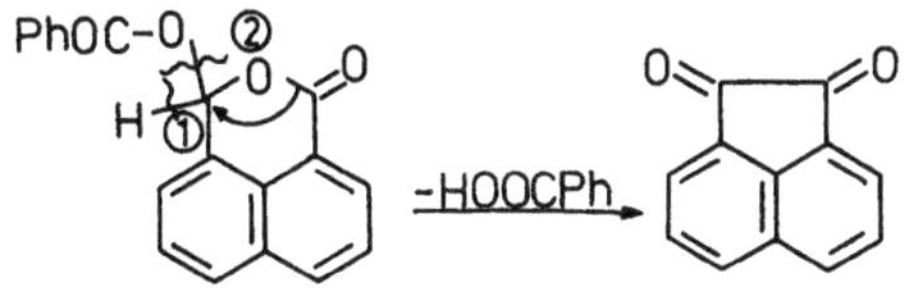

53 54

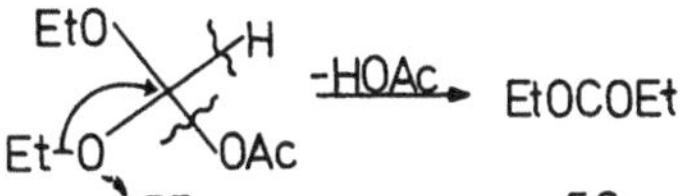

55 56

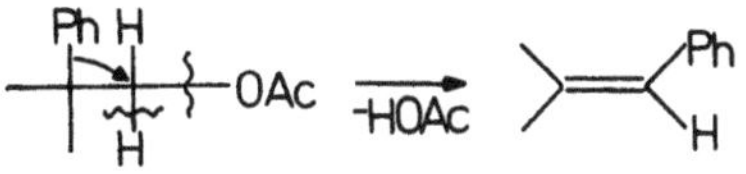

57 58

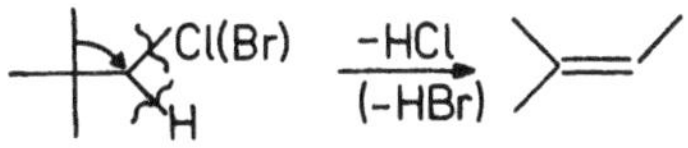

59 60

of considerable synthetic importance [36a, 37]. Further β-alcohol ethers can be transformed, comparably. Correspondingly, the [1,2,1]-elimination of the acetal derivative *53* (560 °C, flash pyrolysis) [38] produces the dione *54* (71 %), the orthoester derivative *55* gives (400 °C) the ester *56* (50–80 %) [39], and the ester *57* (600 °C) the alkene *58* (low conversion) [40]. Also the [1,2,1]-elimination of hydrogen chloride from neopentyl chloride (*59*) in the pulsed shock tube to give the alkene *60* [41], or the pyrolyses of the chloride *59* (445 °C) and the bromide *59* (389–440 °C) with the same result [42] belong here.

Oxidative [1,2,1]-eliminations, in which hydrogen is formally cleaved off, are of preparative importance. Thus, 7-aminotheophylline (*61*) can be oxidized with lead tetraacetate to give the azapterin *62* (74 %) [43], and in the conversion of the 5-O-tosyl-D-glucurone lactone *63* to give *64* under the influence of potassium acetate/crown ether [44] a dehydration takes place (atmospheric oxygen?).

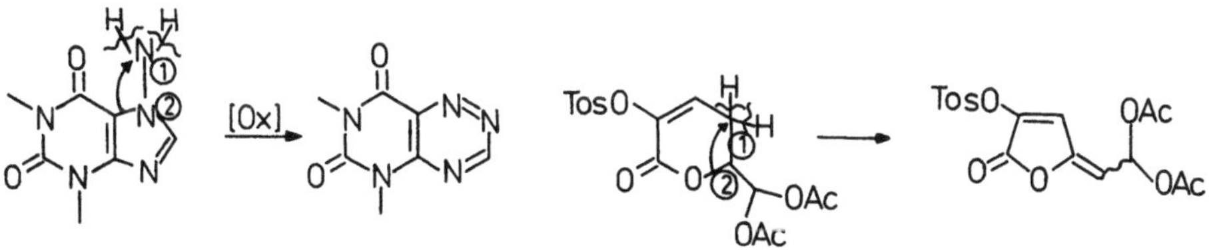

61 62 63 64

3.2 [1,n,1]-Eliminations

The higher members of the [1,n,1]-eliminations are also of preparative importance. Thus, the "Doering allene synthesis" [14] leads to bicyclobutanes (e.g. *66*, 28 %) [45], if bulky substitution as in *65* favours the [1,3,1]-elimination. The related cyclopropene syntheses from *67* [46] and *69* [46], respectively (40 and 6 % *68* in derivatized form: lithiation and carboxylation), are to be classified as [1,3,(2)1]-eliminations of bromine (reductive) as well as of hydrogen chloride. The thermolysis or the photolysis of diazo

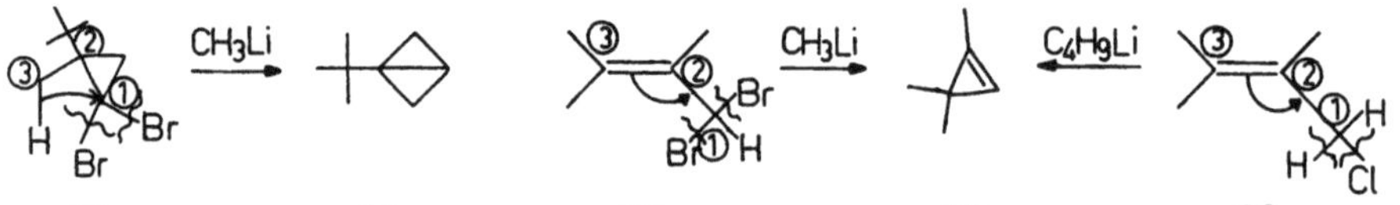

65 66 67 68 69

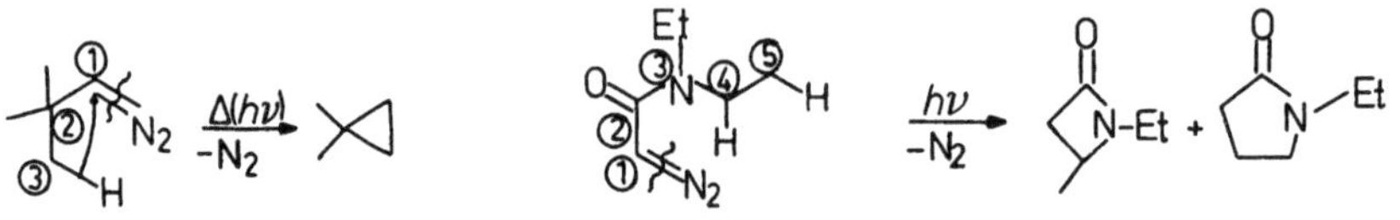

70 71 72 73 74

alkanes with γ-C—H-bonds offer an easy approach to cyclopropanes. Thus, *70* reacts to *71* (yields 41% and 37%)[47] in the sense of a [1,3,1]-elimination, and the photochemical syntheses of *73* (57%) and *74* (43%, dioxane)[48] are classified as [1,4,1]- and [1,5,1]-eliminations. The underlying conformational effects become especially clear in the direct (or with benzophenone sensitization) photolysis of *75* to give *76* ([1,4,1]-; 2.0% (1.7%)), *77* ([1,5,1]-; 31.6% (36.1%)), *78* ([1,6,1]-; 21.5% (20.6%)), and *79* ([1,7,1]-e.; 2.2% (2.8%))[49].

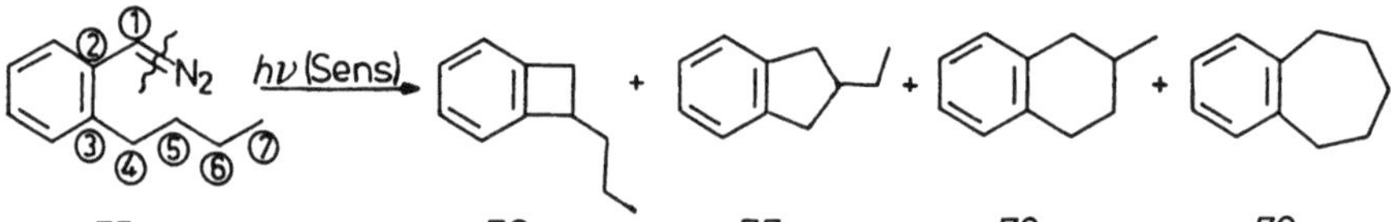

75 76 77 78 79

It is assumed that intermediate carbenes insert into the corresponding C—H-bonds. Accordingly, nitrene insertions in the decomposition of azides are usually assumed. Thus, the thermolysis (180 °C) or photolysis (68–74%) of (substituted) 2-azidobiphenyls *80* lead to carbazoles (*81*)[50] in the sense of a [1,5,1]-elimination.

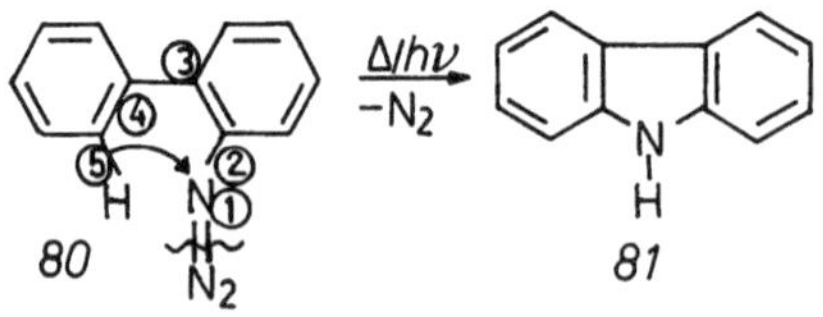

80 81

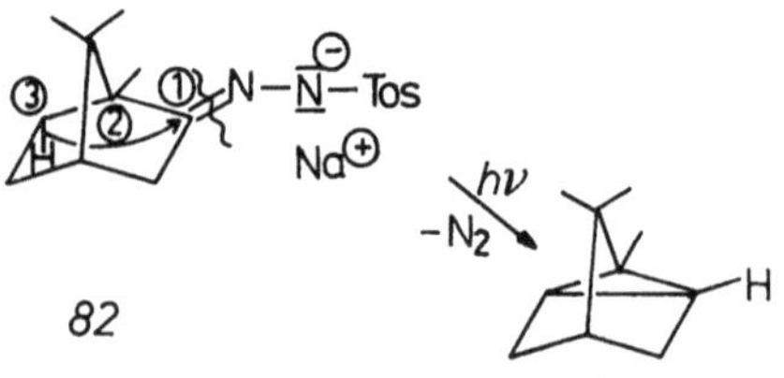

82 83

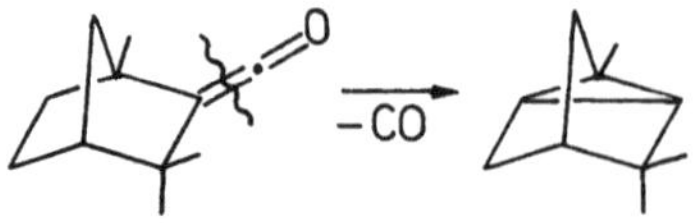

84 85

[1,n,1]-reactions also appear in the thermolysis and photolysis of tosylhydrazone salts, e.g. in the [1,3,1]-elimination of the sodium salt of the camphor-tosylhydrazone (*82*) to give the polycycle *83* (31%)[51,52]. Closely related is the photolytic [1,3,1]-

elimination of carbon monoxide from the ketene *84* in pentane with formation of *85* [32]. If there are double bonds available in a suitable arrangement in the diazo compounds or in the azides, polycyclic compounds will form. Bicyclobutane formations by [1,3,(4)1]-eliminations of nitrogen from *86* (7% *87*) [53] and *88* (3% *89*) [54] as well as the [1,6,(7)1]-elimination of *90*, which leads to *91* (12%, isolated as its hydroxy lactam after addition of water) [55], belong here.

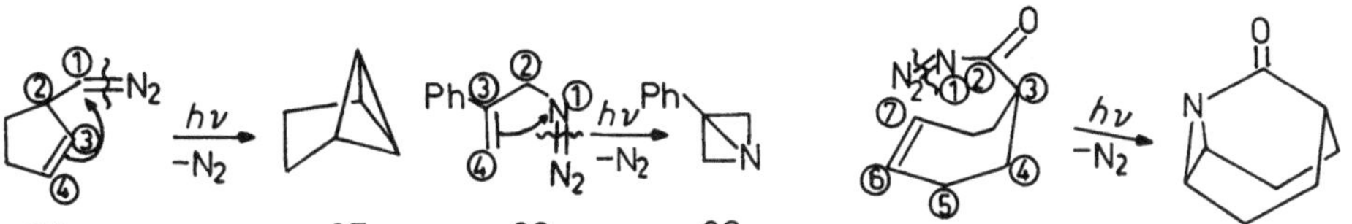

86 *87* *88* *89* *90* *91*

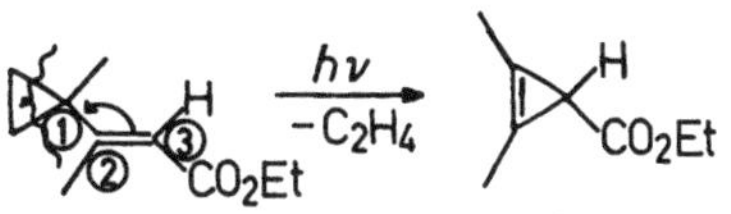

92 *93*

The photolysis of *92* with formation of *93* (40%) [56] is classified as a [1,3,(2)1]-elimination, because the double bond migrates while the three-membered ring is formed (cf. *68*).

Vinylogous eliminations, in which both leaving groups depart from the same centre, are also known. The [1,Δ,4,(5)1]-elimination of ethylene from *94*, which inter alia yields *95* (20%) [56] is typical. Here, the 2/3 double-bond is vinylogously shifted (Δ), while the 4/5 double-bond migrates under formation of a new single-bond (assistant figure for unequivocal classification).

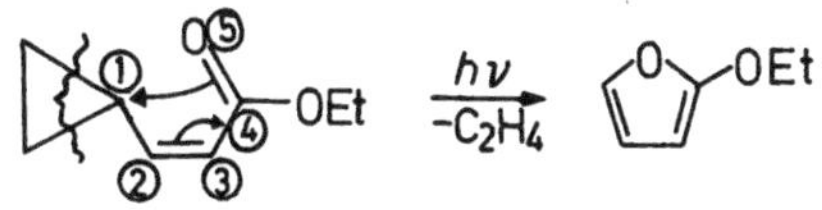

94 *95*

3.3 [1,2,3]-Eliminations

As in the [1,2,1]-eliminations, there are also [1,2]-migrations involved in [1,2,3]-eliminations and both types frequently compete with each other.

In the so called "pinacol(ine)(one)-rearrangement" water is eliminated from 1,2-diols [57]. If it is run under acid-catalysis, it will undoubtedly proceed via a more or less free solvated carbocation, in which a migration ("Wagner-Meerwein-rearrangement") of the substituent occurs from relative position 2 to 3. Therefore, it should be classified as a [1,2,3]-elimination (Scheme 5). The same is true for the Tiffenau-reaction (Scheme 6), which also succeeds in the poly- and acyclic families. By deamination of cyclic β-amino-alcohols [oxidative [1,2,3]-elimination of (formally) ammonia via a possibly not fully free carbocation] the same ring enlarged ketones

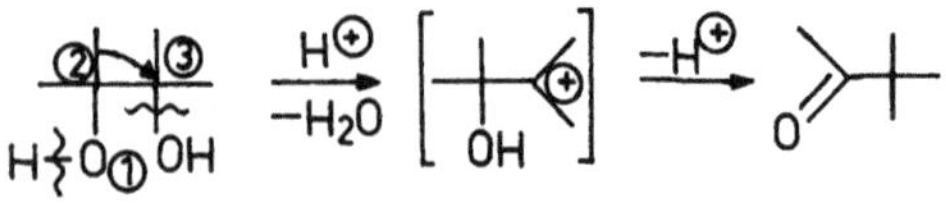

(5)

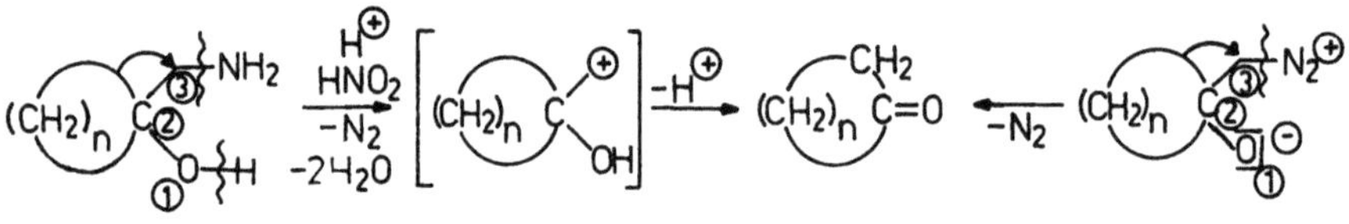

(6)

form (p. 946; 988f. in Refs. [57,58]) as in the corresponding homologizations of cycloalkanones with diazomethane [59] (these via the zwitterion formulated in Scheme 6). α-Halogenoalcohols (p. 995ff. in Ref. [57]) and α-hydroxythioethers [60] react in the same way. A more recent variant on the [1,2,3]-elimination from pinacols converts their monomethanesulphonates such as (*S*)-*96* [−78 °C with triethyl aluminium and thereby phenyl migrates stereospecifically with inversion (ee > 99%)] into (*S*)-*97* (96%) [61]. Here, a carbocation intermediate is excluded.

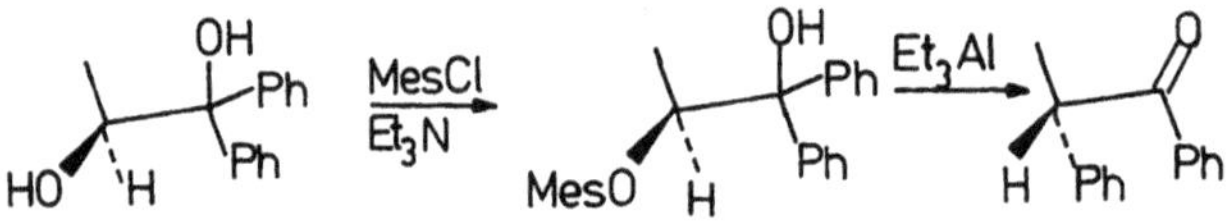

96 *97*

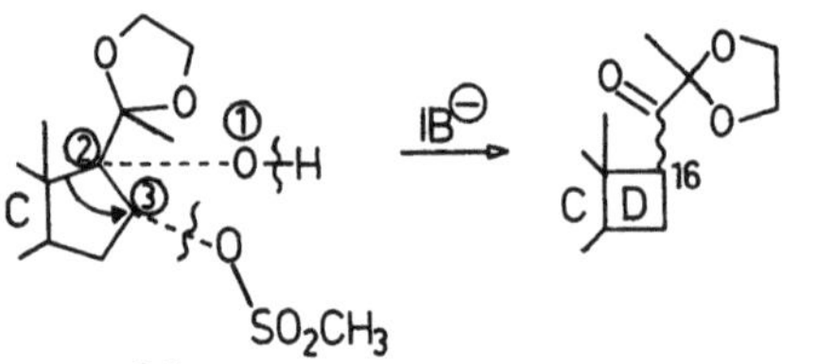

98 *99*

The conversion of the pregnane-derivative *98* into the D-nor-pregnane derivative *99* (16β:16α = 7:3, together 82%) using potassium-t.-butoxide [62] is well known. Alcohols with hydrogen atoms in the β-position such as *100* can be photodehydrated (*101*, 40%) in the sense of the [1,2,3]-elimination by applying the iodine/mercury oxide system [63]. With hydrogen atoms in the γ-position, [1,2,3]-eliminations take place under solvolysis conditions. In this way, upon deprotonation alloxane hydrate (*102*) reacts to the salt of alloxanic acid (*103*, 62%) [64] (no "benzilic-acid rearrangement").

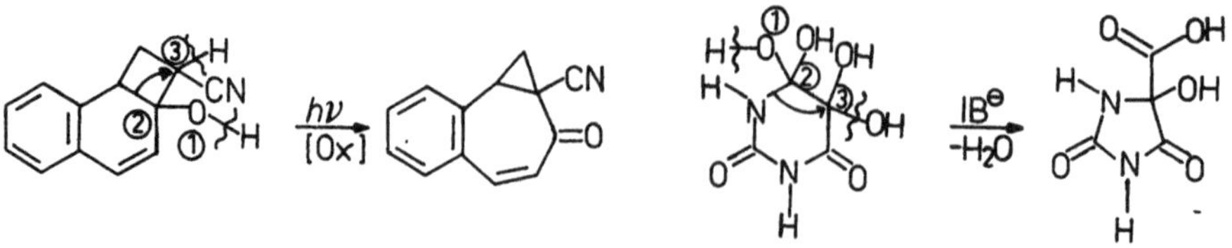

100 *101* *102* *103*

Acetates, benzoates, tosylates, and mesylates exhibit analogous [1,2,3]-eliminations upon (gas phase) pyrolysis. Thus, Patchouli-acetate (*104*) reacts to *105* and *106* (52%, and 46%, resp.) [65] and *57* yields *107* (low conversion) with migration of phenyl [40]. In the pyrolysis of bornyl benzoate (*108*) mainly (54%) camphene (*109*) [66] is formed, the atisine-derivative *110* (α or β) yields stereospecifically *111* (α-epimer, 90%) or *112* (β-epimer, 95%) [67], and the adamantane mesylate (*113*) yields the alkene *114* (38%) [68].

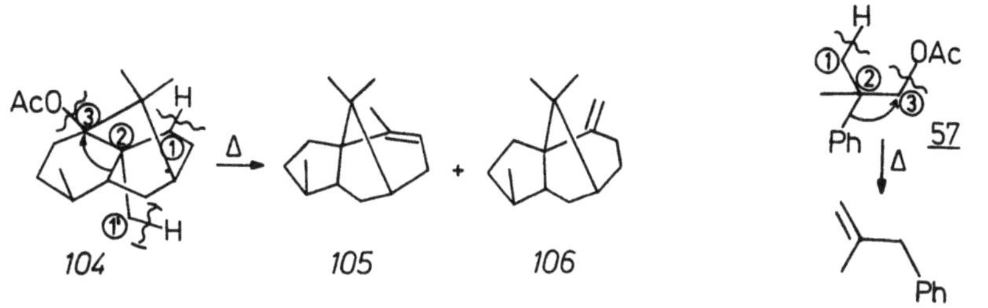

104 *105* *106* *107*

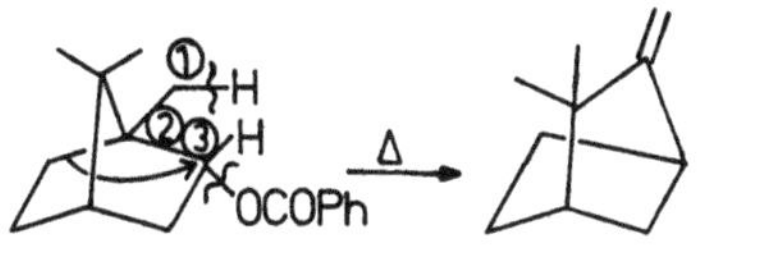

108 *109*

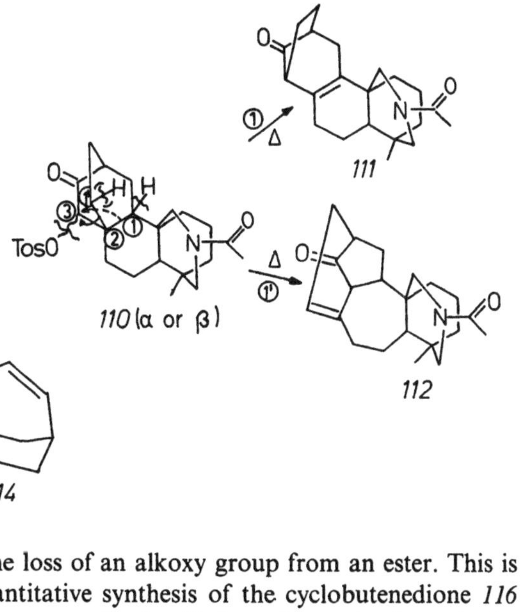

110 (α or β) *111* *112*

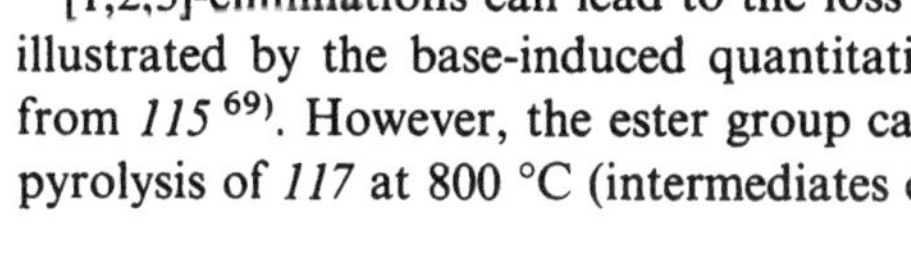

113 *114*

[1,2,3]-eliminations can lead to the loss of an alkoxy group from an ester. This is illustrated by the base-induced quantitative synthesis of the cyclobutenedione *116* from *115* [69]. However, the ester group can also be eliminated completely, as in the pyrolysis of *117* at 800 °C (intermediates could not be detected) [70].

115 *116* *117* *118*

The variability of the [1,2,3]-elimination type is very pronounced. This is shown, e.g., by the pyrolyses of the ortho ester *119* (52%)[71], the (in situ generated) mesylate *121* (quant.)[72], and the (in situ generated) A-nor-androstane derivative *123* (60%)[73]. Furthermore, it appears in the elimination of nitrous acid from *125* (77–90%)[74], in the β-lactone-synthesis from the chloramine *127* (38–65%)[75], and in the anodic oxidation of *129* (45–53%)[76].

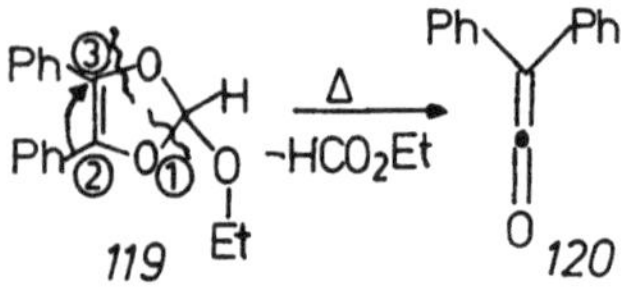

119 *120*

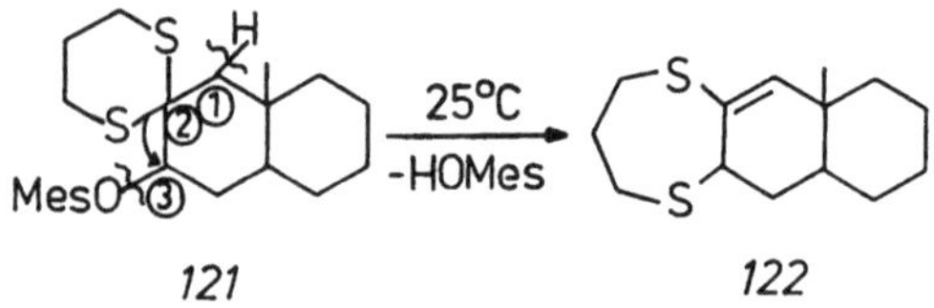

121 *122*

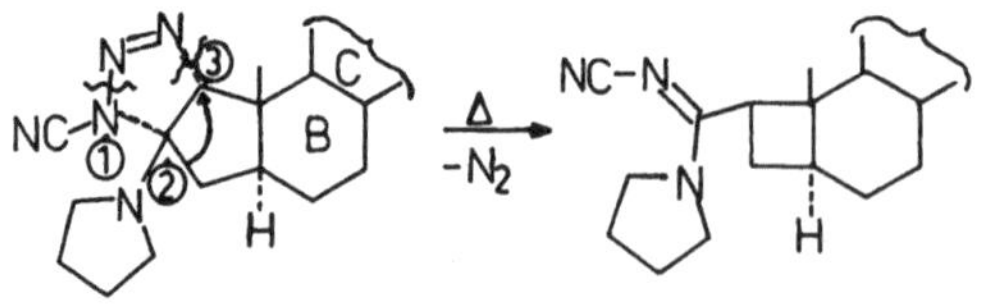

123 *124*

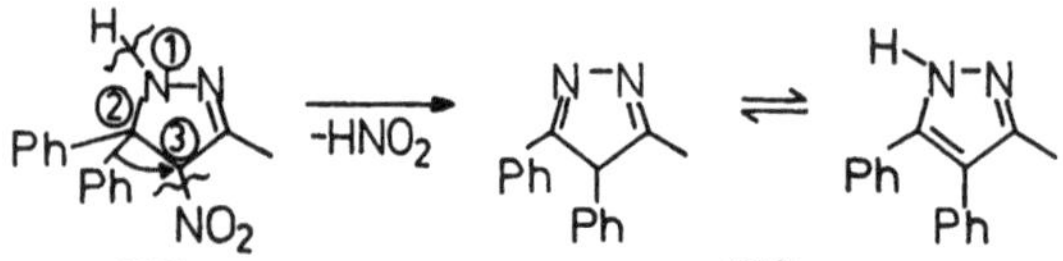

125 *126*

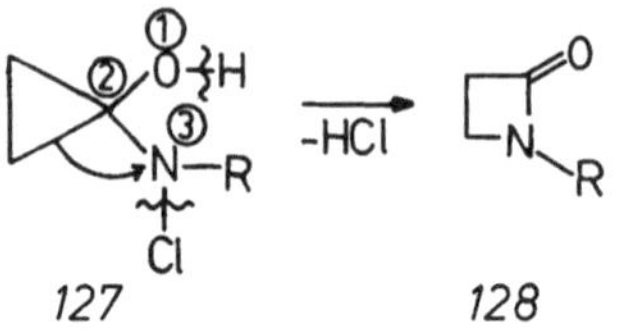

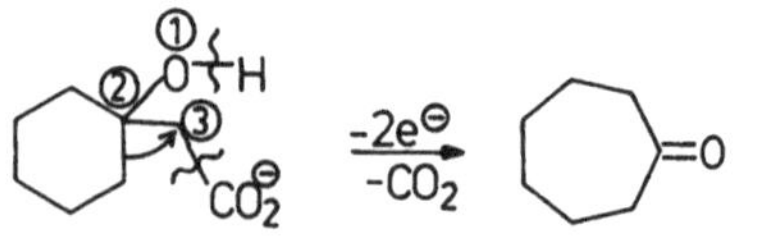

127 *128* *129* *130*

As expected, halogeno compounds also undergo [1,2,3]-eliminations (cf. footnote [68])). Thus, in the gas phase the dehydrochlorination of neopentyl chloride (*59*) inter alia yields the alkene *131*[42]. From gem-dihalogen alkylcyclopropanes such as *132*,

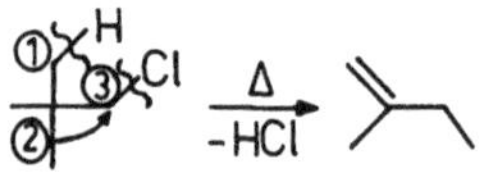

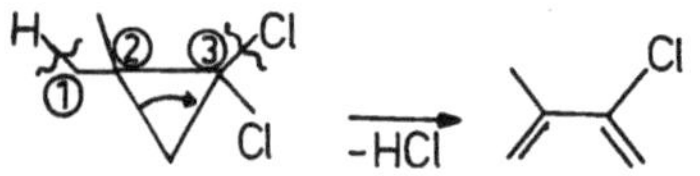

59 *131* *132* *133*

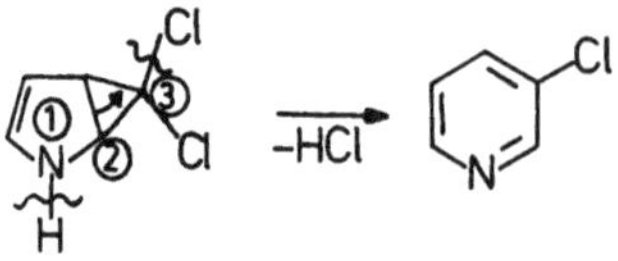

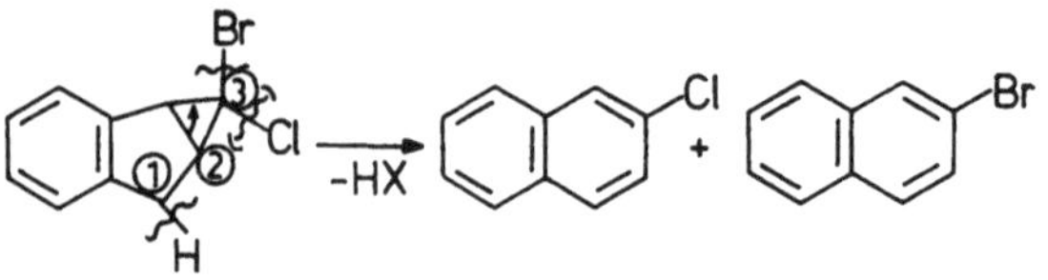

134 *135* *136* *137* *138*

2-halogeno butadienes are generated (*133*, 42%) [77,78] and the synthesis of 3-chloropyridine (*135*) from pyrrole, chloroform, and base (58%) is well known [79]. Undoubtedly, this proceeds via *134* [80] and consequently the possibility of numerous variations arises, e.g., the synthesis of *137* (35%) and *138* (35%) [81]. Under the influence of silver ions, methyl bromide may be [1,2,3]-eliminated from *139* [82]. The

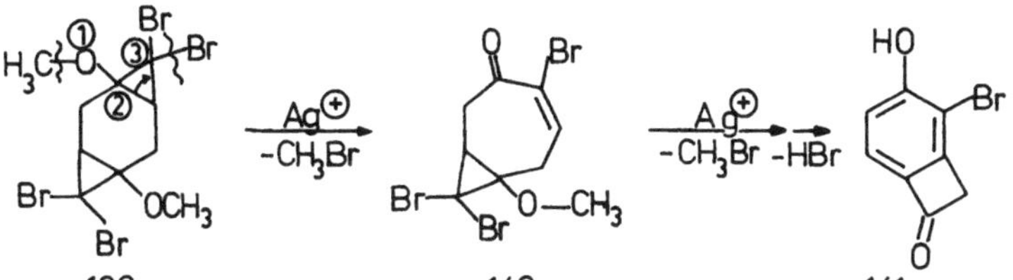

139 *140* *141*

isolable *140* (76%) may undergo a further [1,2,3]-elimination and this yields the benzocyclobutenone *141* (53%) [82] after valence isomerization and [1,2]-elimination of HBr. From 1-methoxynaphthalene and dichlorocarbene the intermediates *142* or *144* are formed, which by [1,2,3]- or vinylogous [1,2,3]-, i.e. [1,Δ,4,5]-elimination give *143* or *145* (12%) [83].

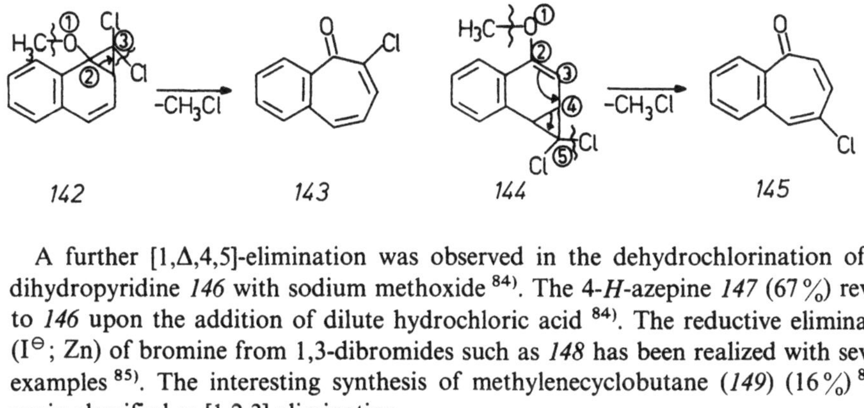

142 *143* *144* *145*

A further [1,Δ,4,5]-elimination was observed in the dehydrochlorination of the dihydropyridine *146* with sodium methoxide [84]. The 4-*H*-azepine *147* (67%) reverts to *146* upon the addition of dilute hydrochloric acid [84]. The reductive elimination ($I^{\ominus}$; Zn) of bromine from 1,3-dibromides such as *148* has been realized with several examples [85]. The interesting synthesis of methylenecyclobutane (*149*) (16%) [85] is again classified as [1,2,3]-elimination.

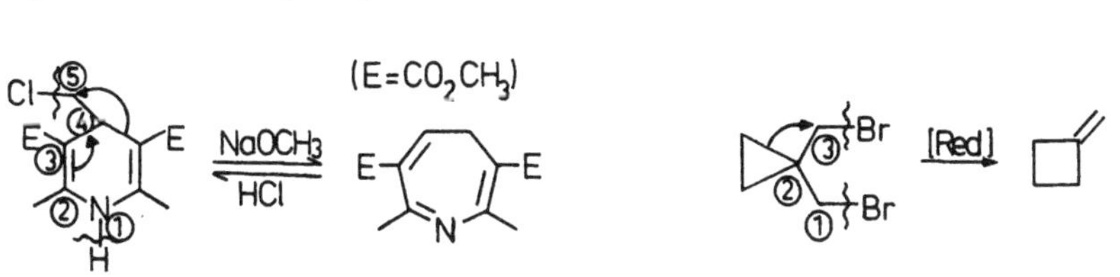

146 *147* *148* *149*

3.4 [1,2,n]-Eliminations

The early observation that dihydroxyacetone (*150*) easily eliminates water in an acidic or weakly basic medium, thus forming pyruvaldehyde *151* [86], was later confirmed [87]. This constitutes a [1,2,4]-elimination, as does the deamination of *152* to propionaldehyde [88]. In addition to the elimination, a hydrogen atom migrates from the relative position 2 to 4.

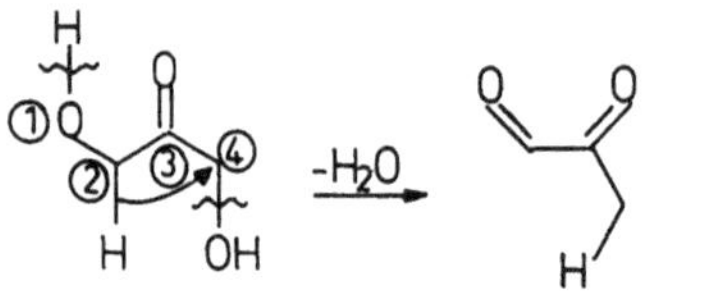

150 *151*

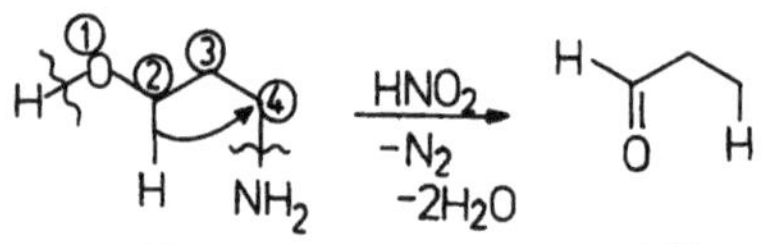

152 *153*

There are totally different reaction conditions in the [1,2,4]-elimination of HBr from isoamyl bromide (*154*), achieved by "electron bombardment", which presumably proceeds via a carbocation $C_5H_{11}^{\oplus}$ as intermediate and yields predominantly *131* [41]. In the solvolysis of *155* an alkyl migration occurs and *156* is formed (17%) [89].

In a [1,2,5]-elimination, the thermal extrusion of ethanol from the aziridine *157* (generated in situ from 2-benzoylaziridine) leads to *158*, which, in addition, undergoes a [1,2]-elimination of water [90].

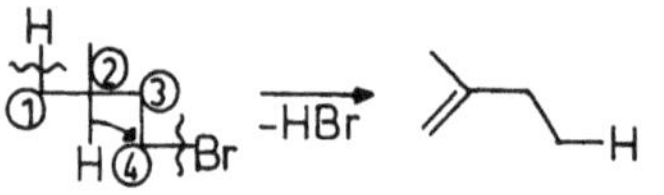

154 *131*

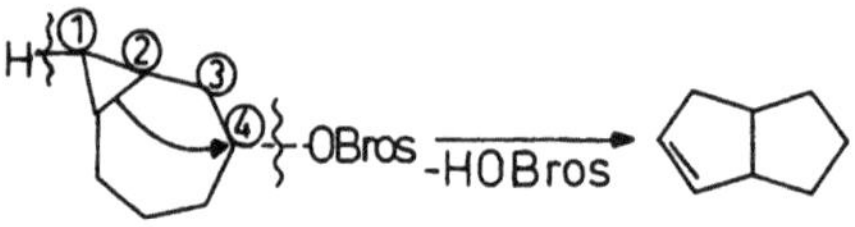

155 *156*

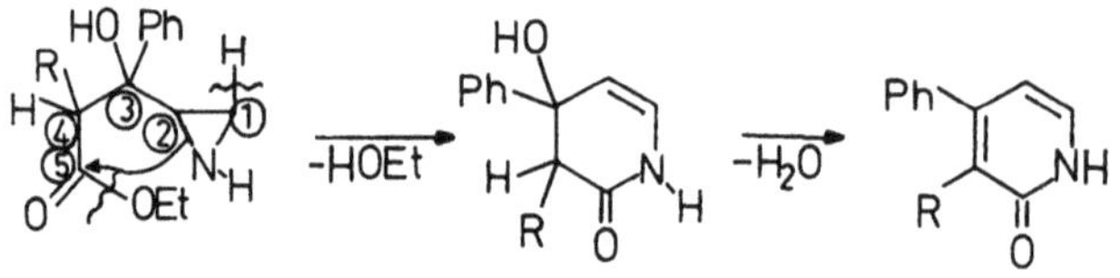

157 *158*

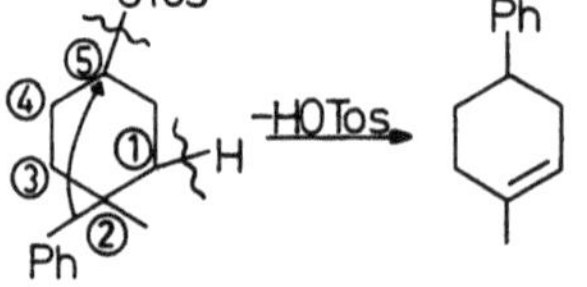

159 *160*

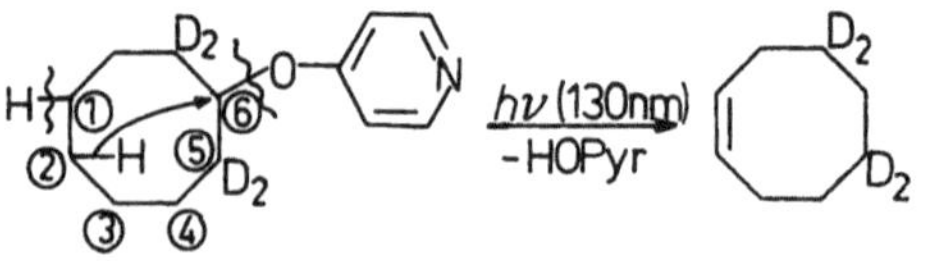

161 *162*

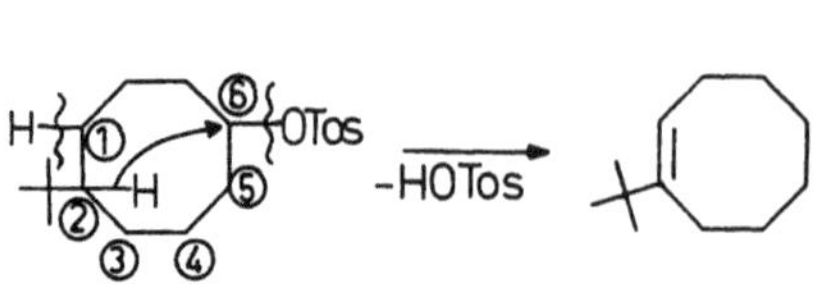

163 *164*

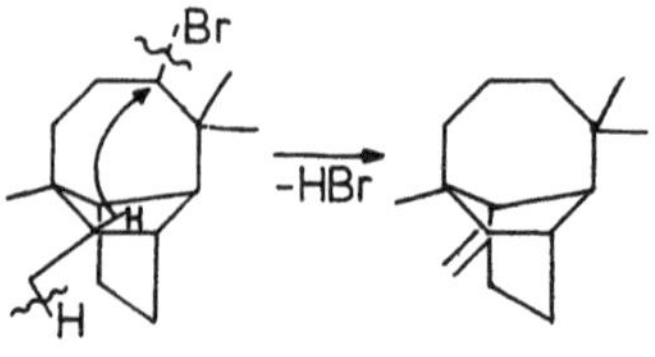

165 *166*

Regressive numbering should be avoided as far as possible. Therefore, the solvolysis of *159* (19%) [91] is classified as [1,2,5]-elimination (not [2,1,4]-e.). The reactions of the cyclooctane derivatives *161* [92], *163* [93], and *165* (quant.) [94] exemplify the [1,2,6]-elimination type.

Of course, favourable conformations in the educts are required for the success of these reactions. The same is valid for the [1,2,6]- or [1,2,7]-eliminations of the cyclodecanes *167* (proof of H-migrations by ^2H- and ^{14}C-labeling experiments) [95] and *170* [96] (*171* forms without loss of deuterium).

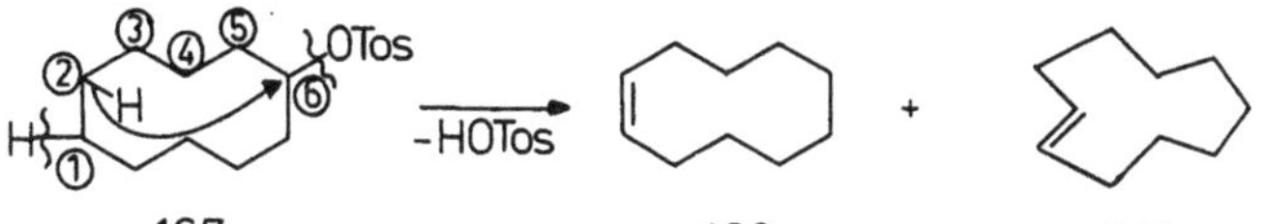

167 *168* *169*

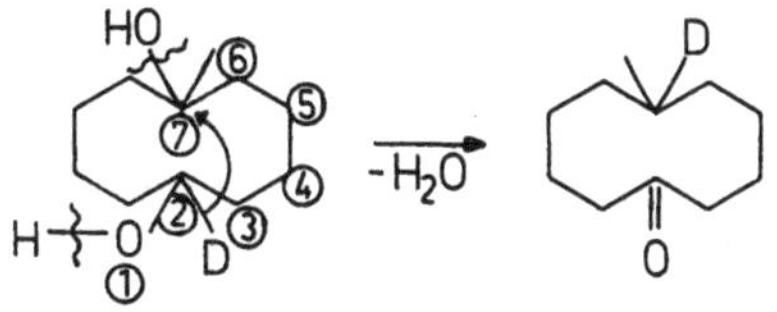

170 *171*

Double bonds can also migrate transannularly if the conformative requirements are fulfilled. Characteristic examples are the [1,2,(3)7]-, and [1,2,(3)8]-eliminations of *172* [97] and *173* [98] to give the bicyclic products *156* (12%) and *174* (14%). The [1,2,(3)5]-elimination type can be realized with cycloalkenes that have a suitable side chain. These complex eliminations succeed thermally, under the influence of bases,

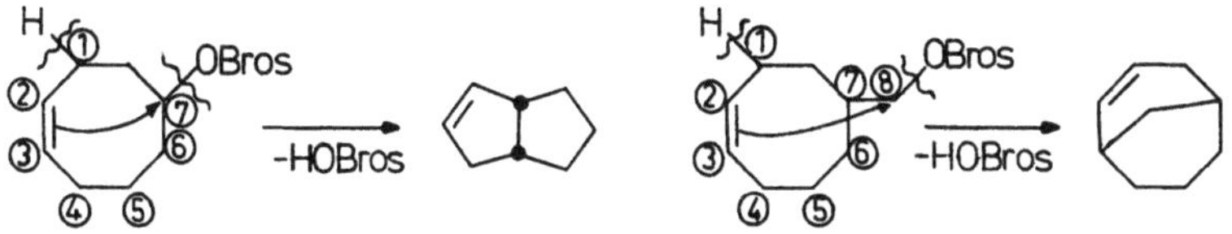

172 *156* *173* *174*

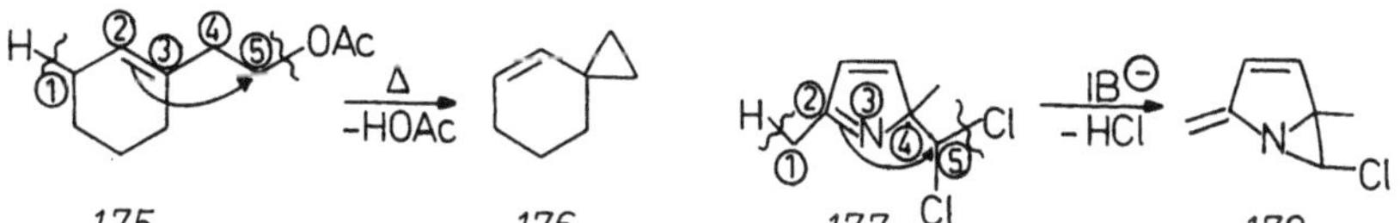

175 *176* *177* *178*

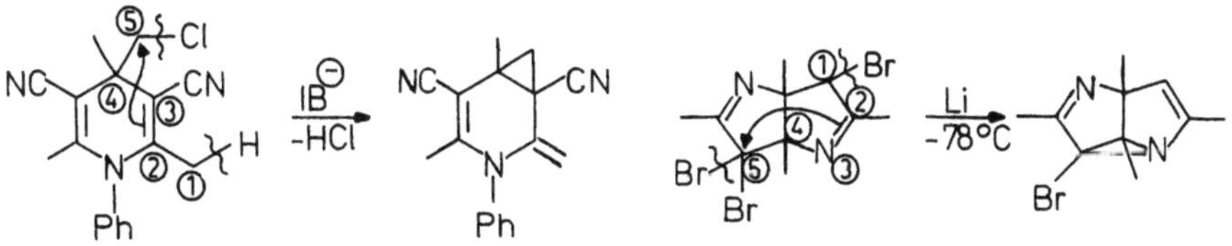

179 *180* *181* *182*

or reductively. They have high synthetic potential. This is shown in the formation of three-membered rings from *175* (26%)[99], *177* (83%)[100], *179* (45%)[101], and *181*[102].

The vinylogous case of a [1,2,(3)5]-, i.e. the [1,Δ,4,(5)7]-elimination, is observed when heating *183*[103]. The 2/3-double-bond is shifted allylicly (Δ), while the 4/5-double-bond migrates (assistant figure). The azanorcaradiene derivative *184* (40–50%) prefers the monocyclic form.

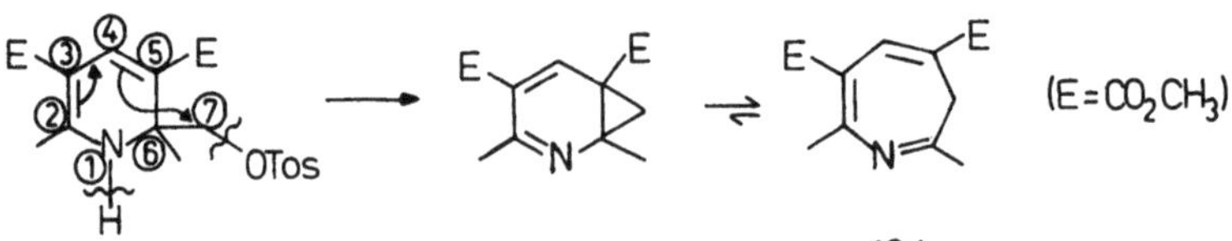

183 *184*

At the end of this section, it is necessary to discuss a special case: If the electronegative leaving group is an inner element of the main chain, its relative position number has to be placed at the end in parentheses in order to reach an unequivocal classification. This procedure is illustrated in the typical [1,2,4(3)]-eliminations of *185* to give urea and thiirane *186*, as well as that of *187* to give acetic acid and *186*[104] (see also *200* → *201*; *253*; *275*; *277*; *279*; *311*; *200* → *292* and *313*; *314* → *313* and *315*; *316*).

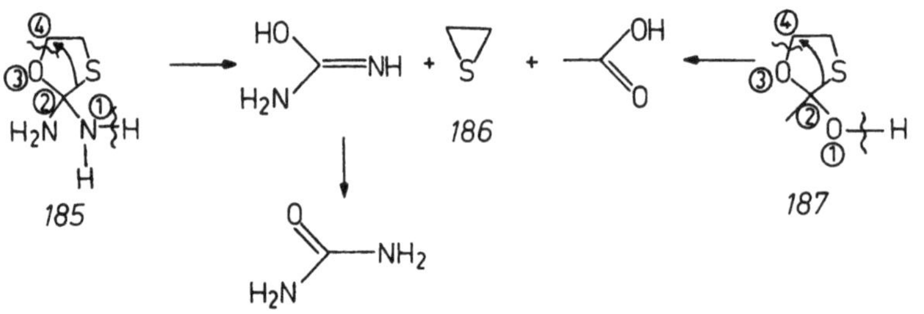

185 *186* *187*

3.5 [1,2,(3)4]-Eliminations and Homologous "Fragmentations"

[1,2,(3)4]-eliminations are only distinguished from the [1,2,4]-type by the fact that it is a bond from the main chain and not a substituent that migrates (from 2–3 to 3–4). This leads to fragmentation (well-known under the name "Grob-fragmentation"; recent literature review in Ref. [105]) and is specified by the assistant figure in parentheses.

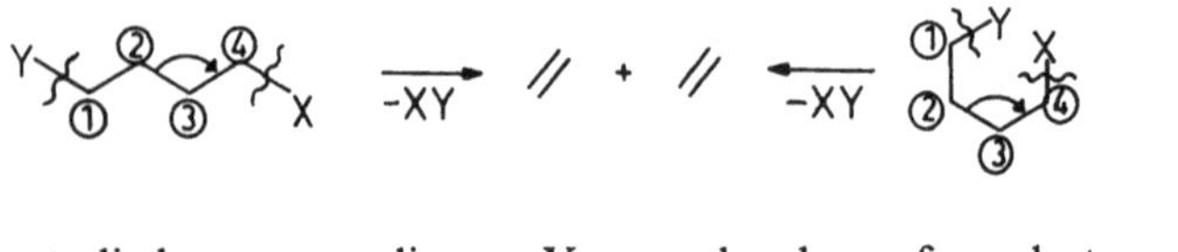

(7)

The most studied cases are linear. Y can also be a free electron pair of an amine (Scheme 7). The chains may be present in extended or cisoid conformation.

There are numerous bridging possibilities. The extraordinary variety can be exemplified by the pyrolysis (700 °C) of *188* to give *190* (40 %) [106], the anodic oxidation of *191* to give *192* (55 %) [107], and the photochemical (mass spectrometrical) "Norrish type-II cleavage" ("McLafferty-rearrangement") of carbonyl compounds with γ-H-atoms (*193*) to give enols and alkenes [108].

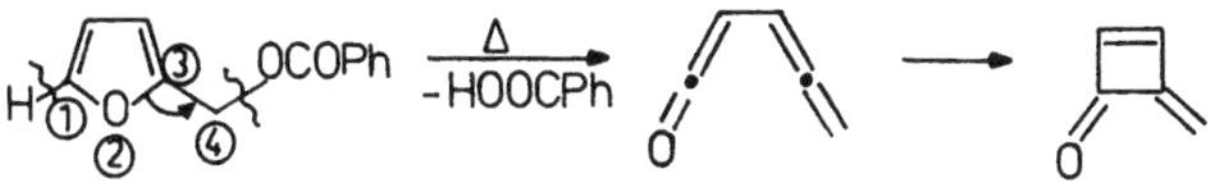

188 *189* *190*

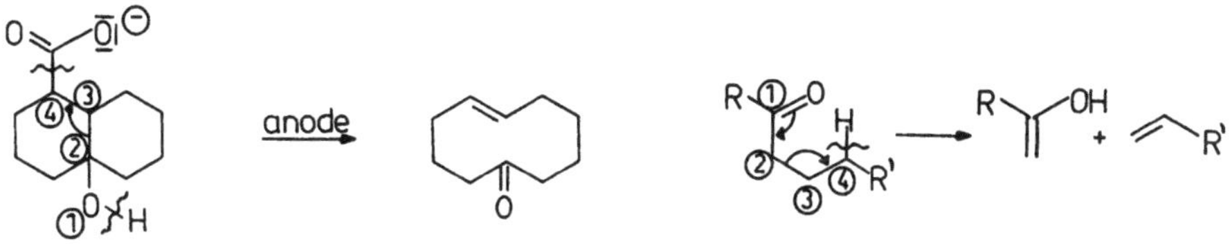

191 *192* *193*

Cyclic exponents of the same elimination type are of particular interest. Thus, numerous cyclopentanones photolytically decarbonylate to give 1,4-dienes (p. 876ff. in Ref. [108]). With thujone (*194*) this [1,2,(3)4]-elimination of carbon monoxide proceeds quantitatively to give *195* [109]. With silver nitrate the norcaradiene *196* yields *197* (95 %) [110] apparently regioselectively, and the [1,2,(3)4]-elimination of methanethiol from *198* to give *199* was realized thermally (16 %), acid-catalyzed (59 %), and photochemically (ca. 5 %) [105]. For the acid-catalyzed reaction (acetic acid, 100 °C) a non-stereospecific process has been proved [105].

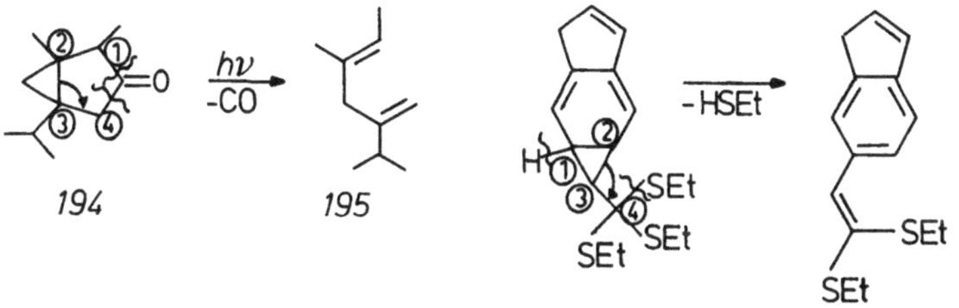

194 *195* *196* *197*

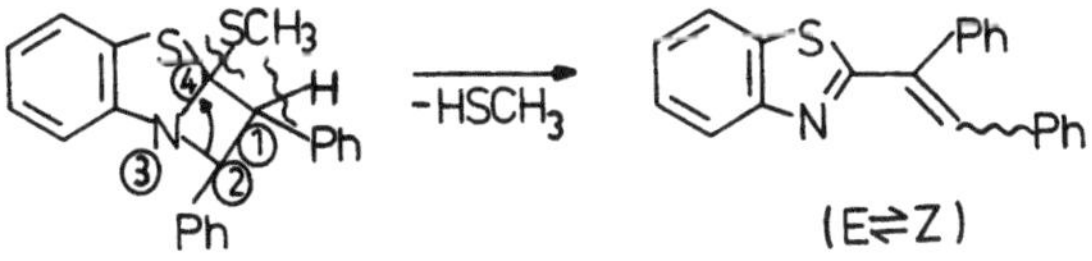

198 *199*

In homologous [1,3,(4)5]-eliminations, three-membered rings are formed in addition to alkenes, and bishomologous [1,3,(4)6]-eliminations lead to two three-membered ring fragments. The reaction of *200* at 220 °C with formation of *201* (18 %) [88b] is a bishomologous [1,3,(4)6(5)]-elimination, in which the electronegative leaving group (carbon atom in relative position 5) is an inner constituent of the main chain (cf. *185*, *187*), and where labile α-lactone [111] forms as the second fragment. The isolated product *201* consists of both leaving groups.

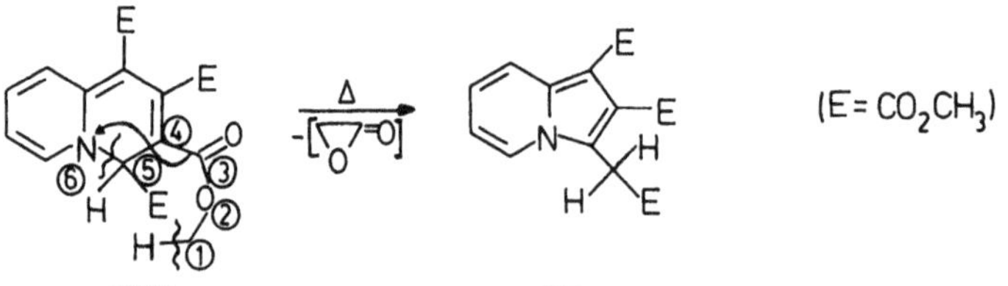

200 *201*

If the reaction type of Scheme (7) is expanded to educts with 2/3-double-bonds, this leads back to the simple (1,Δ,4]-eliminations, with 1,3-butadiene formations [112].

3.6 Higher [1,m,n]-Eliminations

It must be assumed that higher [1,m,n]-eliminations (m ≠ n, m > 2, n ≧ 2) can be found and developed in a large variety of ways. The synthetic potential of those types can be indicated with some examples: Thus, the thermal conversion of *202* to *204* (48%) [113] can be reasonably formulated via initial [1,3,4]-elimination of hydrogen chloride to produce the intermediate bicyclobutane *203* which is stabilized by a [1,2/2,1]-rearrangement. The base-induced reaction of *205* (55%) leads to the displacement of *p*-toluene-sulphonic acid and to the migration of the inner chain bond [114]. Therefore, it must be classified as [1,3,(2)4]-elimination. The dehydroxy-silylation of *207* to *208* (20%) [115] is typical for [1,4,(3)2]-eliminations (migration of the Se—C-chain bond).

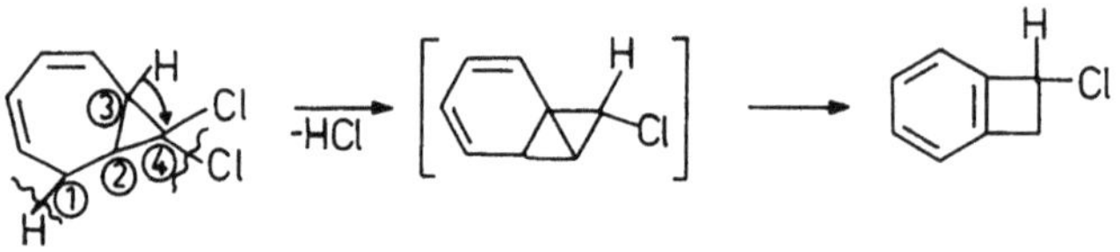

202 *203* *204*

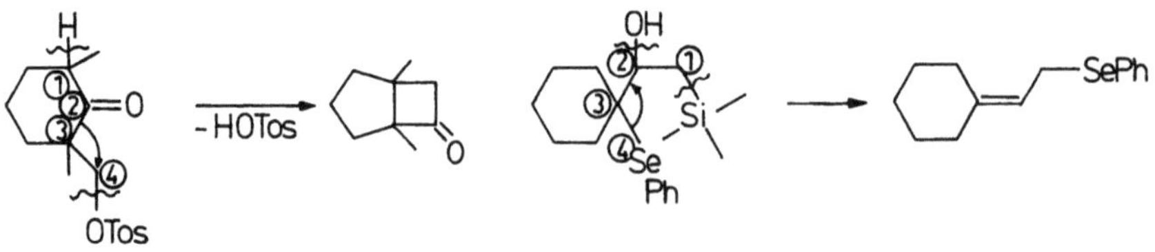

205 *206* *207* *208*

A [1,5,3]-elimination of water is realized in the transformation of *209* into the bicyclic compound *210* (66%) [116]. The formation of imidazole *212* (73%) from *211* [117] is a [1,5,(4)3]- and that of *214* (90%) from *213* [118] a [1,8,(7)3]-elimination of nitrogen.

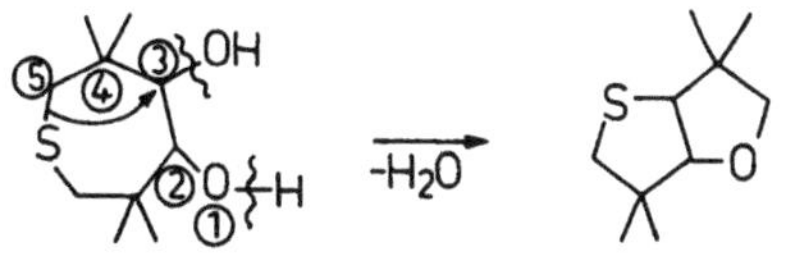

209 210

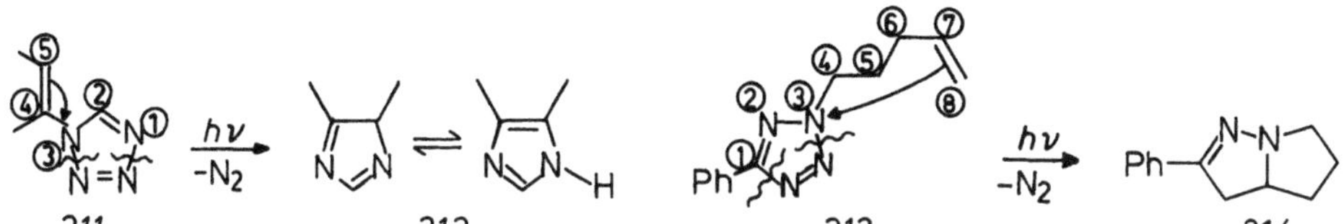

211 212 213 214

3.7 [2,1,3]-Eliminations

In the eliminations in the preceding sections the electropositive leaving group is always situated in the 1-position. However, there are also elimination types in which the electropositive leaving group is situated at an inner chain position. These kinds of reactions give rise to very profound chemical changes, which, nevertheless, can compete with [1,2/2,1]- or [1,3/3,1]-rearrangements.

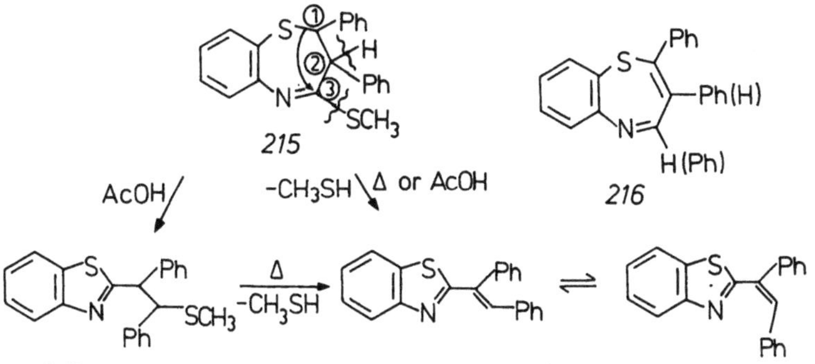

215 216 217 199

The [2,1,3]-elimination of methanethiol from *215* to give *199* proceeds purely thermally at 200 °C (78 %) [119] and also acid-catalyzed (glacial acetic acid, 150 °C, 65 %) [105]. An imagined [1,2,3]-elimination to give *216* is not observed. Only in glacial acetic acid is the product of the [1,3/3,1]-rearrangement *217* (pair of diastereomers, 35 %) formed in addition to *199*. At 200 °C *217* reacts to *199* by a [1,2]-elimination [105]. Yet, *217* is not necessarily an intermediate in the transformations of *215* into *199*, and there are still further possibilities for intermediates (e.g. rearrangement to the heterodiene thiol), as the sequence of bond-cleavages and -formations is not always clarified.

Upon thermolysis (200 °C) of the amide *218*, the [1,2/2,1]-rearrangement to give *220* (16 %, stable to ≥300 °C) competes with the [2,1,3]-elimination of water to give *221* (67 %) [120], which is formulated via the protomer *219*, which must be present in equilibrium with *218* at that temperature (cf. *253*).

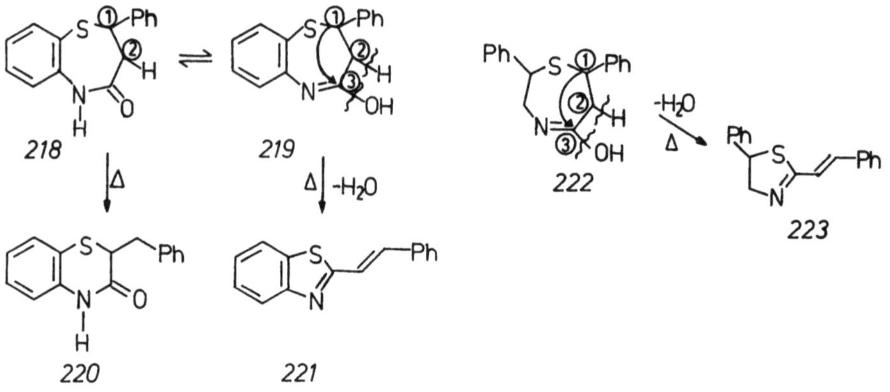

The reaction of *219* can be accelerated by alkylating reagents (triethyloxonium-tetrafluoroborate) in dichloromethane (79%) [121]. Closely related to this is the action of polyphosphoric acid on *222* (enolic form, 135 °C), which yields *223* (27%) [121].

The [2,1,3]-elimination of ethanol from the aziridine *224* is clearly base-induced (mesityl-magnesium bromide) and forms *225* (77%) [122]. In the ester cleavage of homoadamantyl acetate (*226*, gas phase), both possibilities of the [2,1,3]-elimination are observed, i.e.: C—C-migration leads to *227* (30%), C—H-migration to *228* (5%) [123].

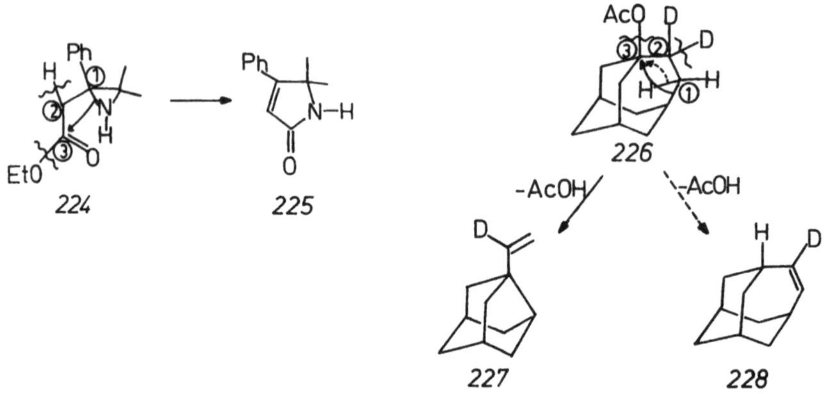

3.8 [2,1,n]- and [m,1,n]-Eliminations

[2,1,n]-eliminations can be observed in displacements of nitrogen or carbon monoxide if the geometrical conditions in the educts are favourable. This is shown by the reactions of *229* ([2,1,4]-elimination to give 43–87% *230*) [124], *231* ([2,1,5]-e. to give 95% *232*) [125], and *233* ([2,1,6]-e. to give *234*) [125]. In these types, the leaving groups may be varied further [126] and one can imagine additional generalizations in terms of [m,1,n]-eliminations [127].

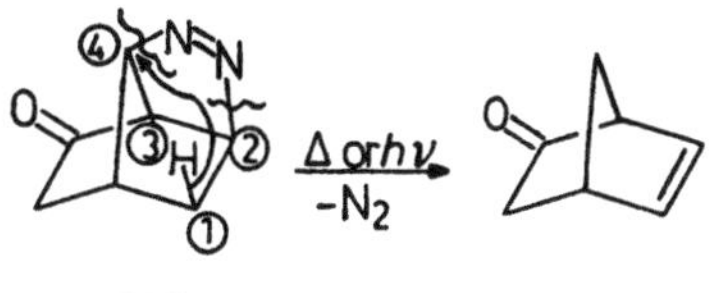

229 230

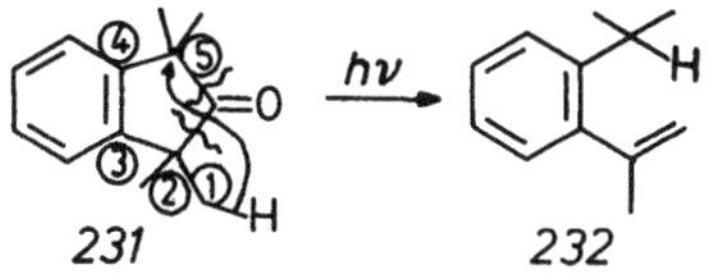

231 232

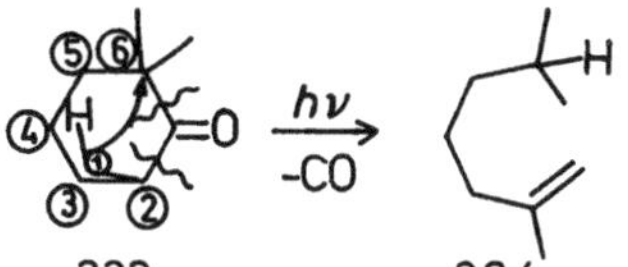

233 234

3.9 Eliminations Involving Branched Chains

It is long known that the displacement of water from *235* yields the spirolactone *236* (90%) in a rather elegant way [128]. As there is a branching in the chains of *235*, it is unavoidable to attribute prime numbers to the relative positions in one

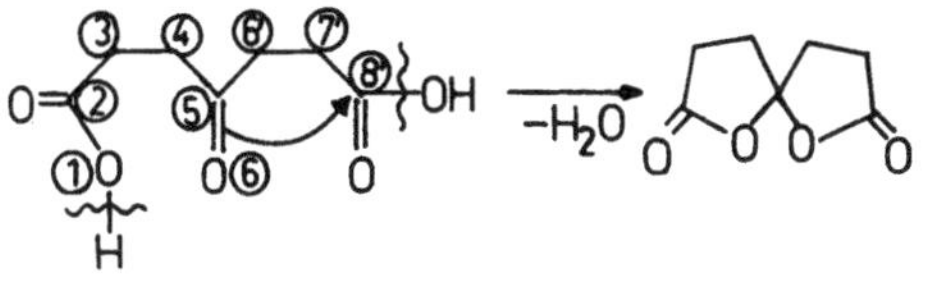

235 236

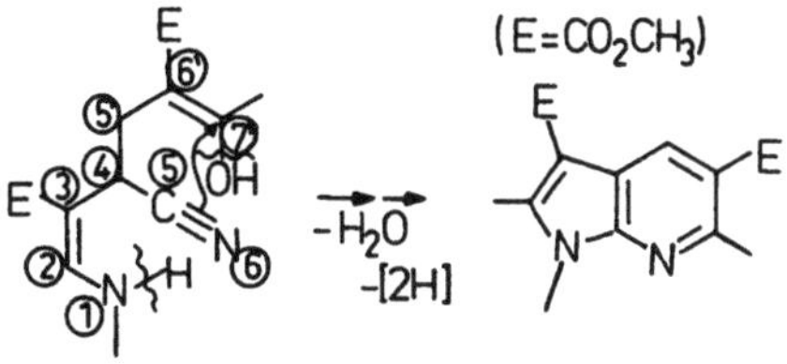

237 238

of the branches. Thus, the formation of *236* is the result of a [1,5,(6)8′]-elimination. Correspondingly, the formation of *238* from the postulated intermediate *237* [129] is an example of a [1,5,(6)7′]-elimination (here followed by a dehydrogenation). Also the 1-H-hexahydrophenalene synthesis of Huisgen (66%) [130] has to be mentioned here. An olefinic intermediate which absorbs at short wavelengths with low ε-values has been isolated upon formolysis of *239* [130]. If the constitution of this were *240*, its formation should be termed a [1,4,(3)4′]-elimination. By "Friedel-Crafts reaction" *240* would consequently form *241*.

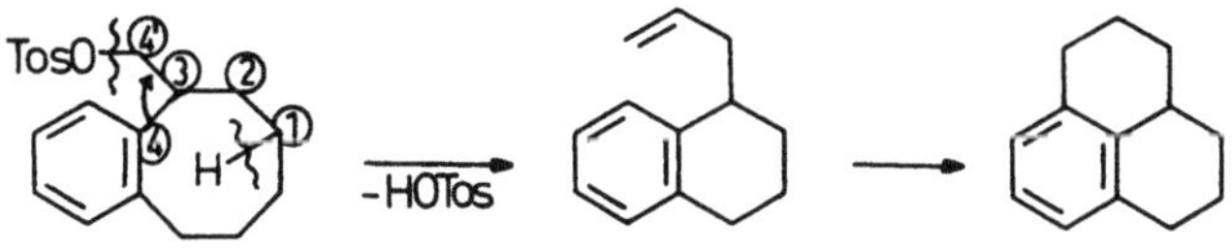

239 240 241

4 Complex Eliminations with Two Migrating Groups (Bonds)

Eliminations with additional migration of two groups or bonds lead to extremely profound chemical changes. For the sake of an easy survey the types with exclusive migration of substituents are treated first. Then follow those types with migration of one substituent and one single- or double-bond (aside from allylic shifts) and finally those types with migration of two bonds from the main chain.

4.1 Migration of Two Substituents

In [1,2,3,4]-eliminations one substituent migrates from 2 to 3 and displaces a substituent from there to 4, from where the electronegative leaving group departs. The acid catalyzed dehydration of *242* (28% + 40% of the acetal of *243* with *242*) [131] corresponds to this type unless a [1,2,3]-elimination would yield intermediately the enol of *243* with a [1,3]-hydrogen-shift to follow.

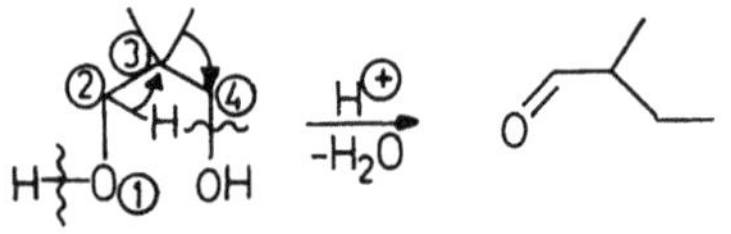

242 *243*

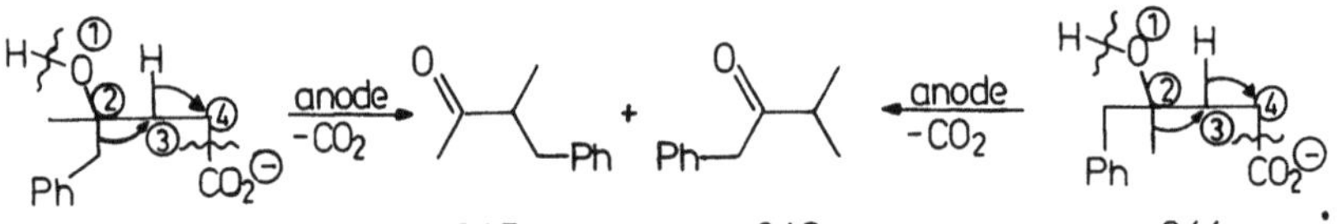

244 *245* *246* *244*

In the anodic oxidation of *244* both of the possible [1,2,3,4]-eliminations are observed. Migration of the benzyl group produces *245* and migration of the methyl group *246* (ratio 2.5:1) [76]. Two homologs constitute the [1,2,3,5]-eliminations of ethanol (induced by bases) from *247* [122] or of hydrogen chloride (photochemically) from *249* [132] which lead to *248* (95%) and *250* (20%). Several mechanistic possibilities may be envisaged, but they are hard to distinguish.

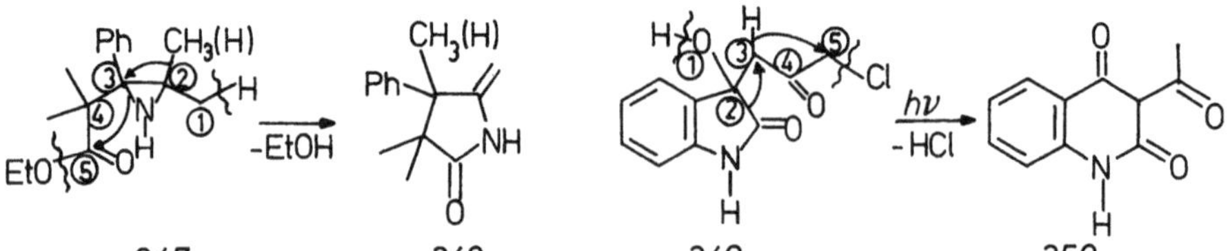

247 *248* *249* *250*

The series may also be extended regressively. Thus, upon heating of *251* [133] the [1,2,3,2]-elimination of water gives *252* (5%) and the [1,2,3,2(3)]-elimination of hydrogen sulfide from *253* (220 °C, one of the chain bonds belongs to the leaving group) produces *254* (55%) [120]. Even [1,2,3,1]-eliminations are known: *255* produces no allene but *256* (50%) instead, upon treatment with sodium [134], and the tosylhydrazonate *257* gives prototropic *258* (48%) upon heating [135].

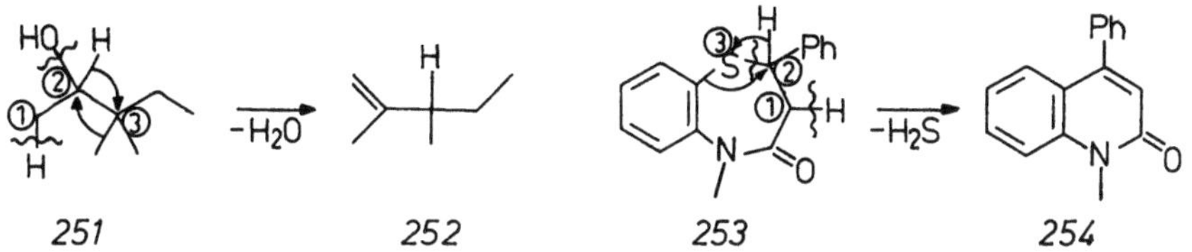

251 *252* *253* *254*

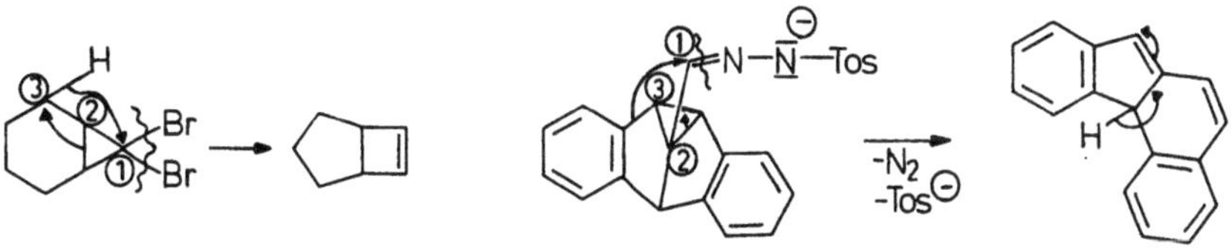

255 *256* *257* *258*

Further types, which may be exemplified, are the [1,2,4,1]-elimination of nitrogen from *259* [136] to give *260* and the solvolytic [1,2,1,6]-elimination of *p*-toluolsulphonic acid from *261*, which yields *262* (40%) [137].

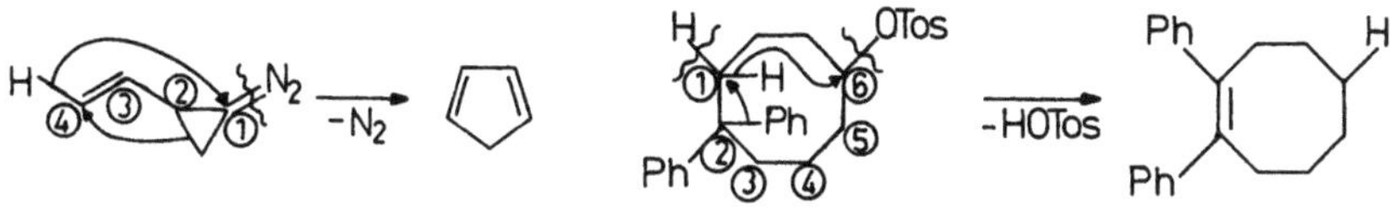

259 *260* *261* *262*

A [1,5,1,6]-elimination of sulphur dioxide has been observed in the transformation of *263* into prototropic *264* (87%, via isolable 4,5-diphenylpyrazol-3-ylmethanesulphinic acid) [138]. It appears, that the precise reaction mechanisms of these complex eliminations, which occur under very different experimental conditions, would be elucidated only with undue efforts, prior to a collection of empirical knowledge by the practical exploitation of these reaction types. The same is true in the following sections.

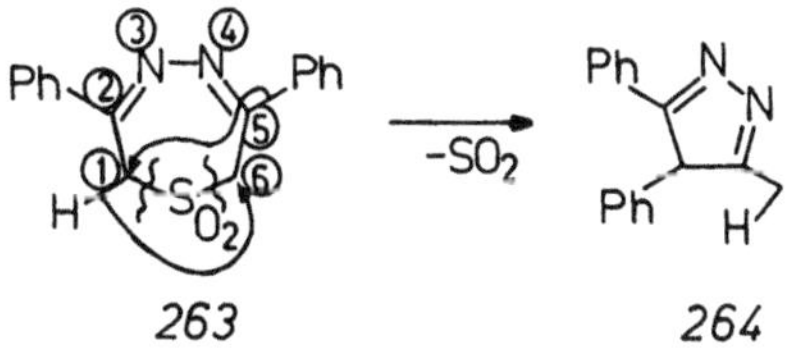

263 *264*

4.2 Migration of One Substituent and One Bond of the Chain

Nitrogen displacement from the tricycle *265* produces the cyclopentene *266* (26%). This complex reaction has been formulated via a photolabile transient intermediate [139].

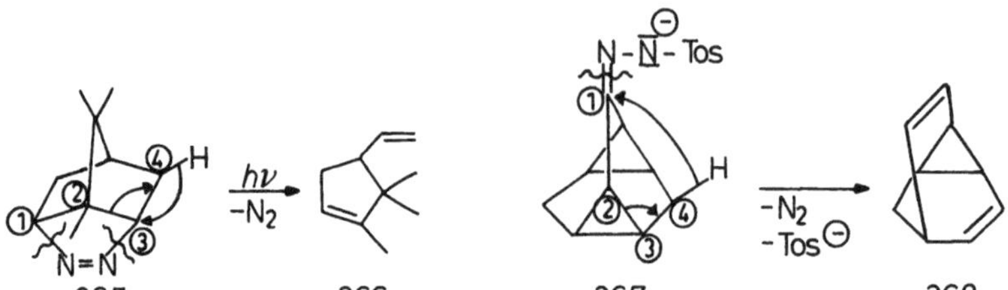

265 266 267 268

Overall, this reaction has to be classified as a [1,2,(3)4,3]-elimination, because nitrogen is liberated from relative positions 1 and 3 (first and last figure), the 2/3-bond migrates to 3/4 (assistant figure = 3) in order to induce the substituent in 4 to migrate to 3. By complete analogy, the synthesis of barbaralane (*268*) from *267* (43%) [140], which has been formulated via a cation, has to be classified as [1,2,(3)4,1]-elimination, because both the 2/3-bond and the 4-H migrate.

Both the thermolysis (700 °C) of phenylazide (*269*) to give *270* (50%) [141] and of the tosylhydrazonate *271* to give the cyclononynone *272* (56%) [142] are [1,2,3,(2)1]-eliminations. For topological reasons the migration of the "substituent" (three-membered-ring bond) in *271* leads to the generation of a double-bond. Consistently, the nitrogen liberation from *273* to give (*E*/*Z*)-*274* (80%, both isomers) [143] is to be classified as [1,2,6,(5)4]-elimination.

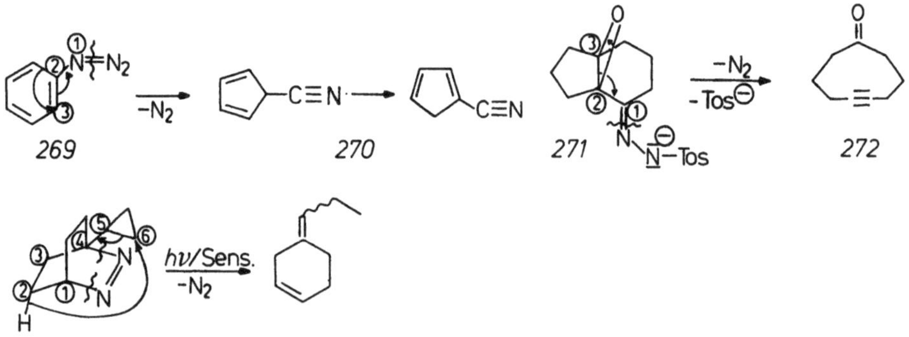

269 270 271 272

273 274

[1,3,1,3(2)]-eliminations occur in the displacements of carbon oxysulfide or carbon dioxide from *275* (81%) [144], *277* (63%) [145], and *279* (71%) [146], which are induced thermally, with tri-*n*-butylphosphine or hydrogen iodide. With respect to the completely different reaction conditions it is to be assumed that these reactions select different but difficult to evaluate reaction mechanisms.

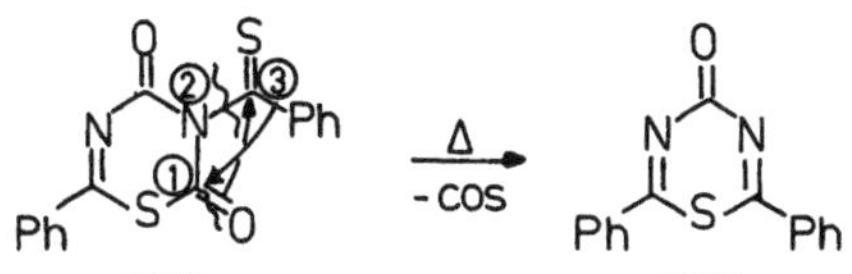

275 276

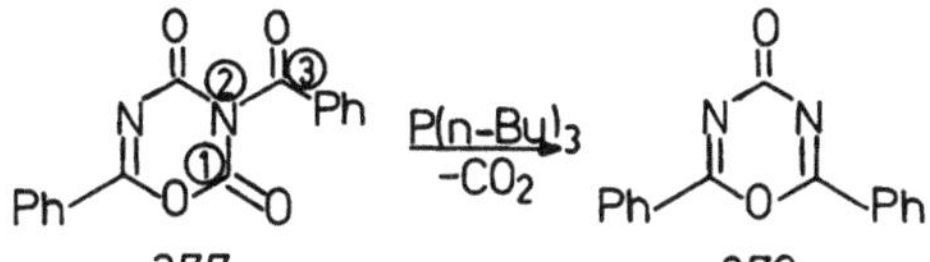

277 278

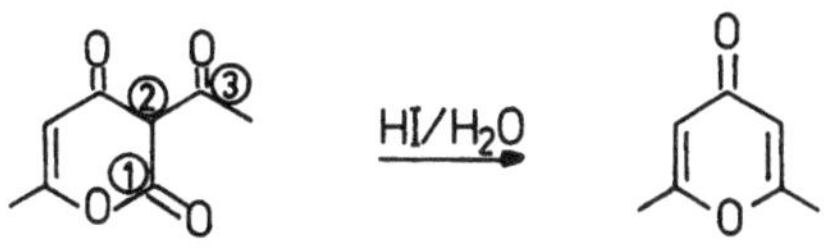

279 280

A subtype of the [1,3,2,3]-elimination is found in the quantitative [1,3,2,(1)3]-elimination of water from oxazepame (*281*) which produces the aldehyde *282* [147]. Here, both the 3-H and the 2/1-bond migrate. The deuteration experiment shows that there is *no* [1,3,(2)3,1]-elimination as depicted in *281'*. Nevertheless, there remain numerous, as yet hardly distinguishable mechanistic possibilities for the process *281* → *282* [147].

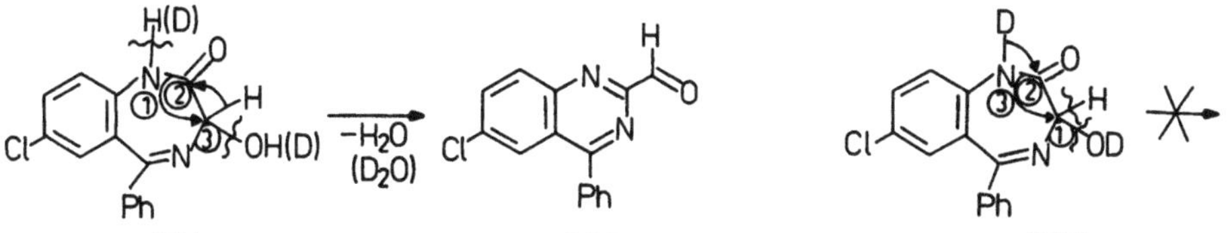

281 282 281'

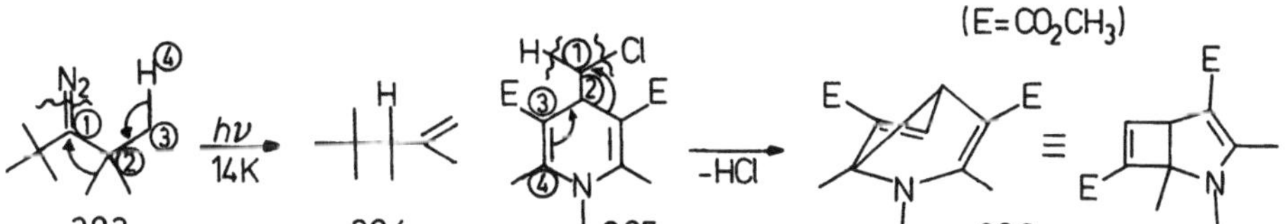

283 284 285 286

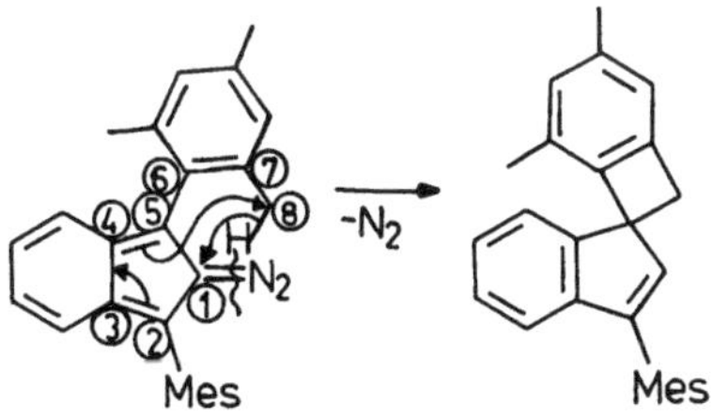

287 288

[1,4,(3)2,1]-eliminations occur in significant proportion beside other reaction modes upon photolysis of *283* [47c)] (the hydrogen is part of the main chain here) and upon heating of *285* (30%) [129)] (here migration of the double bond). The formation of *288* (74%) from *287* [148)] has to be classified as [1,Δ,4,(5)8,1]-elimination (here both allylic shift of the 2/3-double bond and migration of the 4/5-double bond).

Also comparatively high figures are necessary for the classification of elimination types which occur upon photolyses of suitable ketenes, *p*-tosylhydrazonates and diazo compounds. Thus, the compounds *84* (intermediate: fenchenaldehyde) [32)] and *82* (38%) [51a)] react also in terms of the [1,5,(6)7,1]-elimination, thereby producing *289*. Similarly, the diazo compound *290* experiences the photolytic [1,7,(6)5,1]-elimination resulting in *291* (5%) [49)].

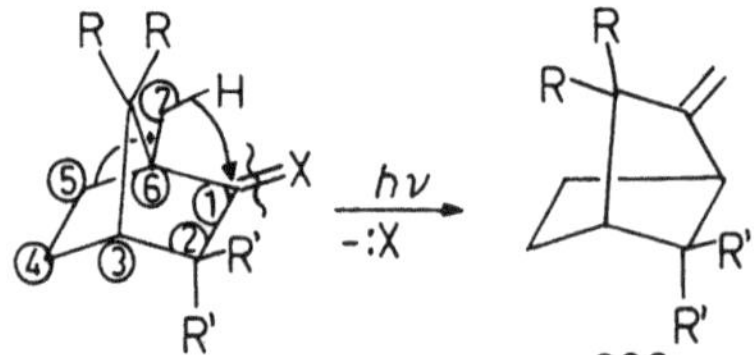

84: X=CO, R=H, R'=CH_3
82: X=NNNaTos, R=CH_3, R'=H

289

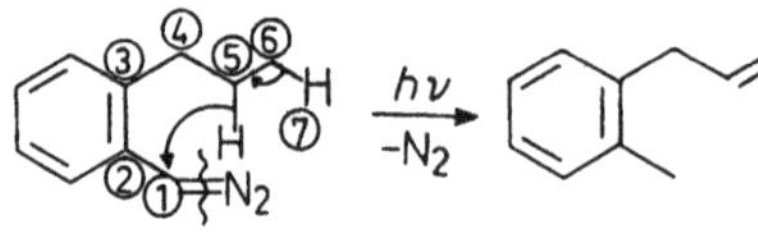

290 *291*

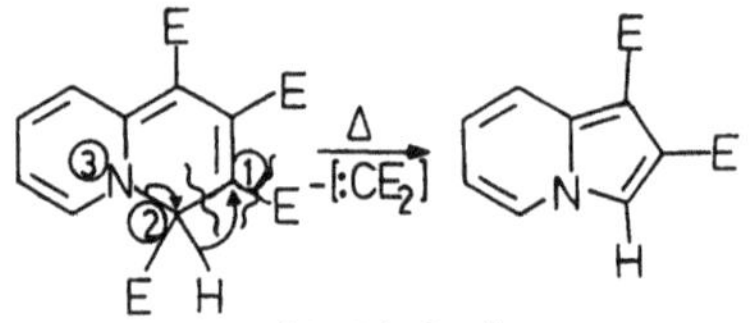

200 ($E=CO_2CH_3$) *292*

Even more complicated is the [1,3,(2)2,1(2)]-elimination of dimethoxycarbene (formally) from *200*, which produces *292* (3%) [88b)]. One of the leaving groups is part of the main chain and this requires therefore the final assistant figure in the unequivocal classification.

The generally formulated decomposition reactions of alkanes, alkenes, and (hetero-) dienes in Scheme 8 are the basic types of [2,3,1,(2)3]-eliminations. Here, the electropositive leaving group (or the double bond to be abandoned) is situated in position 2, i.e. at an inner position of the chain. At first glance in the decomposition

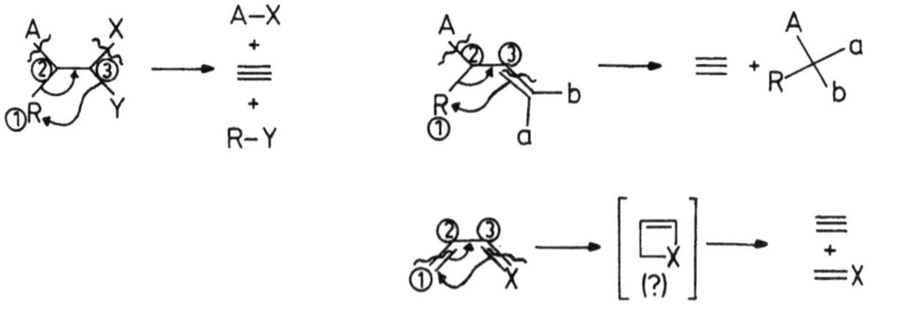

(8)

reactions of the heterodienes, it appears very reasonable that these proceed, as depicted, via heterocyclobutenes. However, such speculation always requires close scrutiny, because there exist also differing mechanistic possibilities. Indeed, it has been shown that the thermal (≧300 °C) [2,3,1,(2)3]-elimination of nitrogen from benzaldazine (*293*) with the formation of *trans*-stilbene (*294*, 61 %) follows a chain reaction, with phenyldiazomethane as chain carrier [150]. Photochemically a further mechanism via initial dimerization of *293* has been observed (*cis*- and *trans-294*, 5 %) [151].

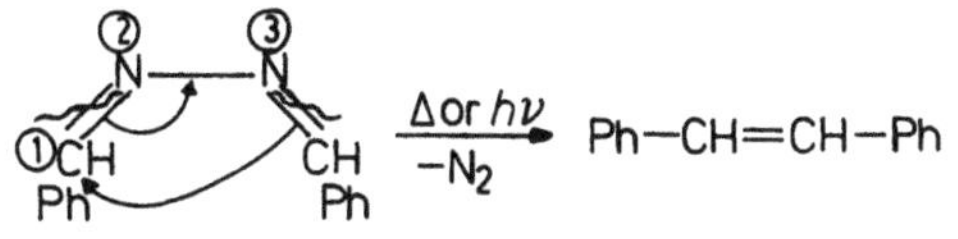

293 *294*

While an (in terms of the chemical result) analogous formation of azo-arenes does not succeed from 1,4-diaryltetraazadienes [152], nitrogen displacements from N-nitrosoimines are well known and have occasionally been used for the synthesis of carbonyl compounds [153]. The early impressive synthesis of *296* (90 %) [154] illustrates this clearly. Closely related are the [2,3,1,(2)3]-eliminations of acylthioketenes such as *297* [155], acylisocyanates such as *299* (quant.) [156], and of thioacylisocyanates such as *301* [157], which produce acetylenes (*298*) and nitriles (*300*). In all these cases hetero-cyclobutenes (cf. Scheme 8) have been postulated as intermediates.

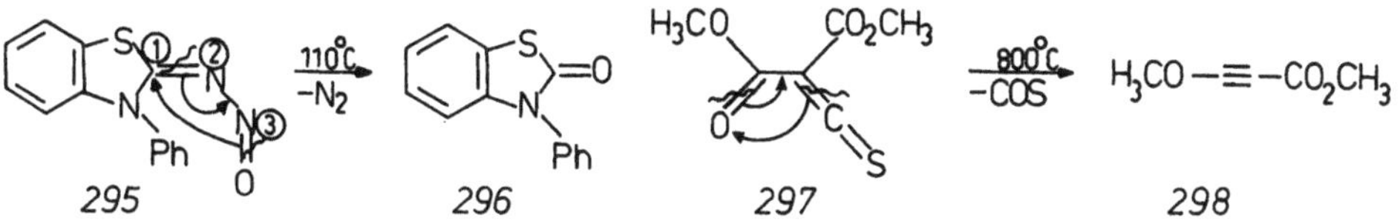

295 *296* *297* *298*

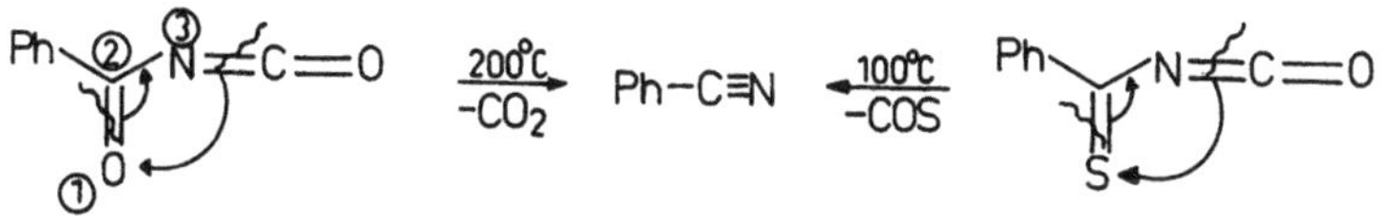

299 *300* *301*

In this section too, a branched type is available for classification, in which a continuous numeration is not possible. Thus, the formation of *303* (8–19 %) from *302* [158] is an example of the [1,3,(2)3′,1]-elimination. There is migration of the 3/2-bond and of a hydrogen, which is situated at position 3′ in *302*.

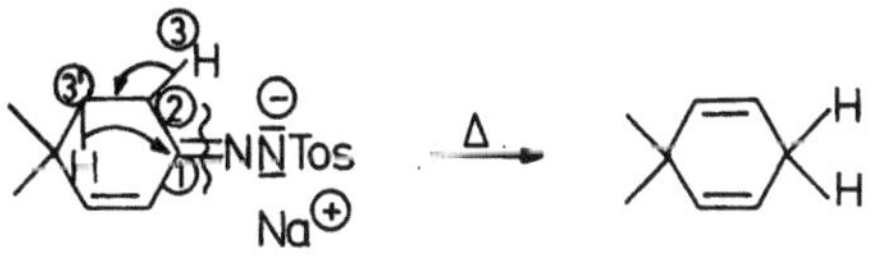

302 *303*

4.3 Migration of Two Bonds of the Chain

There are numerous, even preparatively important, cases of complex eliminations, which do not allow designation of a chain in such a way that at least one of the migrating species is a substituent. Such a reaction may be unequivocally classified if two assistant figures together indicate both of the migrating bonds. The vinylogous (here oxidative) "Grob fragmentation" of *304* to give *305* (34%) [159] is obviously a [1,2,(3)4,(5)6]-elimination, because there is migration both of the 2/3-bond and of the 4/5-bond, while the leaving groups depart from positions 1 and 6. Interesting decarboxylations are found with *119* (to give CO_2, ethanol, and tolane, 24%) [71] and with the saturated orthoester derived from *119* (to give CO_2, ethanol, and stilbene) [71]. These are clearcut [1,2,(3)4,(5)1]-eliminations. The same classification applies in the formation of *306* (12%) from *307* [161], whereas its vinylogous type, i.e. the [1,2,(3)Δ,6,(7)1]-elimination produces *308* (37%) [161].

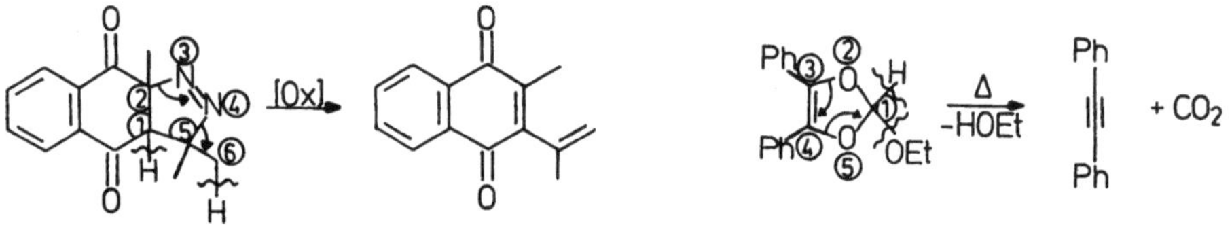

304 *305* *119*

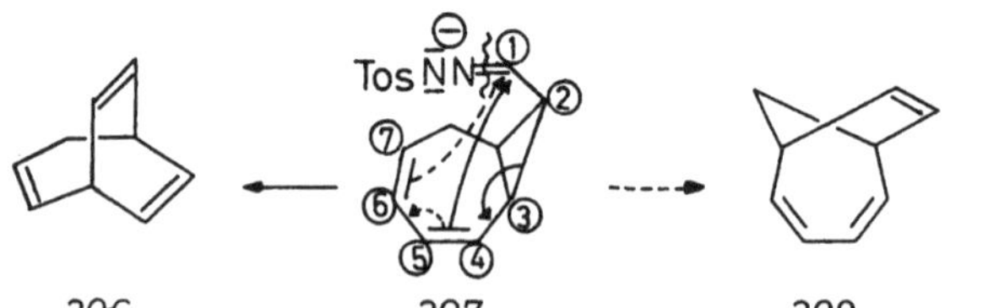

306 *307* *308*

The tetrahedrane synthesis of Maier [162] starting with *309* represents a clearcut [1,3,(4)2,(1)4]-elimination of carbon monoxide. The formation of *310* is an impressive example of the migration of two double bonds.

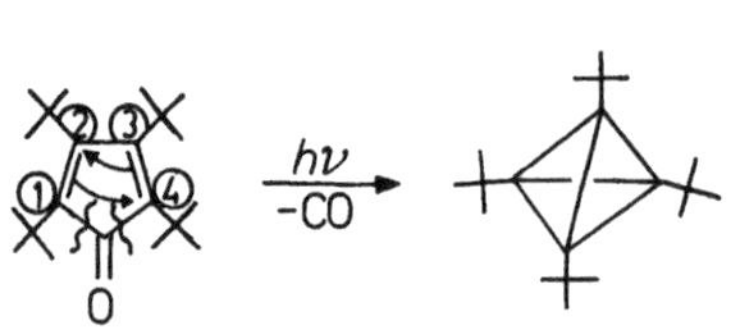

309 *310*

These types are particularly involved whose one of the leaving groups is part of the chain. Thus, in the "abnormal Wolff-Kishner reduction" 4,5-dihydro-1H-pyrazoles are formed (e.g. *311*), whose pyrolysis leads to [2,3,(4)1,(2)3(4)]-eliminations of nitrogen with cyclopropane formation. In this way *311* forms *312* (75%) [163]. The three assistant figures are necessary in order to fix which chain-bonds migrate and where the second leaving group is located. Even the eliminations of carbenes (formally) from *200* ⇌ *314* which yield *313* and *315* (for *292* also the [1,3,(2)2,1(2)]-elimination type applies, see Section 4.2) [88b)] are easily and unequivocally classified. These are obviously [1,3,(2)2,(1)3(2)]-eliminations. The notation fixes the relative position of the leaving groups (one of them being part of the chain) and the migration of the 3/2- and 2/1-bonds. This is also visualized in the depicted formulae.

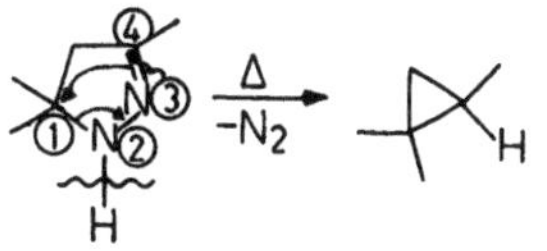

311 *312*

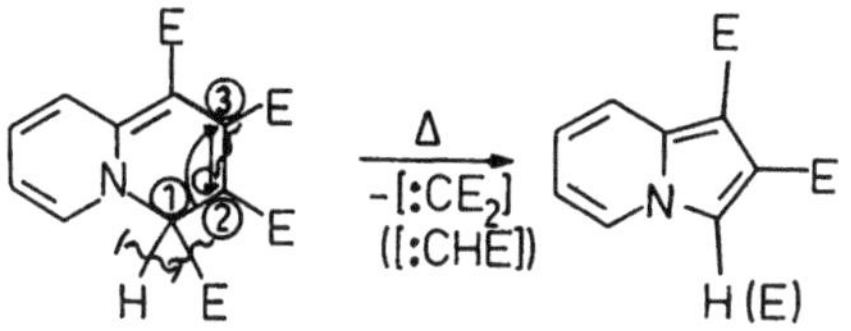

200 *292 (313)*

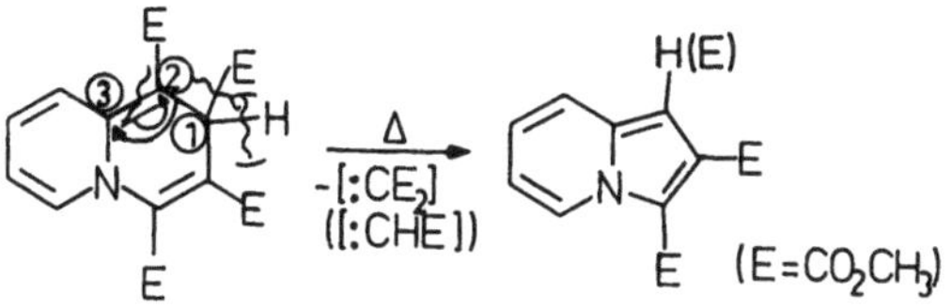

314 *315 (313)*

The formation of enones from isoxazolidines [164)] is to be classified as [1,2,(3)6, (5)4(3)-elimination. This is independent of the question whether there is intermediately a methylation by trimethylphosphate (displacement of dimethylamine and $H^{\oplus}$) or not (displacement of methylamine). An example is shown in the synthesis of *317* (90%) from *316* [164)].

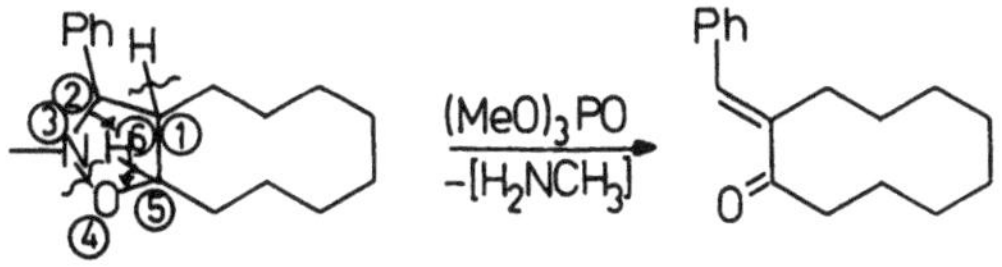

316 *317*

5 Complex Eliminations with More Than Two Migrating Groups

Principally, also the classification of eliminations with multiple migrations is possible according to the developed criteria. Thus, in the so called "Skattebøl rearrangement", which, if there is suitable substitution, results in the migration of

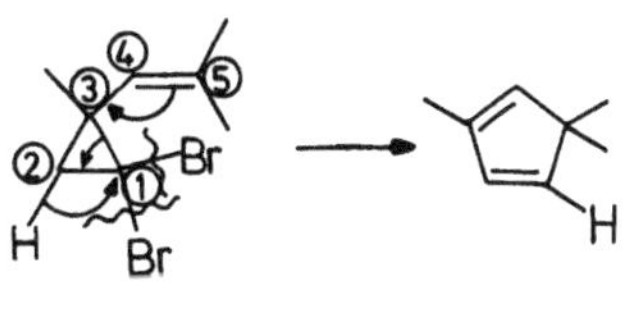

318 *319*

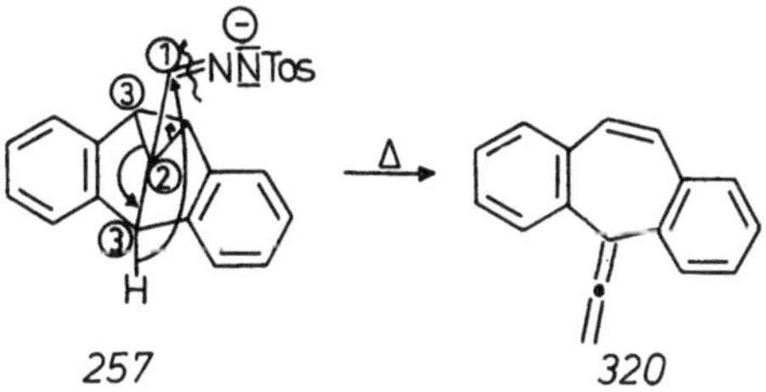

257 *320*

at least three bonds. This is shown in the product analysis upon treatment of *318* with methyllithium at −78 °C [29, 136]. The arrangement of the methyl groups in *319* (85%) proves that a [1,5,(4)3,2,1]-elimination is occurring [165]. Also the branched reaction of *257* with three-fold migration to give *320* (27%) [135] may still be relatively lucidly and comprehensively classified as a [1,2,3,(2)3′,1]-elimination. However, the complexity of the brutto reaction can also be so enormous that any attempt at a closed classification for eliminations with multiple migrations must fail. That is to say, the following examples *321* → *322* (16%) [166], *323* → *324* (90%) [167], *325* → *326* (60%) [168], and *327* → *328* (41%) [169] could be classified. However, such classification would be subject to the mechanism of these reactions. Therefore, it appears unavoidable in these reactions to separate them into reaction steps which can be unambiguously classified according to the Sections 3 and 4. However, such separation into distinct steps does require mechanistic knowledge or mechanistic speculation. Only in a few examples of the preceding sections, simple and clearcut mechanistic terms had to be applied, if there were unavoidable pre- or post-equilibria of the valence isomerization type.

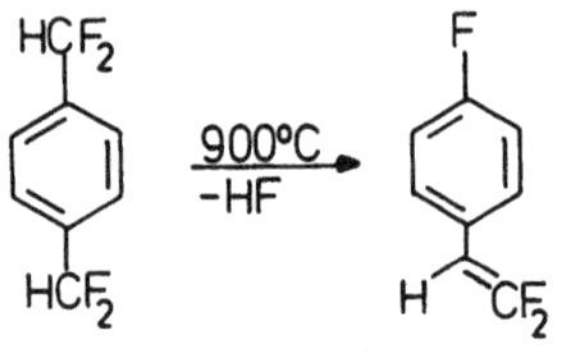

321 *322*

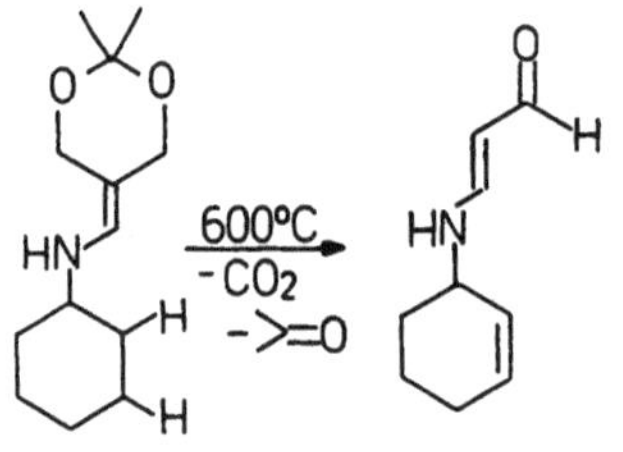

323 *324*

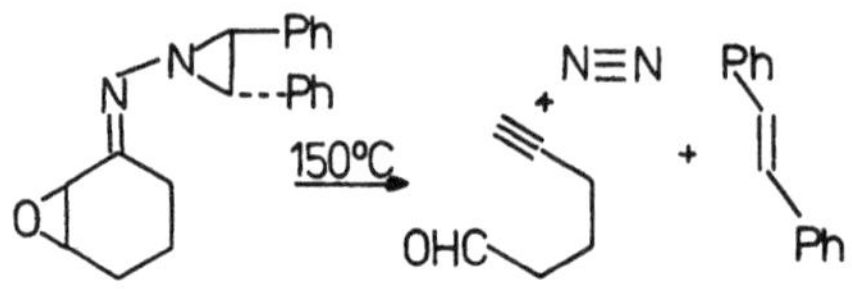

325 *326*

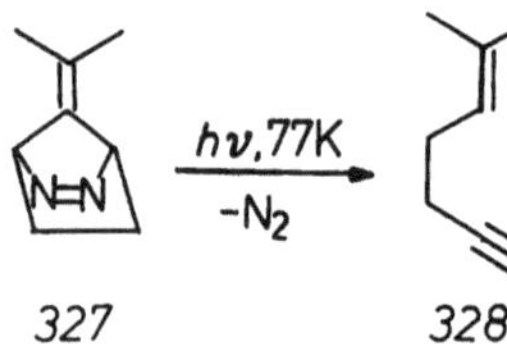

327 *328*

6 Summary and Outlook

The numerous reactions in Sections 3 and 4 have been classified unequivocally without the necessity of invoking detailed mechanistic knowledge or differentiations based on mechanisms. These classifications remain therefore meaningful and valid after more or less complete and detailed mechanistic studies. It is a particular advantage that they embrace all experimental and mechanistic possibilities. In this way, they assist the future mechanistic penetrations, as they create a systematic order of, on first glance, isolated or incoherent reactions independent of the question whether in single cases intermediates had been proven, postulated, or discounted. The practical benefit to the preparative chemist appears to be of great importance. The systematic order now presents a large body of empirical material for a broad extrapolation basis with, as may be already seen, high synthetic potential. Both of them will continually improve as the aspects of reaction mechanisms and reactivity effects are understood more fully.

Further examples and elimination types will undoubtedly be added to the hitherto known ones and it appears highly versatile that our classification system is not limited to complex eliminations, but also covers complex additions (e.g. the so called "benzilic acid rearrangement" [36a)] in terms of a complex [1,2,3]-addition [170)] of water to benzil) and complex substitutions. Thus, the well known technical camphor synthesis starting with α-pinene (Scheme 9) is the result of a sequence of a [1,2,3]-addition [170)], [1,2,3]-elimination, [1,2,3]-addition, and finally saponification and oxidation. This appears to be a highly economic description of the underlying chemical processes.

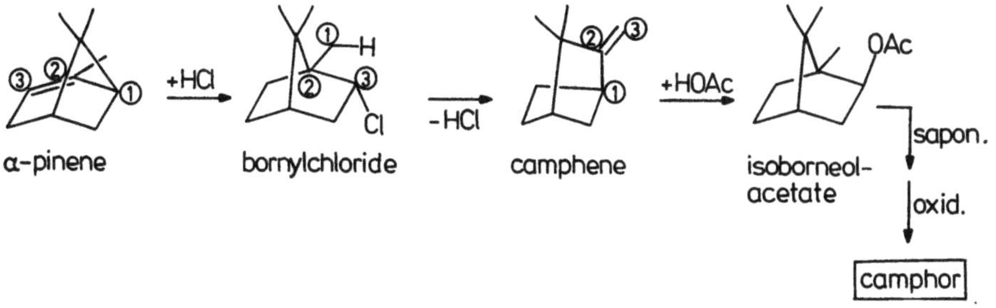

(9)

One of the more important examples of complex substitutions is undoubtedly the synthesis of the much used psychopharmacon chlorodiazepoxide (trade name Librium) (Scheme 10) [171)], which nicely illustrates this classification system. Methylamine reacts in terms of a [1,1,2]-substitution, because it is connected to the carbon in relative position 1, from where the nitrogen migrates to carbon 2 in order to push out the chlorine substituent from that position.

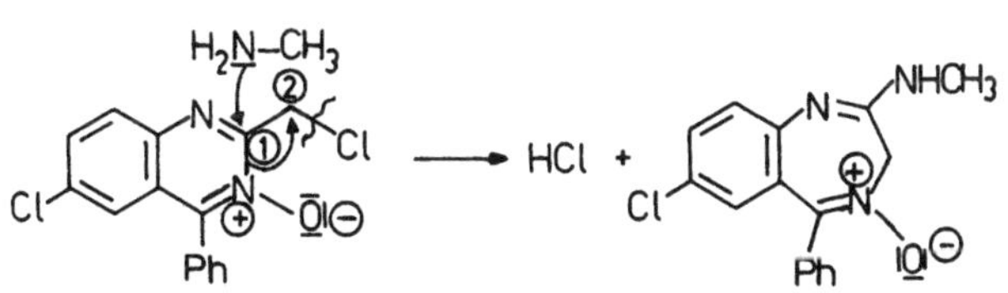

(10)

The comprehensive treatment of complex additions and substitutions may also be done without invoking mechanistic restrictions. Therefore, a unifying consideration of simple and complex reaction classes with a distinct and unambiguous vocabulary is available.

7 Acknowledgement

The author wishes to thank all his coworkers, cited in this review, for their valuable contributions. Support from the Deutsche Forschungsgemeinschaft and the Fonds der Chemischen Industrie is highly acknowledged.

8 References

1. This provision does not implicate any mechanistic assumptions. It takes into account that the more electropositive leaving group will tend to leave without the electron pair, whereas the more electronegative group will tend to take up this electron pair. If both leaving groups are identical, the figures should be chosen such, that the sequence of smaller digits obtains
2. See e.g. Becker, K. B., Labhart, M. P.: Helv. Chim. Acta *66*, 1090 (1983); Nakayama, J., Ohshima, E., Ishii, A., Hoshino, M.: J. Org. Chem. *48*, 60 (1983); Org. Reactions *25*, 1 (1977)
3. Alkhader, M. A., Smalley, R. K.: J. Chem. Res. (S) *1982*, 295; J. Chem. Res. (M) *1982*, 3283
4. Komppa, G., Nyman, G. A.: Ber. Dtsch. Chem. Ges. *69*, 334 (1936)
5. Gukasyan, A. O., Shakhnazaryan, G. M.: Arm. Khim. Zh. *36*, 374 (1983), C.A. *99*, 157907 (1983)
6. Jain, A. C., Rohatgi, V. K., Seshadri, T. R.: Tetrahedron *23*, 2499 (1967)
7. Kugita, H., Inoue, H., Ikezaki, M., Takeo, S.: Chem. Pharm. Bull. (Tokyo) *18*, 2028 (1970)
8. Berti, C.: Gazz. Chim. Ital. *90*, 559 (1960); photochemical example with acyl migration: Tortajada, J., van Hemelrych, B., Morizur, J. P.: Tetrahedron *40*, 613 (1984)
9. Starting points are cyclohexenes, α-chloronitrones, and KCN: DasGupta, T. K., Felix, D., Kempe, U. M., Eschenmoser, A.: Helv. Chim. Acta *55*, 2198 (1972)
10. Vystrěil, A., Křeček, V., Buděšinský, M.: Coll. Czech. Chem. Commun. *48*, 1499 (1983)
11. Pirrung, M. C., Hwu, J. R.: Tetrahedron Lett. *24*, 565 (1983)
12. Fitjer, L., Modaressi, S.: Tetrahedron Lett. *24*, 5495 (1983); also 1,1-diaryl- and 1,1-dicyclopropyl-2-haloethenes react with bases to give acetylenes in terms of [1,2,1]-eliminations: Köbrich, G., Merkel, D.: Angew. Chem. *82*, 257 (1970); Angew. Chem. Int. Ed. Engl. *9*, 243 (1970); mechanistic discussion: Köbrich, G.: ibid. *77*, 75 (1965) and *4*, 49 (1965)
13. Merz, A., Thumm, G.: Liebigs Ann. Chem. *1978*, 1526
14. Review: Murray, M.: Methoden der Organischen Chemie, Houben-Weyl-Müller, Vol. 5/2a, p. 985, Thieme Verlag, Stuttgart, 1977
15. a. Marquis, E. T., Gardner, P. D.: Tetrahedron Lett. *1966*, 2793; b. optical active 1,2-cyclononadiene: Moore, W. R., Bach, R. D.: J. Am. Chem. Soc. *94*, 3148 (1972)
16. Blickle, P., Hopf, H.: Tetrahedron Lett. *1978*, 449; cf. Kleveland, K., Skattebøl, L.: J. Chem. Soc. Chem. Commun. *1973*, 432
17. Jefford, C. W., Rossier, J. C., Zuber, J. A., Kennard, O., Cruse, W. B. T.: Tetrahedron Lett. *24*, 181 (1983); additional reactions of 1,1-dibromocyclobutenes: Brinker, U. H., Schenker, G.: J. Chem. Soc. Chem. Commun. *1982*, 679
18. Jones, W. M., Grasley, M. H.: Tetrahedron Lett. *1962*, 927; cf. the synthesis of highly reactive ketene imines from α-diazoamino compounds: Arnold, B., Regitz, M.: Angew. Chem. *91*, 337 (1979); Angew. Chem. Int. Ed. Engl. *18*, 320 (1979)
19. Tomioka, H., Hayashi, N., Izawa, Y., Senthilnathan, V. P., Platz, M. S.: J. Am. Chem. Soc. *105*, 5053 (1983) and lit. cit. therein; cyclohepta-1,2,4,6-tetraene from phenylcarbene: West, P. R., Chapman, O. L., LeRoux, J.-P.: ibid. *104*, 1779 (1982); 1-azacyclohepta-1,2,4,6-tetraene from phenylazide: Takenchi, H., Koyama, K.: J. Chem. Soc. Perkin I *1982*, 1269
20. Review: Dürr, H.: Methoden der Organischen Chemie, Houben-Weyl-Müller, vol. IV/5b, p. 1158ff., Thieme Verlag, Stuttgart, 1975; diaziridines react correspondingly (Δ and hν), e.g. with migration of hydrogen: Liu, M. T. H., Chishti, N. T., Tencer, M., Tomioka, H., Izawa, Y.: Tetrahedron *40*, 887 (1984); in monolayers: Holden, D. A., Ringsdorf, H., Haubs, M.: J. Am. Chem. Soc. *106*, 4531 (1984)
21. Masamune, S., Nakamura, N., Suda, M., Ona, H.: J. Am. Chem. Soc. *95*, 8481 (1973)
22. Eisenbarth, P., Regitz, M.: Chem. Ber. *115*, 3796 (1982); Angew. Chem. *94*, 935 (1982); Angew. Chem. Int. Ed. Engl. *21*, 913 (1982); Angew. Chem. Suppl. *1982*, 2016
23. Wolff, L.: Liebigs Ann. Chem. *325*, 129 (1902); *394*, 25 (1912)
24. Blaustein, M. A., Berson, J. A.: Tetrahedron Lett. *1981*, 1081
25. Dauben, W. G., Walker, D. M.: Tetrahedron Lett. *23*, 711 (1982); related example: Wiberg, K. B., Olli, L. K., Golembeski, N., Adams, R. D.: J. Am. Chem. Soc. *102*, 7467 (1980); paddlanes: Eaton, P. E., Leipzig, B. D.: ibid. *105*, 1656 (1983)
26. Stork, G., Szajewski, R. P.: J. Am. Chem. Soc. *96*, 5787 (1974)
27. Süs, O.: Liebigs Ann. Chem. *556*, 85 (1944); Pape, M.: Pure Appl. Chem. *41*, 535 (1975)

28. Linke, S., Tisue, G. T., Lwowski, W.: J. Am. Chem. Soc. *89*, 6308 (1967); Eibler, E., Sauer, J.: Tetrahedron Lett. *1974*, 2569 and lit. cit. therein; correspondingly 3-azidonoradamantane: Radziszewski, J. G., Downing, J. W., Wentrup, C., Kaszynski, P., Jawdosiuk, M., Kovacic, P., Michl, J.: J. Am. Chem. Soc. *106*, 7996 (1984); Sasaki, T., Equchi, S., Okano, T.: Tetrahedron Lett. *23*, 4969 (1982); 1-azidobicyclo[3.2.1]octane: Becker, K. H., Gabutti, C. A.: ibid. *23*, 1883 (1982); tris-dimethylaminoazidophosphonium salts reacting to give iminophosphonium salts: Mulliez, M., Majoral, J. P., Bertrand, G.: J. Chem. Soc. Chem. Commun. *1984*, 284; diphenylphosphinic acid azide: Harger, M. J. P., Westlake, S.: J. Chem. Soc. Perkin I *1984*, 2351
29. Jones, W. M., Brinker, U. H.: Pericyclic Reactions, chap. 3, p. 109ff., Academic Press, New York, 1977
30. Review: Prinzbach, H., Auge, W.: Methoden der Organischen Chemie, Houben-Weyl-Müller, vol. V/1b (1972), p. 686ff.; syntheses of cyclobutenes: Seebach, D.: ibid., vol. IV/4, p. 112ff., 1971; photochemical syntheses of cyclopropanes: Brinker, U. H., König, L.: Chem. Ber. *116*, 882 (1983)
31. Review: Adlington, R. M., Barrett, A. G. M.: Acc. Chem. Res. *1983*, 55; Shapiro, R. H.: Org. React. *23*, 405 (1975)
32. Kirmse, W., Spaleck, W.: Angew. Chem. *93*, 791 (1981); Angew. Chem. Int. Ed. Engl. *20*, 776 (1981)
33. El'Cov, A. V., Krivzejko, K. M.: Žh. Org. Chim. *6*, 635 (1970); Cheminform *26*, 158 (1970)
34. Trost, B. M.: J. Am. Chem. Soc. *89*, 138 (1967)
35. Bestmann, H. J., Pfohl, S.: Liebigs Ann. Chem. *1974*, 1688; Bestmann, H. J., Häberlein, H., Pils, I.: Tetrahedron *20*, 2079 (1974)
36. a) Autorenkollektiv: Organicum, 15th edit., VEB Deutscher Verlag der Wissenschaften, Berlin 1981,
    b) Isocyanates from pyridinium 1-acylimines: Katritzky, A. R., Lewis, J., Nie, P.-L.: J. Chem. Soc. Perkin I *1979*, 447
37. Organic Synthesis, Coll. Vol. III, 276 (1955)
38. Brown R. F. C., Eastwood, F. W., Lim, S. T., McMullen, G. L.: Aust. J. Chem. *29*, 1705 (1976)
39. Oele, P. C., Louw, R.: Tetrahedron Lett. *1972*, 4941
40. Kwart, H., Hoster, D. P.: Chem. Commun. *1967*, 1155
41. Morton, Th.: Tetrahedron *38*, 3195 (1982)
42. Also [1,2,3]-elimination (cf. Section 3.3): Failes, R. L., Mollah, Y. M. A., Shapiro, J. S.: Int. J. Chem. Kinet. *11*, 1271 (1979); *13*, 7 (1981); neopentyl phenylether behaves similarly upon electron bombardment [41]
43. Pozharskii, A. F., Kuz'menko, V. V., Nanavyan, I. M.: Khim. Geterotsikl. Soedin. *11*, 1564 (1983); C.A. *100*, 85661x (1984); N-aminooxindole: Atkinson, R. S., Rees, C. W., J. Chem. Soc. C *1969*, 772
44. Itoh, H.: Noguchi Kenkyusho Jiho *25*, 44 (1982); C.A. *98*, 126529w (1983)
45. Also [1,2,1]-elimination: Brown, D. W., Hendrick, M. E., Jones Jr., M.: Tetrahedron Lett. *1973*, 3951; [1,5,1]-elimination of 8,8-dibromobicyclo[5.1.0]octane: lit. [159]; methylbicyclobutanes (e.g. from 1,1-dibromotetramethylcyclopropane) may rearrange subsequently [methyl-1-(2-propenyl-)cyclopropane]: Skattebøl, L.: Acta Chem. Scand. *17*, 1683 (1963)
46. Closs, G. L., Closs, L. E.: J. Am. Chem. Soc. *85*, 99 (1963)
47. a) Kirmse, W., v. Scholz, H. D., Arold, H.: Liebigs Ann. Chem. *711*, 22 (1968, lit. cit. therein; review in lit. [20])
    b) more recent examples: Erni, B., Khovana, H. G.: J. Am. Chem. Soc. *102*, 3888 (1980)
    c) Gano, J. E., Wettach, R. H., Platz, M. S., Senthilnathan, V. P.: ibid. *104*, 2326 (1982)
48. Rando, R. R.: J. Am. Chem. Soc. *94*, 1629 (1972); Tomioka, H., Kondo, M., Izawa, Y.: J. Org. Chem. *46*, 1090 (1981); corresponding synthesis of penicillanic acid: Corey, E. J., Felix, A. M.: J. Am. Chem. Soc. *87*, 2518 (1965)
49. Bear, T. A., Gutsche, C. D.: J. Am. Chem. Soc. *93*, 5180 (1971); there detailed mechanistic discussion; for a further example of a [1,5,1]-elimination: Platz, M. S., Carroll, G., Pierrat, F., Zayas, J., Auster, S.: Tetrahedron *38*, 777 (1982); [1,6,1]-elimination: Banerjee, A. K., Caraballo, P. C., Hurtado, H. S., Carrasco, M. C., Rivas, C.: Tetrahedron *37*, 2749 (1981);

superphanes by [1,8,1]-elimination: Schirch, P. F. T., Boekelheide, V.: J. Am. Chem. Soc. *103*, 6873 (1981)

50. Swenton, J. S., Ikeler, T. J., Williams, B. H.: J. Am. Chem. Soc. *92*, 3103 (1970), lit. cit. therein; analogously fluorenes from 2-biphenyldiazomethanes: Denny, D. B., Klemchuk, P. P.: ibid. *80*, 3289 (1958)
51. a) Kirmse, W., Arend, G.: Chem. Ber. *105*, 2738 (1972), lit. cit. therein
    b) Review of further three-membered ring syntheses: Wendisch, D.: Methoden der Organischen Chemie, Houben-Weyl-Müller, vol. IV/3, p. 333ff. ([1,3,1]-e.); p. 697ff. ([1,(2)3,1]-e.), Thieme Verlag, Stuttgart, 1971
    c) [1,5,1]- and [1,6,1]-e. with tosylhydrazones: Cope, A. C., Martin, M. M., McKervey, M. A.: Quart. Rev. *20*, 119 (1966)
52. Thermally *82* (Li-salt) does not produce tricyclene (*83*), but via a [1,2,1]-e. bornylene (2-bornene) instead: Shapiro, R. H., Heath, M. J.: J. Am. Chem. Soc. *89*, 5734 (1967)
53. Lemal, D. M., Shim, K. S.: Tetrahedron Lett. *1964*, 3231; tricyclo[2.1.0.0^{2,5}]pentane-3-one by [1,4,(5)1]-e.: Maier, G., Hoppe, M., Reisenauer, H. P.: Angew. Chem. *95*, 1009 (1983); Angew. Chem. Int. Ed. Engl. *22*, 990 (1983), lit. cit. therein; cyclopropene formation via photolysis of conjugated tosylhydrazonates: Klett, M. W., Johnson, R. P.: Tetrahedron Lett. *24*, 2523 (1983); Farnum, D. G., Ghandi, M., Raghu, S., Reitz, T.: J. Org. Chem. *47*, 2598 (1982)
54. Hortmann, A. G., Martinelli, J. E.: Tetrahedron Lett. *1968*, 6205
55. Brown, I., Edwards, O. E., McIntosh, J. M., Vocelle, D.: Can. J. Chem. *47*, 2751 (1969); [1,6,(5)1]-e. of acyclic halogenides: Garst, J. F., Hines, J. B.: J. Am. Chem. Soc. *106*, 6443 (1984)
56. Jorgenson, M. J.: J. Am. Chem. Soc. *91*, 6432 (1966); applications in the synthesis of natural products: Balmain, A.: J. Chem. Soc. Perkin II *1975*, 1253; more recent example (azide): Montevecchi, P. C., Spagnolo, P.: J. Org. Chem. *47*, 1996 (1982)
57. Dieterich, D.: Methoden der Organischen Chemie, Houben-Weyl-Müller, vol. VII/2a, p. 927 ff.; 1016ff., Thieme Verlag, Stuttgart, 1973
58. More recent examples: Della, E. W., Pigou, P. E.: Austr. J. Chem. *36*, 2261 (1983); ring contraction: Ebsiu, K., Batty, L. B., Higaki, J. M., Larson, H. O.: J. Am. Chem. Soc. *88*, 1995 (1966); phenyl migration with partial racemization: Martin, J. C., Bentrude, W. G.: J. Org. Chem. *24*, 1902 (1959)
59. Eistert, B., Regitz, M.: Methoden der Organischen Chemie, Houben-Weyl-Müller, vol. VII/2b, p. 1852ff.; 1855ff., Thieme Verlag, Stuttgart, 1976; cyclobutanones; Turro, N. J., Gagosian, R.B.: J. Am. Chem. Soc. *92*, 2036 (1970)
60. Cohen, T., Kuhn, D., Falck, J. R.: J. Am. Chem. Soc. *97*, 4749 (1975); Knapp, S., Trope, A. F., Theodore, M. S., Hirata, N., Barchi, J. J.: J. Org. Chem. *49*, 608 (1984)
61. Suzuki, K., Katayama, E., Tsuchihashi, G.: Tetrahedron Lett. *24*, 4997 (1983)
62. Ghera, E.: Tetrahedron Lett. *1965*, 4181
63. Suginome, H., Liu, C. F., Tokuda, M.: J. Chem. Soc. Chem. Commun. *1984*, 334
64. Clark-Lewis, J. W., Edgar, J. A.: J. Chem. Soc. *1962*, 3887; Wöhler, F., Liebig, J.: Liebigs. Ann. Chem. *26*, 241 (1838)
65. Büchi, G., MacLeod, W. D., Padilla, J.: J. Am. Chem. Soc. *86*, 4438 (1964); Dobler, M., Dunitz, J. D., Gubler, B., Weber, H. P.: Proc. Chem. Soc. *1963*, 383; more recent [1,2,3]-eliminations of acetates with H-migration: Chuchani, G., Rotinov, R., Dominguez, R. M., Gonzales, N.: J. Org. Chem. *49*, 4157 (1984)
66. Also tricyclene (*83*) by [1,3,1]-e. (21.5%) and bornylene by [1,2]-e. (24.5%); similar behaviour of bornylisobenzoate and bornylmethylxanthate: Bunton, C. A., Khaleeluddin, K., Whittaker, D.: Nature *190*, 715 (1961)
67. Johnston, J. P., Overton, K. H.: J.C.S. Perkin Trans. I *1972*, 1490
68. Boyd, J., Overton, K. H.: J.C.S. Perkin Trans. I *1972*, 2533; photolysis of 2-iodoadamantane to give *114*: Kropp, P. J., Sawyer, J. A., Snyder, J. J.: J. Org. Chem. *49*, 1583 (1984); corresponding reactions of vinyltriflates to give butadienes: Martinez, A. G., Hanack, M., Summerville, R. H., von R. Schleyer, P., Stang, P. J.: Angew. Chem. *82*, 323 (1970); Angew. Chem. Int. Ed. Engl. *9*, 302 (1970); Pfeifer, W. D., Bahn, C. A., von R. Schleyer, P., Bocher, S., Harding, C. E., Hummel, K., Hanack, M., Stang, P. J.: J. Am. Chem. Soc. *93*, 1513 (1971); [1,2,3]-e. of quarternary ammonium hydroxides: Cocker, W., Geraghty, N. W. A., McMurry, T. B. H., Shannon, P. V. R.: J. Chem. Soc. Perkin I *1984*, 2245

69. DeBoer, C. D.: Chem. Commun. *1972*, 377
70. Mamer, O. A., Rutherford, R. G., Seidewand, R. J.: Can. J. Chem. *52*, 1983 (1974); naphthalene series: Grützmacher, H. F., Hübner, J.: Liebigs Ann. Chem. *1973*, 793; cyclopentadienothioketene and -selenoketene via [1,2,3]-e. of N_2 from 1,2,3-benzothio(seleno-)diazole: Schulz, R., Schweig, A.: Tetrahedron Lett. *25*, 2337 (1984)
71. Moss, G. I., Crank, G., Eastwood, F. W.: J. Chem. Soc. Chem. Commun. *1970*, 206
72. Marshall, J. A., Roebke, H.: J. Org. Chem. *34*, 4188 (1969); cf. Coffeu, D. L., Lee, M. L.: ibid. *35*, 2077 (1970)
73. Scibner, R. M.: Tetrahedron Lett. *1967*, 4737; photochemical example with succeeding hydrogen migration: Hirakawa, K., Tanabiki, T.: J. Org. Chem. *47*, 280 (1982); more recent thermal reaction: Kato, N., Hamada, Y., Shioiri, T.: Chem. Pharm. Bull. *32*, 2496 (1984)
74. Parham, W. E., Braxton, H. G., O'Connor, P. R.: J. Org. Chem. *26*, 1805 (1961)
75. Wassermann, H. H., Adickes, H. W., deOchoa, O. E.: J. Am. Chem. Soc. *93*, 5586 (1971)
76. Corey, E. J., Bauld, N. L., LaLonde, R. T., Casanova, J., Kaiser, E. T.: J. Am. Chem. Soc. *82*, 2645 (1960)
77. Weyerstall, P., Klamann, D., Finger, C., Fligge, M., Nerdel, F., Buddrus, J.: Chem. Ber. *101*, 1303 (1968); in-situ production of *132*: Engelsma, J. W.: Rec. Trav. Chim. Pays-Bas *84*, 187 (1965)
78. Further examples: Barlet, R., Vo-Quang, Y.: Bull. Soc. Chim. France *1969*, 3729; sulfonyl groups: Griffiths, G., Hughes, S., Stirling, C. J. M.: J. Chem. Soc. Chem. Commun. *1982*, 236; 658
79. Ciamician, G., Dennstedt, M.: Ber. Dtsch. Chem. Ges. *14*, 1153 (1881); pyrimidines via oxidation of 1,3-diazabicyclo[3.1.0]hexenes: Heine, H. W., Weese, R. H., Cooper, R. A., Durbetaki, A. J.: J. Org. Chem. *32*, 2708 (1967)
80. Parham, W. E., Schweizer, E. E.: Org. Reactions *13*, 55 (1963)
81. Parham, W. E., Twelves, R. R.: J. Org. Chem. *22*, 730 (1957); analogously 2-chloropyrazine and 5-chloropyrimidine: Busby, R. E., Iqbal, M., Parrick, J., Shaw, C. J. G.: J. Chem. Soc. Chem. Commun. *1969*, 1344; 3-bromocycloheptatriene: Ketley, A. D., Berlin A. J., Gorman, E., Fischer, L. P.: J. Org. Chem. *31*, 305 (1966); 2-chloro-3-ethoxy-(Z-)1,(E-)3-cyclotridecadiene: Parham, W. E., Sperley, R. J.: ibid. *32*, 927 (1967); with hydrogen migration: Davalian, D., Garratt, P. J.: Tetrahedron Lett. *1976*, 2815
82. Birch, A. J., Brown, J. M., Stansfield, F.: J. Chem. Soc. *1964*, 5343
83. Parham, W. E., Bolon, D. A., Schweizer, E. E.: J. Am. Chem. Soc. *83*, 603 (1961)
84. Anderson, M., Johnson, A. W.: J. Chem. Soc. *1965*, 2411
85. Schubert, W. M., Leahy, S. M.: J. Am. Chem. Soc. *79*, 381 (1957)
86. Fischer, H. O. L., Feldmann, L.: Ber. Dtsch. Chem. Ges. *62*, 854 (1929), Fischer, H. O. L., Taube, C.: ibid. *57*, 1502 (1924); Dakin, H. D., Dudley, H. W.: J. Biol. Chem. *15*, 127 (1913)
87. Reich, H., Samuels, B. K.: J. Org. Chem. *21*, 68 (1956)
88. a) Saavedra, J. E.: J. Org. Chem. *46*, 2610 (1981); additionally [1,2,(3)4]-elimination (cf. Section 3.5); N-migration has been realized in the [1,2,4]-elimination of methanol in lit. [88b)] (there from *5* to give *4*)
    b) Kaupp, G., Hunkler, D., Zimmermann, I.: Chem. Ber. *115*, 2467 (1982)
89. Cope, A. C., Moon, S., Park, C. H.: J. Am. Chem. Soc. *84*, 4850 (1962); additionally [1,1,3]- and [1,1,(2)4]-substitutions
90. Bartnik, R., Laurent, A., Lesniak, S.: Compt.rend. C *288*, 505 (1979)
91. Abdun-Nur, A. R., Bordwell, F. G.: J. Am. Chem. Soc. *86*, 5695 (1964)
92. Biermann, H. W., Freemann, W. P., Morton, T. H.: J. Am. Chem. Soc. *104*, 2307 (1982)
93. Allinger, N. L., Greenberg, S.: J. Am. Chem. Soc. *84*, 2394 (1962)
94. Ourisson, G.: Proc. Chem. Soc. *1964*, 274
95. Heck, R., Prelog, V.: Helv. Chim. Acta *38*, 184 (1955); Urech, H. J., Prelog, V.: ibid. *40*, 477 (1957); Prelog, V., Küng, W., Tomljenović, T.: ibid. *45*, 1352 (1962); related examples with 9-, 10- and 11-membered rings: Prelog, V., Kägi, H. H., White, E. H.: ibid. *45*, 1658 (1962); King, J. F., de Mayo, P. in "Molecular Rearrangements", vol. 2, p. 771ff., Interscience Publishers, New York, 1963; lit. [51c)]
96. Prelog, V., Küng, W.: Helv. Chim. Acta *39*, 1395 (1956)

97. Cope, A. C., Peterson, P. E.: J. Am. Chem. Soc. *81*, 1643 (1959); additionally [1,2,6]-substitutions to give *exo*- and *endo*-2-acetoxybicyclo[3.3.0]octane; open-chain examples: Lochead, A. W., Proctor, C. R., Caton, M. P. L.: J. Chem. Soc. Perkin I *1984*, 2477
98. Cope, A. C., Nealy, D. L., Scheiner, P., Wood, G.: J. Am. Chem. Soc. *87*, 3130 (1965); yield in *156*: 14%; main reaction in acetolysis is the [1,2,7]-substitution to give 2-acetoxy-bicyclo[3.3.1]nonane; cf. Hanack, M., Kaiser, W.: Angew. Chem. *76*, 572 (1964); Angew. Chem. Int. Ed. Engl. *3*, 583 (1964)
99. Hanack, M., Schneider, H.-J., Schneider-Bernlöhr, H.: Tetrahedron *23*, 2195 (1967)
100. Jones, R. L., Rees, C. W.: J. Chem. Soc. C *1969*, 2255
101. Gill, G. B., Harper, D. J., Johnson, A. W.: J. Chem. Soc. C *1968*, 1675
102. Schnieders, C., Altenbach, H. J., Müllen, K.: Angew. Chem. *94*, 638 (1982); Angew. Chem. Int. Ed. Engl. *21*, 637 (1982); Angew. Chem. Suppl. *1982*, 1353
103. van Bergen, T. J., Kellogg, R. M.: J. Org. Chem. *36*, 978 (1971)
104. Culvenor, C. C., Davies, W., Savige, W. E.: J. Chem. Soc. *1952*, 4480 and lit. cit. therein; if the N—H(O—H)bonds in *185* (*187*) do not in fact participate, these would be simple [1,3]-eliminations; cf. Calo, V., Lopez, L., Marchese, L., Pesce, G.: J. Chem. Soc. Chem. Commun. *1975*, 621; [1,2,5(4)]- and [1,2,7(6)]-elimination of water from 2- and 4-hydroxymethylpyridine N-oxide: Chilton, W. S., Butler, A. K.: J. Org. Chem. *32*, 1270 (1967); related [1,2,6(5)]-eliminations: Haddadin, M. J., Issidorides, C. H.: Tetrahedron Lett. *1968*, 4609; Kaupp, G., Voss, H.: to be published
105. Kaupp, G.: Chem. Ber. *118*, 4271 (1985)
106. Trahanovsky, W. S., Alexander, D. L.: J. Am. Chem. Soc. *101*, 142 (1979) and lit. cit. therein; deuterium which may be introduced in relative position 4 is not lost during the reaction
107. Wharton, P. S., Hiegel, G. A., Coombs, R. V.: J. Org. Chem. *28*, 3217 (1963)
108. Review of literature: Heinrich, P., Methoden der Organischen Chemie, Houben-Weyl-Müller, vol. IV/5b, p. 891ff., Thieme Verlag, Stuttgart, 1975; "McLafferty-rearrangement" of ketones and numerous further classes of compounds: see e.g. Rose, M. E., Johnstone, R. A. W.: Mass spectrometry for chemists and biochemists, Cambridge University Press, Cambridge, p. 222ff. (1982)
109. Cooke, R. S., Lyon, G. D.: J. Am. Chem. Soc. *93*, 3840 (1971); here also decarbonylations of five-membered ring anhydrides, like e.g. phthalic acid anhydride (further products: CO_2 and dehydrobenzene) should be mentioned: Dunkin, I. R., MacDonald, J. G.: Tetrahedron Lett. *23*, 4839 (1982) and lit. cit. therein
110. Hünig, S., Ort, B.: Angew. Chem. *96*, 231 (1984); Angew. Chem. Int. Ed. Engl. *23*, 237 (1984)
111. See e.g. Adam, W., Cadiz, C., Mazenod, F.: Tetrahedron Lett. *1981*, 1203 and lit. cit. therein
112. This type should be distinguished from simple [1,4]-eliminations which provide cyclobutanes
113. TerBorg, A. P., Bickel, A. F.: Rec. trav. chim. PaysBas *80*, 1217 (1961); 2-chloro-2,3-epoxy-norbornane→nortricyclanone (via α-oxocarbene by [1,1,2]-elimination?): MacDonald, R.N., Steppel, R. N., Cousins, R. C.: J. Org. Chem. *40*, 1694 (1975)
114. Nerdel, F., Frank, D., Marschall, H.: Chem. Ber. *100*, 720 (1967); further example: Nerdel, F., Frank, D., Rehse, K.: ibid. *100*, 2978 (1967)
115. Nishiyama, H., Kitajima, T., Yamamoto, A., Itoh, K.: J. Chem. Soc. Chem. Commun. *1982*, 1232
116. de Groot, A., Boerma, J. A., Wynberg, H.: Tetrahedron Lett. *1968*, 2365
117. Casey, M., Moody, C. J., Rees, C. W.: J. Chem. Soc. Chem. Commun. *1982*, 714
118. Sato, E., Kanaoka, Y., Padwa, A.: J. Org. Chem. *47*, 4256 (1982)
119. Kaupp, G.: Chem. Ber. 117, 1643 (1984)
120. Kaupp, G., Gründken, E., Matthies, D.: Chem. Ber. *119*, 3109 (1986)
121. Wilhelm, M., Schmidt, P.: Helv. Chim. Acta *53*, 1697 (1970)
122. Kryczka, B., Laurent, A., Marquet, B.: Tetrahedron *34*, 3291 (1978)
123. Kwart, H., Slutsky, J.: J. Org. Chem. *41*, 1429 (1976); Adams, B. L., Kovacic, P.: J. Am. Chem. Soc. *97*, 2829 (1975)
124. Adam, W., DeLucchi, O., Hill, K.: J. Am. Chem. Soc. *104*, 2934 (1982)
125. Starr, J. E., Eastmann, R. H.: J. Org. Chem. *31*, 1393 (1966)
126. Cf. say N-pyrrolidyl- and N-piperidylnitrene: Schultz, P. G., Dervan, P. D.: J. Am. Chem. Soc. *104*, 6660 (1982)

127. Cf. e.g. eliminations of sulfur from unstable intermediates which are represented as [5,1,8]- and [6,1,9]-eliminations, resp.: Brown, J. P.: J. Chem. Soc. Perkin I *1974*, 869; [2,1,5]-, [6,(2)1,9]-, and [7,(2)1,10]-e. of carbon dioxide: Hodgetts, I., Noyce, S. J., Storr, R. C.: Tetrahedron Lett. *25*, 5435 (1984)
128. Volhard, J.: Liebigs Ann. Chem. *253*, 206 (1889); Katritzky, A. R., Robinson, R.: J. Chem. Soc. *1955*, 2481; tricyclic dilactone: Sladkov, V. I., Anisimova, O. S., Turchin, K. F., Sheinker, Y. N., Suvorov, N. N.: Zh. Org. Khim. *12*, 52 (1976); C.A. *84*, 121229z (1976); homologous [1,6,(7)10']-elim.: Hoye, T. R., Peck, D. R., Trumper, P. K.: J. Am. Chem. Soc. *103*, 5618 (1981); correspondingly spiroketals by oxidative cyclodehydration: Burgstahler, A. W., Widiger, G. N.: J. Org. Chem. *38*, 3652 (1973)
129. Childs, R. F., Johnson, A. W.: J. Chem. Soc. C *1966*, 1950
130. Huisgen, R., Seidl, G.: Tetrahedron *20*, 231 (1964); in which there is a proposal of a mechanism via benzocyclonones, which after independent synthesis [Marshall, P. A., Prager, R. H.: Austr. J. Chem. *32*, 1251 (1979)] reacted to give *241*; [1,3,(2)3']-elimination of nitrogen from 4-methylene pyrazolines: Dolbier, W. R., Burkholder, C. R.: J. Am. Chem. Soc. *106*, 2139 (1984)
131. Yvernault, T., Mazet, M.: Bull. Soc. Chim. France *1969*, 638; further examples: Mazet, M.: ibid. *1969*, 4309 and lit. cit. therein
132. Folz, C. M., Kondo, Y.: Tetrahedron Lett. *1970*, 3163
133. Additional [1,2]-, [1,3]-, [1,2,3]-, [1,2,1]-eliminations: Pillai, C. N., Pines, H.: J. Am. Chem. Soc. *83*, 3274 (1961); correspondingly 1-methylcyclopentene from bromocyclohexene by "electron bombardment" in the presence of triethylamine [41)]
134. Boll, W. J., Landor, S. R.: Proc. Chem. Soc. *1961*, 143
135. Paquette, L. A., Meehan, G. V.: J. Am. Chem. Soc. *92*, 3039; further examples in lit. [30)]
136. Kirmse, W., IUPAC, Chemistry for the Future, Pergamon Press, Oxford, 1984, p. 225; a mechanism via vinylcyclopropylcarbene → cyclopentenylcarbene is proposed ("Skattebol-rearrangement"). Then, however, there is a migration of three bonds, not of two (see Sect. 5), and the reaction of *259* should then (with changed topology) be classified as a [1,5,(4)3,2,1]-elimination. A series of substituted 1,1-dibromo-2-vinylcyclopropanes undoubtedly react in accordance with the more complicated topology, as is shown by the analysis of the products: e.g. Holm, K. H., Skattebøl, L.: Acta Chem. Scand. B *38*, 783 (1984) and lit. cit. therein
137. Cope, A. C., Burton, P. E., Caspar, M. L.: J. Am. Chem. Soc. *84*, 4855 (1962)
138. London, J. D., Young, L. B.: J. Chem. Soc. *1963*, 5496
139. Adam, W., Gillaspey, W. D.: Tetrahedron Lett. *24*, 1699 (1983)
140. Biethan, U., Klusacek, H., Musso, H.: Angew. Chem. *79*, 152 (1967); Angew. Chem. Int. Ed. Engl. *6*, 176 (1967)
141. Hedaya, E., Kent, M. E., McNeil, D. W., Lossing, F. P., McAllister, T.: Tetrahedron Lett. *1968*, 3415; [1,2,3,(2)1]-e. of N_2 from 2-fluorenyltetrazole followed by [1,6,Δ,2,1]-e. of N_2 and [1,2/2,1]-rearrangement to give benz[*a*]azulene: Wentrup, C., Becker, J.: J. Am. Chem. Soc. *106*, 3705 (1984); analogously 1-azaazulene (followed by [1,7,Δ,3,(2)1]-e. and [1,2/2,1]-r.)
142. Reese, C. B., Sanders, H. P.: Synthesis *1981*, 276 and lit. cit. therein; further examples with cyclopropane instead of the oxirane ring: Cristol, S. J., Harrington, J. K.: J. Org. Chem. *28*, 1413 (1962); Moss, R. A., Wetter, W. P.: Tetrahedron Lett. *22*, 997 (1981)
143. Engels, P. S., Keys, D. E.: J. Am. Chem. Soc. *104*, 6860 (1982); [1,5,6,(5)1]-elimination of an azide: Schultz, A. G., McMahon, W. G.: J. Org. Chem. *49*, 1676 (1984)
144. Goerdeler, J. G., Schenk, H.: Chem. Ber. *98*, 2954 (1965)
145. Etienne, A., Bonte, B., Druet, B.: Bull. Soc. Chim. France *1972*, 251
146. Feist, F.: Liebigs Ann. Chem. *257*, 253 (1890)
147. Kaupp, G., Knichala, B.: Chem. Ber. *118*, 462 (1985)
148. Jones, D. W., Pomfret, A.: J. Chem. Soc. Chem. Commun. *1982*, 919
149. Curtius, T., Jay, R.: J. pr. Chem. [2] *39*, 27 (1889)
150. Zimmerman, H. E., Somasekhara, S.: J. Am. Chem. Soc. *82*, 5865 (1960)
151. Binkley, R. W.: J. Org. Chem. *34*, 931 (1969)
152. Brede, O., Mehnert, R., Naumann, W., Becker, H. G. O.: Ber. Bunsenges. Phys. Chem. *84*, 666 (1980)

153. Akiba, K., Inamoto, N.: Heterocycles 7, 1131 (1977); Challis, B. C., Challis, J. A.: Supplement F (Part. 2): The chemistry of amino, nitroso, and nitro compounds and their derivatives (ed.) Patai, S., p. 1209ff., Wiley, New York, 1982
154. Passing, H.: J. pr. Chem. *153*, 1 (1939); cf. also Besthorn, E.: Ber. dtsch. chem. Ges. *43*, 1519 (1910)
155. Seybold, G., Heibel, C.: Chem. Ber. *110*, 1225 (1977)
156. Sheludyakov, V. D., Tkachev, A. S., Sheludyakova, S. V., Kozyukov, V., Mironov, V. F.: Zh. Obshch. Khim. *47*, 2259 (1977); C.A. *88*: 23041v (1979); it is possible that *299* is an intermediate in the formation of *300* from sodium benzoate and cyanobromide: Douglas, D. E., Eccles, J., Almond, A. E.: Can. J. Chem. *31*, 1127 (1953); the decomposition of methylisocyanate at 500 °C via a radical chain which produces hydrogen, hydrogen cyanide and carbon monoxide (Blake, P. G., Ijadi-Maghsoodi, S.: Int. J. Chem. Kinetics *14*, 945 (1982); here no rearrangement) is to be classified as a [1,2/2,1]-elimination
157. Goerdeler, J., Nandi, K.: Chem. Ber. *114*, 549 (1981); corresponding decomposition of an α-iminoisocyanate as a postulated intermediate: Goerdeler, J., Sappelt, R.: ibid. *100*, 2064 (1967)
158. Kirmse, W., Ruetz, L.: Liebigs Ann. Chem. *726*, 30 (1969)
159. Aldersley, M. F., Dean, F. M.: J. Chem. Soc. Chem. Commun. *1983*, 331
160. Crank, G., Eastwood, F. W.: Austr. J. Chem. *17*, 1392 (1964)
161. Jones, M., Reich, S. D.: J. Am. Chem. Soc. *89*, 2935 (1967); additionally [1,2,1]- (bicyclobutene-derivative) and [1,2,3,(2)1]-e. (tropilidene and presumably acetylene); cf. lit. [142]; corresponding reactions with the tosylhydrazonate of bicyclo[6,1,0]nona-2,4,6-triene-9-carbaldehyde (there is no distinction between the [1,(2)3,(4)5,1]-e. and the [1,(2)3,Δ,(6)7,1]-e. possibilities): Jones, M., Scott, L. T.: ibid. *89*, 150 (1967); linear [1,2,(3)4,(5)1]-e. of diazocarbonyl compounds ("vinylogous Wolff-rearrangement"): Smith, A. B., Toder, B. H., Branca, S. J.: ibid. *106*, 3995 (1984); Smith, A. B., Toder, B. H., Richmond, R. E., Branca, S. J.: ibid. *106*, 4001 (1984)
162. Maier, G., Pfriem, S., Schäfer, U., Malsch, K.-D., Matusch, R.: Chem. Ber. *114*, 3965 (1981)
163. Review: Wendisch, D.: Methoden der Organischen Chemie, Houben-Weyl-Müller, vol. IV/3, p. 71 ff., Thieme Verlag, Stuttgart, 1971
164. Liguori, A., Sindona, G., Uccella, N.: Tetrahedron *40*, 1901 (1984)
165. 3-Methylenecyclobutenes are stable compounds (e.g. p. 168, 425, 435 in Methoden der Organischen Chemie, Houben-Weyl-Müller, vol. IV/4) and therefore a [1,4,(3)2,1]-e. to give 2,4,4-trimethyl-3-methylenecyclobutene followed by a [1,2/2,1]-r. to give *319* does not apply
166. Fugua, S. A., Parkhurst, R. M., Silverstein, R. M.: Tetrahedron *20*, 1625 (1964)
167. Gordon, H. J., Martin, J. C., McNab, H.: J. Chem. Soc. Perkin I *1984*, 2129
168. Felix, D., Müller, R. K., Horn, U., Joos, R., Schreiber, J., Eschenmoser, E.: Helv. Chim. Acta *55*, 1276 (1972); cf. p. 247ff. in lit. [14]
169. Schmidt, S. P., Pinhas, A. R., Hammons, J. H., Berson, J. A.: J. Am. Chem. Soc. *104*, 6822 (1982)
170. The first and the last figure specify the relative positions of the addition, the intermediate figure specifies — for reasons of microscopic reversibility — the final point of migration of that group, which is displaced from the first named position; examples for [1,3,4]- and [1,4,5]-additions: McCabe, P., Stewart, A.: J. Chem. Soc. Chem. Commun. *1980*, 100
171. Sternberg, L. H., Reeder, E.: J. Org. Chem. *26*, 1111 (1961); further [1,1,2]-substitutions: "Favorskij-rearrangement": (Organic Reactions *11*, 261 (1960)); Jäckel, K.-P., Hanack, M.: Liebigs Ann. Chem. *1975*, 2305; Kirmse, W., Siegfried, R., Wroblowsky, H.-J.: Chem. Ber. *116*, 1880 (1983); [1,1,3]-substitutions: Reutov, O. A.: Pure Appl. Chem. *7*, 203 (1963); Akiyama, T., Yoshida, Y., Hanawa, T., Sugimori, A.: Bull. Chem. Soc. Japan *56*, 1795 (1983); [1,1,(2)3]-substitutions: lit. [105,120]; Skattebøl, L.: J. Org. Chem. *35*, 3200 (1970); [1,1,(2)5]-substitutions (reductively): Sauers, R. R., Parent, R. A., Damle, S. B.: J. Am. Chem. Soc. *88*, 2257 (1966); [1,2,6]-substitution: lit. [97]; [1,2,7]-substitutions: lit. [51c,98]; Hanack, M., Kaiser, W.: Angew. Chem. *76*, 572 (1964); Angew. Chem. Int. Ed. Engl. *3*, 583 (1964); [1,1,2,3]-substitution: Kirmse, W., Wroblowsky, H. J.: Chem. Ber. *116*, 1118 (1983)

Polycyclic Anions: From Doubly to Highly Charged π-Conjugated Systems

Mordecai Rabinovitz

Department of Organic Chemistry, The Hebrew University of Jerusalem, Jerusalem 91904, Israel

Table of Contents

Topics in Current Chemistry, Vol. 146
© Springer-Verlag, Berlin Heidelberg 1988

1 Introduction

The carbanions represent the first group of reactive intermediates which was studied in organic chemistry [1,2,3]. Nevertheless, the research concerning the nature of these species is still in the frontier of organic chemistry. A major group of great importance in carbanion chemistry consists of π-conjugated anions, known since the turn of the century, exemplified by Thiele's study of the cyclopentadienyl anion [4]. This early study opened one major avenue of generation of π-conjugated anions, i.e. deprotonation (proton abstraction) of a sp^3 hybridized carbon. The second major avenue, i.e. alkali metal reduction of conjugated systems, was discovered fourteen years later [5,6,7]. This chapter will concentrate on cyclic carbanionic systems containing more than one ring and bearing more than one negative charge. All the systems that will be discussed are fully conjugated polycyclic systems, benzenoid and nonbenzenoid. Monocyclic carbanions have recently been reviewed very thoroughly [8], and will be omitted, unless required for comparison with a relevant polycyclic system.

The study of the polycyclic-multicharged systems represents a meeting point between the synthetic, the physical-organic and the theoretical chemist. These systems are ideal for confronting experiment and theory and afford an insight to such problems as the relationships between charge distribution, magnetic properties and electronic structure of conjugated molecules. It is therefore not surprising that in the last ten years there has been a revival of interest in multicharged systems.

2 Generation of Polycyclic Anions

2.1 Electron Transfer Reactions

In the late 20's, Schlenck and his group demonstrated that the reduction of π-conjugated systems can be carried out by alkali metals and that the reaction affords dianions [5]. It was also realized that a one electron transfer process occurs in an alternating fashion [1,5] to form radical anions and polyanions. Ever since, this reaction has become the main process of preparation of polycyclic anions [10-13], (Eq. 1):

$$A \underset{-e}{\overset{M(+e)}{\rightleftharpoons}} A^{\overline{\cdot}} \cdot \underset{-e}{\overset{M(+e)}{\rightleftharpoons}} A^{2-} \underset{-c}{\overset{M(+e)}{\rightleftharpoons}} A^{3\overline{\cdot}} \cdot \underset{-e}{\overset{M(+e)}{\rightleftharpoons}} A^{4-} \text{ etc. } M = \text{Li, Na, K, Cs} \tag{1}$$

The same set of events will occur in electrochemical processes [14-17], (Eq. 2):

$$A \overset{+e}{\rightleftharpoons} A^{\overline{\cdot}} \cdot \overset{+e}{\rightleftharpoons} A^{2-} \overset{+e}{\rightleftharpoons} A^{3\overline{\cdot}} \cdot \overset{+e}{\rightleftharpoons} A^{4-} \tag{2}$$

The reverse process can also be achieved by a photochemical process which can eject an electron from a dianion and form a radical-anion [18,19], (Eq. 3):

$$A^{2-} \overset{h}{\longrightarrow} A^{\overline{\cdot}} \cdot + e \tag{3}$$

The formation of anions can be studied by UV-Vis spectroscopy [20] and by magnetic resonance techniques. The alternate formation of odd-electron and even-electron species is followed by ESR and NMR spectroscopies.

The metal reduction of the polycyclic system is usually carried out in an ether solvent and by an alkali metal at low temperature (—78 °C). When potassium metal is applied it is best to prepare a metal mirror. Sodium and lithium react, either directly in the form of a metal wire, or after treatment in an induction furnace. Cesium is prepared by thermolysis of cesium azide. It has recently been found that the application of an ultrasonic bath facilitates the reaction and avoids side reactions. The reaction can be carried out in a modified NMR tube or in an ESR cavity. Diamagnetic ions are prepared in extended NMR tubes to which the metal is extruded and sealed under vacuum. Reaction rates are difficult to compare as the electron-transfer process depends on various experimental conditions such as concentration, temperature, the presence of impurities, the solvent and the nature of the metal surface. It may take from minutes to days to form the first radical-anion; the second step then follows and can sometimes be rather slow [10-13].

The process of reduction can be monitored by recording ^1H NMR and ESR spectra. The first reaction product, the radical-anion, is detected by ESR spectroscopy. In the NMR experiment, the NMR signals of the protons of the starting material start to broaden due to an intermolecular electron-excharge process between the starting material and the radical-anion. This process affects the relaxation process of the protons of the neutral hydrocarbon. As soon as the diatropic species (dianion or tetraanion) is formed, its NMR signals can be detected. Such an exchange process may also affect the appearance of the diatropic dianion. When the reduction proceeds beyond the dianion stage and a tetraanion is produced, there is a second stage of disappearance of the spectrum. This phenomenon is due to the formation of a paramagnetic trianion-radical prior to the formation of a diamagnetic tetraanion.

2.2 Deprotonation of Hydrocarbons

The application of strong bases to hydrocarbons which contain a sp^3 hybridized carbon and a conjugated system may abstract a proton and form a fully conjugated carbanion. This reaction has several synonyms such as deprotonation, proton abstraction or metalation reactions. The most common bases are alkyl lithium derivatives, e.g., butyl lithium [21-24]. Sometimes the addition of tetramethylethylenediamine (TMEDA) or potassium *tert*-butoxide is required especially when dianions and polyanions are prepared [24e,f]. Ether solvents or hydrocarbon solvents are most common. This process can be demonstrated by the preparation of anthraene dianion 2^{2-} from 9,10-dihydroanthracene (1) [25]. The metalation reaction can also be carried out in the NMR tube.

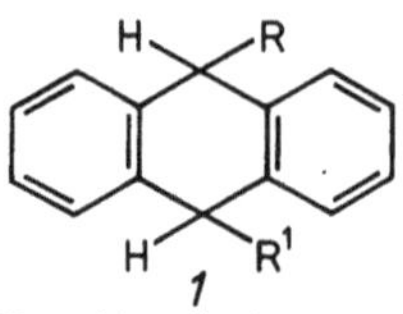

1

R = H, alkyl
R¹ = alkyl

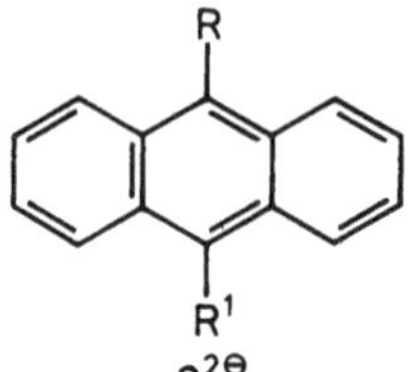

$2^{2\ominus}$

3 Methods of Study

3.1 Magnetic Resonance Spectroscopies — NMR and ESR

NMR features of charged systems have been a subject of wide interest [26–28]. Apart from basic structure elucidation of the molecule, NMR chemical shifts afford further information on π-charge densities [26–32], charge delocalization patterns [8–12, 33–37], the anisotropy of the system and the state of solvation [38–45]. Vicinal H—H coupling constants were correlated with the carbon-carbon bond length and therefore with the bond order of the M.O. theory [46], (vide infra). The development of sophisticated techniques of 2D-NMR enable unambiguous assignments of carbon chemical shifts of polycyclic multicharged systems and hence the estimation of their π-charge densities. The anisotropic effects and the state of solvation, i.e., the ion solvation equilibria, can be deduced from the proton and carbon chemical shifts, and will be discussed later.

In earlier studies, whenever a radical-anion was obtained, the π-charge densities were estimated from ESR parameters [12–13]. This is due to the well known assumption that the local spin densities of a radical-ion which can be estimated from the π charge distribution obtained from ESR can then be compared to the charge distributions obtained from NMR chemical shifts. The exact estimation of empirical parameters of polycyclic ions enables the confrontation of experiment and theory (see Sect. 3.1.3).

3.1.1 NMR Chemical Shifts, Charge Densities and Degree of Charging

Early in the development of NMR spectroscopy, it was noticed that the chemical shift of an atom reflects the electron density in its vicinity. Empirically, it has been shown that shielding of hydrogen or carbon atoms in charged conjugated systems varies linearly with the corresponding π-electron density [28].

Pioneering studies of G. Fraenkel [29] demonstrated a linear relationship between π-charges and proton chemical shifts. In later studies, such correlations were applied to proton [30] and carbon [31] spectroscopies. These relationships can be formulated for protons and carbons in the following general forms (Eqs. 4 and 5).

$$\delta_H = \delta_N - K_H \Delta q_\pi \tag{4}$$

$$\delta_C = \delta_N - K_C \Delta q_\pi \tag{5}$$

δ_H, chemical shift of the proton of the charged species.
δ_C, chemical shift of the carbon of the charged species.
δ_N, the corresponding chemical shift for the neutral precursor.
q_π, the extent of π-charge.
K, proportionality factor (K_H — for protons, K_C — for carbons).

It should be noted that the proportionality factors depend strongly on the molecular structure, on hybridization of the various carbon atoms and on the method of calculation. Actually, each family of charged species would require a modified factor. The two most commonly used correlations for cyclic anions show a proportionality constant of 10.7 ppm per electron for protons (K_H) and 160 ppm per electron for carbons

(K_C) [29–31]. These correlations were derived from chemical shift data of monocyclic anions and cations [32]. In π-conjugated anions the factor of 10.7 ppm per electron for protons seems to fit many diatropic systems. Paratropic systems may, however, show severe deviations. The proton and especially the carbon correlations enable estimation of the charge at the individual atom ($\Delta\delta_H = \delta_N - \delta_H$; $\Delta\delta_C = \delta_N - \delta_C$). On the other hand, summing up of the total shift observed at all the atoms ($\Sigma\,\Delta\delta_H$ and $\Sigma\,\Delta\delta_H$) would give the estimation of the total charge residing on the entire charged molecule. Proton shifts are influenced by induced magnetic fields (diatropic and paratropic) and $K_H = 10.7$ ppm would apply only if these effects are taken into account. Carbon shifts are preferred for this purpose as they are usually free from the influence of anisotropy effects, e.g. ring currents or induced magnetic fields. It was shown that the carbon correlation can be applied to families of compounds having a similar anisotropic effect even with nonuniform charge distribution. The local π-charged density is indeed the main factor influencing the individual ^{13}C chemical shifts (δ_C) [27] within a family of anions. However, ion-pairing, ring current effects and paratropicity should be considered (vide infra). The development of solid state NMR enabled the study of the tensorial features of the ^{13}C shielding constant [33]. This research was carried out on monocyclic neutral and charged molecules (benzene, cyclopentadienyl anion and tropylium cation). Based on single crystal studies, σ^{11} value is assigned along the C—H axis, σ^{22} lies in the molecular plane (perpendicular to C—H axis) and σ^{33} perpendicular to the molecular plane. The isotropic shift dependence on the π-electron density results primarily from variations in σ^{22} and σ^{11} while the dependence on π-electron density of σ^{33} is negligible. When the average of σ^{11} is plotted against q_π (π-electron density), the relationship is linear with a slope of 218 ppm/e [33, 34]. A problem, however, remains: What is the best choice of the proportionality constant K_C? It will be shown how the relationship between K_C and the type of compound studied may varry [27]. This is called by Müllen "compound specific" proportionality constants [8].

As already mentioned, the application of ^1H chemical shifts for the estimation charge density (K_H) is not straightforward. The chemical shifts, (δ_N), considered should represent values which have been corrected for the influence of the induced magnetic fields (ring-current effect). From carbon chemical shifts, which are insensitive to ring current effects along the perimeter of the molecule, the local charge densities, q_π can be estimated (Eq. 6):

$$\delta_C = K_C \cdot \Delta q_\pi \qquad (6)$$

The estimation of charge density at the local sites of the charged molecules and the degree of charging on the entire charged species of the conjugated ($4n + 2$) π-electron systems did not represent serious problems. However, the $4n\pi$-electron species which show high-field (paratropic) shifts presented difficulties. It has been demonstrated [35, 36] that the outstanding feature of the $4n\pi$ series is the narrow energy gap between their HOMO and LUMO energy levels (ΔE). The electronic structure affects, inter alia, the chemical shifts [36] and line shapes [37], and is characteristic of paratropic "antiaromatic" [38, 39] systems. As will be shown later, the involvement of high energy states in the ground state determines the systems' properties [39]. In view of these ideas [40], Edlund and Müllen [8, 41] found that the energy of the mixing of the electrons from the ground state (HOMO) to low lying excited states, should be taken into account when K_C of 4nπ anions is considered. It is suggested that the total shift of carbons $\Sigma \Delta\delta_C$ is the sum of a charge term and an anisotropy term of the form (Eq. 7) [41]:

$$\Sigma \Delta\delta_C = \varrho_\pi F_C + n_C\chi_C \tag{7}$$

χ_C = anisotropy term
n_C = total number of carbon atoms
F_C = "pure" chemical shift/charge factor
ϱ_π = π-charge change

When this equation is divided by the total π-charge change, ϱ_π (as $\chi_C = a\chi_H$), an expression for K_C is obtained (Eq. 8) [41]:

$$K_C = F_C + (n_C/\varrho_\pi)a\chi_H \tag{8}$$

From Equation 8 the value of K_C can be obtained and it varies from 238 to —1. The value of F_C, which is the meaningful term, is 134 ppm per electron. It can therefore be seen that the shift/charge relationship (K_C) in the 4nπ systems is less sensitive to charge than the K_C value of the diatropic (4n + 2) π-systems, and reflects the special electronic properties of the 4nπ-anion series. An intrinsic property of the (4n + 2) π-series is the wide HOMO-LUMO gap (ΔE), while the 4nπ-series are characterized by a narrow HOMO-LUMO gap [40]. These differences should always be taken into account when the spectrospic behaviour of the different series is considered [35, 36].

The structure elucidation by NMR methods will be discussed in detail in relevance to the various systems.

3.1.2 Assessment of Anisotropic Effects as a Criterion for Aromaticity

Aromatic compounds are defined as cyclic or polycyclic systems which sustain a diamagnetic ring current and consequently exhibit a total diatropic lowfield ^{1}H NMR chemical shift relative to that of vinylic protons. Paramagnetic ring current is expected and was shown to be induced in antiaromatic species and to result in a total highfield ^{1}H NMR band displacement, while non aromatic compounds give rise to characteristic vinylic ^{1}H patterns.

The induced magnetic field on protons inside and outside a cyclic fully-conjugated system is described in terms of the "ring current model" [42–52]. Experimental observations have shown that protons inside and outside a (4n + 2) π-conjugated ring system appear at highfield and lowfield, respectively. In alternant polycyclic hydrocarbons the diatropic or ring current effect can be calculated (Eq. 9) where the values of e, m_e and μo are well known:

$$\sigma = \frac{\mu_0}{4\pi}\frac{e^2r^2}{2m_0}\sum_i R_i^{-3} \tag{9}$$

r = 1.4 Å, R is the distance of the various protons, from the center of the ring.

These systems (when the outer protons are considered) are known as diatropic systems. In (4n + 2) π-conjugated polycyclic systems where no inner protons are present, only the low field shift (diatropic effect) is observed. A reversed effect in 4nπ systems namely a shift to highfield which is synonymous with a paratropic effect is observed. While the diatropicity of a system can be described in terms of the "ring current effect" the paratropic effect arises from the involvement of high energy states in the ground state, as will be discussed later. In charged systems the effect of the charge on the chemical shift should be added to the diatropic or paratropic effect. However, the addition of two or four electrons may render a (4n + 2) π-system into a 4nπ system and *vice versa*. The estimation of the "pure" anisotropic effect is assisted by the charge/shift correlations which enable the evaluation of the net charge effect and determination of the net anisotropic effect.

While the diatropic shift is consistent with all aromatic systems, there is no satisfactory experimental criterion for "antiaromaticity" [38]. The antiaromatic systems are all paratropic but the "degree" of paratropicity is confusing. It has therefore been suggested that the extent of the HOMO-LUMO gap (ΔE) correlates with the paratropicity of the system (see section 6.1) [35, 36]. A generalized approach can be obtained from the Ramsey shielding formula (Eq. 10) [53]:

$$\sigma = \sigma_d + \sigma_p + \sigma' \tag{10}$$

According to this relationship the net shielding at an atom is composed of a diamagnetic term σ_d, a paramagnetic term σ_p and a term which includes anisotropic effects and the effect of neighboring atoms.

The diamagnetic term σ_d is fairly easy to estimate theoretically, since it depends only on the electron distribution in the electronic ground state. The paramagnetic contribution depends also on the excited states. It vanishes for electrons in σ orbitals, which have zero angular momentum; but it may become very significant when there is an asymmetric distribution of p and d electrons close to the nucleus and when these electrons have low-lying excited states. This may be the origin of the effect of the HOMO-LUMO gap ΔE on the correlation of charge-carbon chemical shifts in 4nπ series as shown in Equation 7. The term σ′ contains anisotropy factors. A system will reveal a diatropic low field shift when ΔE is large, and a paratropic (high field) shift when the HOMO-LUMO gap ΔE is narrow. Therefore, the relative importance of the diamagnetic and paramagnetic components which compose σ′ will determine the net anisotropy of the system. This link between σd, σp and σ′ has been pointed out by Mallion [39], and it emphasizes the effect of σ′ on proton spectra as a function of ΔE and the effect of both σp and σ′ on the carbon spectra, taking into account the involvement of ΔE.

3.1.3 ESR Studies

The first step in a one-electron transfer process is the formation of the respective radical anion. Local spin density distributions in radical ions can be obtained from the ESR spectrum [54, 55]. The estimation of π-charge densities is based on the assumption that the doubly occupied M.O. of a diamagnetic anion (dianion or tetraanion) is the same as the singly occupied M.O. in the respective radical-anion. They belong to the same point group symmetry. Based on this assumption it is concluded that the spin density of a radical-anion enables the prediction of the π-charge density of a dianion.

The hyperfine proton coupling constants a_H are used for estimating local spin densities ϱ_μ (at proton bearing centers μ). The McConnell Equation [56,57] (Eq. 11) relates a_H and ϱ_μ [57]. The proportionality constant Q was chosen for benzene radical-anion as −2.25 mT [56]. For larger π-systems higher values of Q were used. An agreement between the local spin densities and the total one should be sought. The McLach-lan procedure is used to interrelate between the spin density at a particular π-center of the square of the atomic orbital coefficient of the singly occupied molecular orbital.

$$a_H = Q\varrho \qquad (11)$$

The significance of the ESR studies stems from the possibility of predicting the symmetry of the singly occupied molecular orbital of the paramagnetic species [59–61],

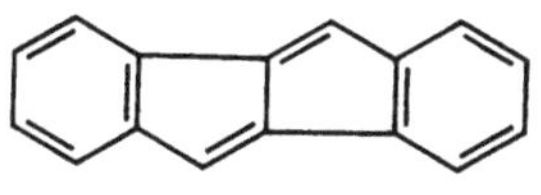

3

by correlating ESR spectroscopy and M.O. theoretical parameters. A specific example is the tetracyclic system, viz., dibenzo[b,f]pentalene radical anion $3^{\overline{\cdot}}$ [60], in which the parameter Q = −2.4 mT was adopted.

3.2 X-Ray Studies of Polycyclic Anions

A very powerful tool for the research of carbanions is their X-ray study in the solid state [62]. Crystal structures of polycyclic polyanions were studied extensively by Stucky et al. [63,64]. He demonstrated that alkali metal anion salts of polycyclic dianions are stabilized by the addition of an amine complexing agent and crystallize as N,N,N′,N′-tetramethylethyldiamine (TMEDA) complexed salts in hydrocarbon solvents. For

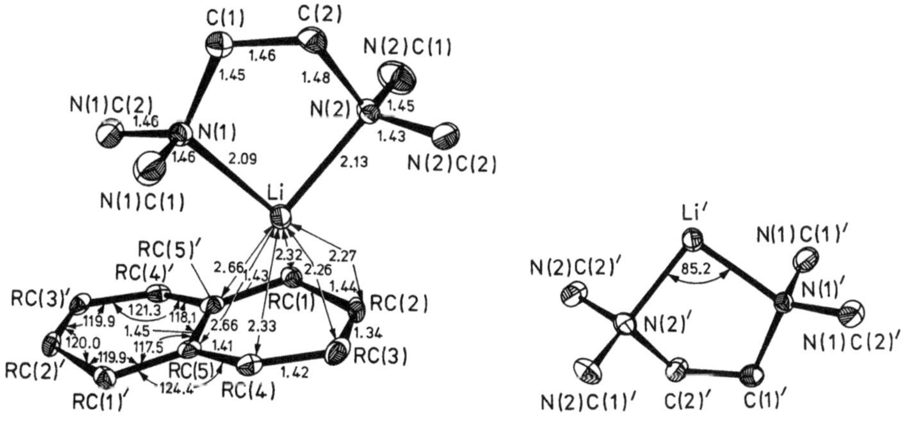

Fig. 1. The molecular geometry of $(Li\text{-}TMEDA)_2C_{10}H_8^{2-}$ (*4*) [63]

illustration, the structure of $(Li\text{-}TMEDA)_2C_{10}H_8^{2-}$ (*4*) consists of the $C_{10}H_8$ moiety bound to two lithium atoms each of which is coordinated to an N,N,N',N'-tetramethyl-ethylenediamine group. The molecule is situated on an inversion center so that only half of the naphthalene ring atoms, one lithium atom and one TMEDA molecule are unique. The coordination sphere of the lithium atom contains two tertiary amine nitrogen atoms and one unsaturated organic group as in the monoanions studied by Stucky. It was shown that the carbanion geometry is consistent with the symmetry of the HOMO (highest occupied M.O.) and that the lithium cation assumed a symmetrical position with respect to the ring. This anion (as well as anthracene dianion [64]) is chosen to demonstrate the potential of X-ray determination of complexed polycyclic dianions. These studies in the solid state are complementary to NMR studies in solution.

3.3 Electrochemical Studies

The intensive electrochemical studies of polycyclic systems, especially cyclic voltametry (CV) are now at a stage which justifies naming cyclic voltametry an "electrochemical spectroscopy" as was suggested by Heinze [65]. Early electrochemical studies referred only to the thermodynamic parameters while CV studies provide direct insight into the kinetics of electrode reactions. These include both heterogeneous and homogeneous electron-transfer steps, as well as chemical reactions which are coupled with the electrochemical process. The kinetic analysis enables the determination of reactive intermediates in the same sense as spectroscopic methods do. As already mentioned, electron transfer processes occur in both the electrochemical and metal reduction reactions.

The essential parameters which determine the electrochemical process are the electron affinity of the neutral compound, which correlates with the energy of the LUMO, the energies of interaction with the solvent and counterions, the electron-electron repulsion energies and stereochemical factors. A precondition for an electrochemical study is that the chemical reaction which may occur, e.g. with the solvent, is much slower than the electron transfer process, and that the electrochemical reaction is reversible [66]. Correlation of half-wave potentials with the energies of Hückel LUMO's has been one of the early successes of the Hückel model [8,20,67,68]. The power of the electrochemical method in the study of polycyclic anions has been demonstrated recently [69a]. Studies on reactions occurring during electrochemical reductions report reductive alkylations of polycyclic systems and their mechanism [70,69b].

3.4 Calorimetric Measurements: Thermodynamic Stabilities of Dianion Salts

Calorimetric measurements of the heat evolved from the reactions of sodium polyacene dianion salts with water afforded the determination of thermodynamic parameters [70]. For all the polyacene dianion salts, the heats of formation were found to be negative. Crystal lattice energies between 400 and 440 kcal/mol were found. These studies add a very important insight into the properties of $4n\pi$ conjugated polycyclic dianions (Sect. 6.4). The measurements were carried out on solvent free dianions in a colimeter in sealed bulbs which were broken under the water in the calorimeter. The change of the temperature of the calorimeter was monitored and the contents of the calorimeter were titrated with standardized hydrochloric acid [70].

3.5 Reaction with Electrophiles

The reaction of electrophiles with anions is often used for synthetic purposes as well as for the elucidation of the structure of ions. The common reactions are alkylation, protonation and acylation. There have been many successful reaction paths which started with a polyanion and an electrophile. The alkylation of cyclooctatetraene dianion 18^{2-} is an instructive example [71]. When the degree of charging of a multi-charged system is deduced from the degree of alkylation, a successive process may occur. Consequently, great care should be taken when the degree of charging is deduced from such experiments. This reaction will be discussed later (s. Sect. 6.3).

4 Ion Solvation Equilibrium of Polycyclic Anions

In the early stages of the study of polycyclic anions, they were referred to as "bare" π-systems, ignoring the role of the counter cation and the solvent. This approach was successful in the understanding of bulk properties of the π-conjugated components. The limited ability of the early theoretical approaches, avoided the counter cations as well. In reality, the anionic structure is a system in which an interplay of the cation, the solvent and the anion (in our case the π-conjugated one) occurs and should not be overlooked.

In the early 50's, an ion pair model was introduced by Winstein to rationalize the mechanism and stereochemistry of solvolysis of sulfonates [72]. This research of carbocationic intermediates and the role of ion solvation equilibrium in reaction mechanisms represents a landmark in the study of charged species. These thermodynamically different ionic species were coined as "free" ions, contact ion-pairs (c.i.p.), and solvent-separated ion pairs (s.s.i.p.). The ion pair situation can be described as an equilibrium between thermodynamically distinct contact (c.i.p.) and solvent-separated ion pairs (s.s.i.p.) [2,73–76]. The situation should be represented by a continuum of ion-solvation equilibria states in which the two extreme states are the c.i.p. and the s.s.i.p. [2,76] (Eq. 12)

$$RX \rightleftharpoons R^+X^- \rightleftharpoons R^+\|X^- \rightleftharpoons R^+ \quad X^- \tag{12}$$

R^+X^- = Contact ion pair (c.i.p.)
$R^+\|X^-$ = Solvent-separated ion pair (s.s.i.p.)
$R^+ \ X^-$ = "free" ions

The equilibrium between the thermodynamically distinct states will be governed by a delicate balance between the solvation requirements of the ions and the Coulombic interactions [73]. One can illustrate two ions, of which at least one has a solvation shell, that approach each other. The potential energy of the system is reduced as the solvation shell disintegrates. As the inter-ionic distance diminishes the potential energy increases due to a partial rupture of the solvation shell which separates the ions. A further decrease of the inter-ionic distance would bring the system to a state of tight ion pairing, which means a complete disappearance of the solvation shell. It should, however, be noted that even a contact ion-pair is solvated, mainly the cation. So that the effective cation radius will vary even for a contact ion-pair depending on the ability of the solvent to form "externally solvated contact ion-pairs". As pointed out by Szwarc [2], this description applies only to those systems in which the energy barrier between s.s.i.p. and c.i.p. is greater than kT. It therefore follows that solvent-separated ion-pairs are favoured by smaller cations, lower temperature, solvents of higher polarity (higher cation solvating ability) and highly π-delocalized anions of large π-delocalized perimeter [8,20,80–82]. The last factor could only be deduced in large ring annulene type of dianions. Due to the increased anion delocalization pattern, a "looser" ion pair prevails. The s.s.i.p./c.i.p. ratio normally increases with a change of the cation in the order Cs^+, Rb^+, K^+, Na^+ and Li^+ due to an increase of the cation-medium electrostatic interaction. The situation can be even more complex when aggregation effects and the relative position of the ions within the ion-pair are taken into account [76]. An important conclusion from ion-pairing studies is that it is essential to know the actual ion-pair state before deriving experimental charge patterns. It has also been pointed out that the nature of the ion-pair influences reaction rates. If the same reaction mechanism applies to a contact ion-pair or a solvent separated ion-pair, it is very difficult to isolate the effect of ion-pairing on the reaction rate [76]. However, in electron transfer reactions, it has been shown that only where understanding the thermodynamics and kinetics of the formation of the ion-pair exists, can a full account of the reaction be achieved [77]. The study of the disproportionation of cyclooctatetraene dianion in THF and ammonia shows that the equilibrium depends on the ion-pairing [77–79]. In ammonia, the equilibrium is shifted to the dianion (NH_4^+ large cation → c.i.p.). It has also been shown that the reaction rate is accelerated with the increase of cation size. These results agree with previous ESR studies and demonstrate that in electron transfer processes the cation acts as a bridge in the course of the electron transfer. It seems that the cation may have an important role in decreasing the electronic and steric repulsions during the formation of the anion. Detailed studies using spectrophotometric and conductometric methods were carried out [26,79d,f] with the aim of studying the equilibrium between the two extreme states, viz. the c.i.p. and the s.s.i.p. The s.s.i.p. state represents a complete shielding of the organic anion [2]. NMR studies [26,79–84] have demonstrated that in c.i.p. the carbon atom near the cation, (α-carbon), accommodates a higher charge density and undergoes a shift to highfield. However, it should be noted that protons are affected both by indirect charge effects and direct field effects causing polarization of the carbon-hydrogen bond. So, by going to contact ion-pairs, direct and indirect effects oppose each other. This condition limits the usefulness of proton shieldings. Carbon atoms (and protons) which are more remote will be characterized by a low density of negative charge and will be shifted to low field. On the other hand, the more the cation is shielded by solvent molecules (solvated), the contribution of charge polarization, will be reduced, and consequently, the NMR shifts, will appear at lower field.

In the polycyclic multicharged anions, a detailed study of ion-pairing was carried out by Edlund [80–83]. This series of studies started with monoanions and extended later into the polycyclic dianionic series. Compounds such as the fluorenyllithium (*5*) and andenyl lithium (*6*) were chosen because of the rigid, well defined structure in

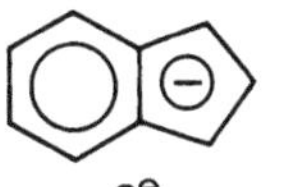

$6^{\ominus}$

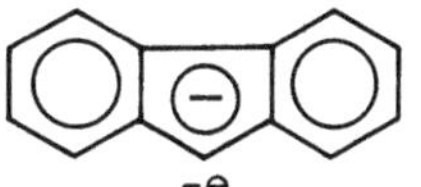

$5^{\ominus}$

which no conformational changes could occur in the course of ion formation [80,81]. The advantage of using ^{13}C NMR for carbanion characterization is that carbon atoms remote from the site of events are still very sensitive to varying charge polarization due to the different equilibrium situations. Charge distribution, on the other hand, is dependent on the effective cation field. It can be concluded from these studies of the indenyl (*6*) and fluorenyl (*5*) anions that there is (a) no cation effect at low temperatures, (b) no concentration dependence, i.e. no aggregation occurs at typical NMR concentrations (<0.5 M), (c) lithium chemical shifts are independent of the anion at s.s.i.p. conditions, (d) total carbon shifts are practically solvent independent, although large individual changes may occur at individual carbons. At low temperature the ion solvation equilibrium favors solvent separated ion pairs (s.s.i.p.) [80,81].

Polycyclic systems bearing more than one negative charge present two problems. (a) The early method of studying ion-pairing of monoanions failed as the UV-Vis measurements have limitations which arise from difficulties in distinguishing between discrete ion-pair states in systems with extended π-conjugation [1,2]. In such cases a situation may occur where one cation is solvated while the other one appears as a c.i.p. Due to the limited time scale of the optical methods, no distinction could be made between the two species. (b) The mutual location of the cations presents an additional problem. The description of the system itself is complicated due to Coulomb repulsions between the cations. The ion-solvation equilibria of dibenzo[b,f]pentalene dianion (3^{2-}) (dibenzopentalenide) [82], its 5,10-dimethyl derivative (7^{2-}), and acenaphthylene dianion (8^{2-}) [83] have been studied in detail.

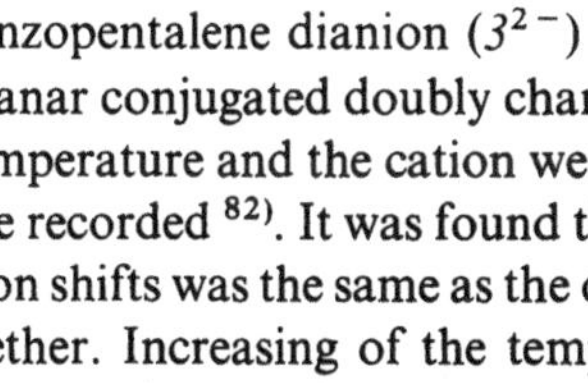

R = CH_3 $7^{2\ominus}$

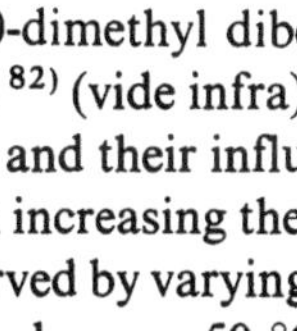

$8^{2\ominus}$

The dibenzopentalene dianion (3^{2-}) and 5,10-dimethyl dibenzopentalene dianion (7^{2-}) are planar conjugated doubly charged ions [82] (vide infra). Variations in the solvent, the temperature and the cation were made, and their influence on the ^{13}C NMR spectra were recorded [82]. It was found that upon increasing the temperature the effect on the carbon shifts was the same as the one observed by varying the solvent from THF to diethyl ether. Increasing of the temperature above —50 °C polarizes the charge towards the pentalene moiety. No levelling out of the chemical shifts was observed in THF, which may indicate a c.i.p. $\rightleftharpoons$ s.s.i.p. equilibrium. It is concluded that the lithium salt exists at r.t. in 2-methyl-THF and diethyl-ether mainly as a contact ion-pair (c.i.p.). In THF, which is a better cation solvating medium the equilibrium is

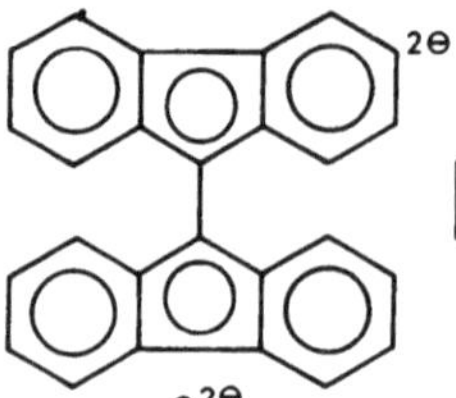

$9^{2\ominus}$

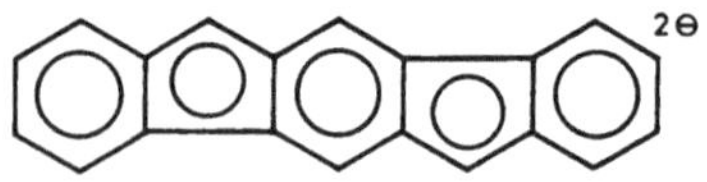

$10^{2\ominus}$

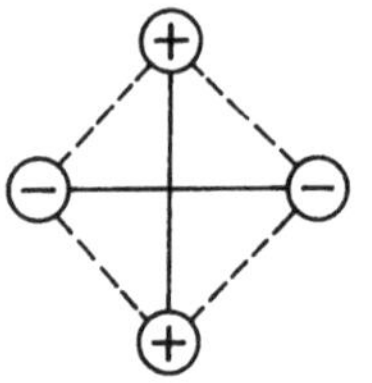

Fig. 2. Point Charge Coulomb Interaction of a Dication Salt of a dicarbanion [84a)]

shifted towards a solvent separated ion pair (s.s.i.p.), but is still in favour of tight ion-pairs. The ^{7}Li NMR was also studied and it was also very valuable for the chracterization of the ion pairs. The preferred lithium positions are close to the pentalene moiety where most of the negative charge is located. Proceeding to contact ion-pairs (c.i.p.), a significant charge localization is observed. In the dicesium salts of 9,9-bifluorenyl (*9*) and indenofluorene (*10*$^{2-}$) Streitwieser and Swanson [84a)] found that the ion pair structure is described by a "triplet ion pair". This term represents a structure in which each cation is close to the anionic center but relatively far from the other cation (s. Fig. 2). Such a structure affords a relief of the repulsion between the cations especially in c.i.p. It seems that Edlund's considerations are also in line with the "triplet ion pair" structure and the two approaches afford a nice description of the situation in doubly charged conjugated systems.

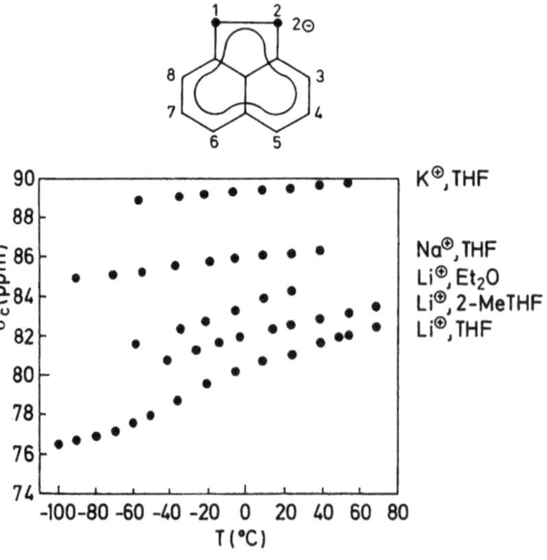

Fig. 3. Variable-temperature ^{13}C NMR chemical shifts at C-1 and C-2 of acenaphthylene dianion (8^{2-}) [83a)]

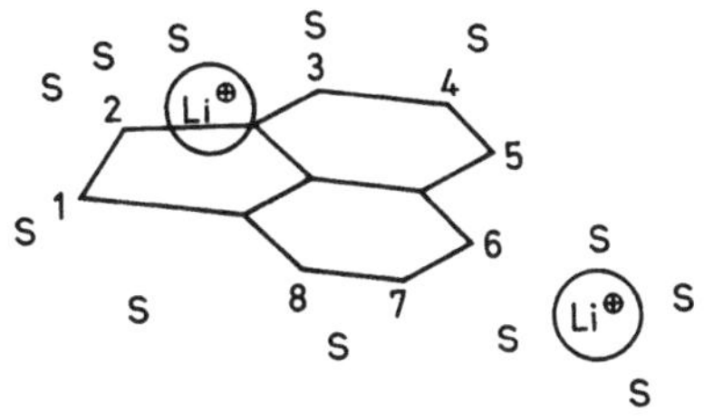

Fig. 4. The location of the lithium cations around 8^{2-} [83]

The dianion of acenaphthylene (8^{2-}) shows a structure that cannot be rationalized in terms of a triplet ion pair [83]. From its NMR data, it seems that the two negative charges are delocalized over the periphery (see section 5.3.1 for a detailed discussion). This structure was rationalized in terms of "minimum antiaromatic contribution" [84b]. A detailed ion-pairing study of this system based on ^{1}H, ^{13}C, ^{7}Li NMR and optical spectroscopy was carried out by Edlund [83]. It can be seen (Fig. 3) that the carbon chemical shifts of the sodium and potassium salts are almost unaffected by temperature variations, and therefore it is concluded that these are contact ion pairs. The lithium salt shows a much more significant change of the chemical shifts especially for carbon atoms C1, C2, C5, and C6. These shifts point to an unsymmetrical location of the cations. One cation seems to be located at the five-membered ring while the second one is located on the other edge of the molecule. This system also shows the following characteristics: (a) A significant amount of charge is located at positions C1, C2. (b) Complexing the cations with TMEDA shows (X-ray studies) that both cations Li-TMEDA appear at the five-membered ring [120]. (c) The agreement between ^{13}C chemical shifts and calculations are better than those for the highly solvated lithium salts. All these observations and considerations are depicted in Fig. 4. According to this presentation, one lithium cation is linked to the five-membered ring while the other lithium cation appears as a solvated ion at positions C5 and C6. Variable temperature experiments and solvents of high solvation power confirmed this conclusion [83]. The solvated cation does not influence the charge distribution, while the other lithium, which is a partner of a tight ion pair, helps to concentrate the charge at the five-membered ring. Changing of the cation from lithium to potassium causes in the ^{1}H NMR spectrum a deshielding of ca. 9 ppm at C1 and C2 and in the optical spectrum a red shift of ca. 65 nm of the long wavelength absorption maximum. Preliminary MNDO studies of acenaphthylene dianion 8^{2-} have shown that the potential energy surface is quite shallow and that the preference for a given structure could be altered by the mode of cation solvation. This could explain why solution studies and X-ray studies using TMEDA-complexed species may disagree [83d].

The detailed understanding of such a system also allows the assessment of the positions to be attacked, most likely by electrophiles. Moreover, the degree of the system's diatropicity may also be dependent on ion-solvation equilibrium. In this case it has already been shown that the system is not diatropic even if its total number of electrons may predict diatropicity [85]. Müllen studied independently a family of anions including the dianion of acenaphthylene (8^{2-}) and he also concludes that in THF solutions the lithium and potassium salts exist predominantly as contact ion pairs [83b].

It is also found that these systems behave in a different fashion than the charged annulenes [8]. In another π-conjugated systems, viz. 11^{2-} tight ion pairing was concluded from a NMR solvent and temperature dependency study [83b].

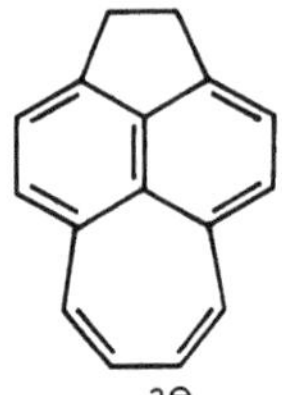

$11^{2\ominus}$

The most pronounced difference of chemical shifts is in the seven-membered ring moiety and therefore it can be assumed that the cation resides at this ring as a contact ion pair.

Diindeno[cd,lm]perylene (*12*) forms a diatropic dianion [85]. This dianion 12^{2-}, has building blocks of the previously mentioned acenaphthylene dianion (8^{2-}). The

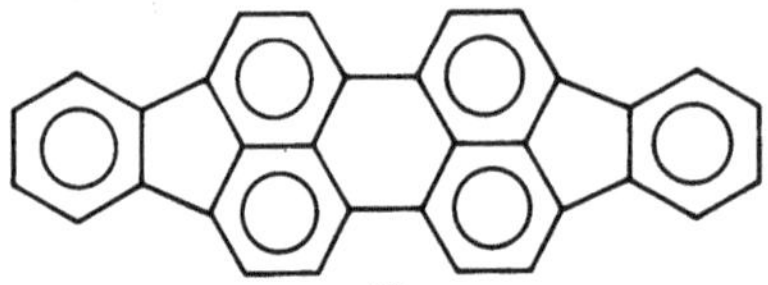

12

system represents a π-conjugated dianion of a rather extended periphery (vide infra). Here, too, the conclusion from ^1H and ^{13}C NMR studies is that the dianion forms contact ion pairs with sodium and lithium and probably in a triplet ion mode [85b]. It seems, therefore, that a mutual interaction exists between the π-conjugated system and the ion solvation equilibrium of each salt. In conclusion, one should not forget the role of the cation when carbanions are considered, both in chemical and spectroscopic studies.

5 Nonbenzenoid Anions: From Doubly to Quadruply Charged Systems

5.1 Bicyclic Polyanions

The study of monocyclic nonbenzenoid Hückeloid series established a field of challenging synthetic and theoretical effort [86]. The aromaticity of bicyclic and polycyclic π-conjugated frameworks required further investigation. Conjugated polyfused $(4n + 2)\pi$ dianions can be visualized either as peripheral systems in which the perimeter of the charged molecule accounts for its properties [87] (the Platt approach) or as an assembly of contributing fractions or rings where each contributor participates in forming the character of the entire system [88] (the Randić-Aihara approach). Conjugated polycyclic systems can be classified according to three different modes of fusion [89]:

(a) The fusion of two $(4n + 2)\pi$ fractions which results in a polycycle that includes a $(4n + 2)\pi$ periphery, as exemplified by the octalene tetraanion (14^{-4}) [90] prepared by Müllen and Vogel [90] and by the pentalene dianion (13^{-2}) prepared by Katz [91]. (b) The fusion of a $(4n + 2)\pi$ ring with an antiaromatic $4n\pi$ ring which results in a peripheral $4n\pi$ delocalization. This system is exemplifed by Edlund's azulene dianion

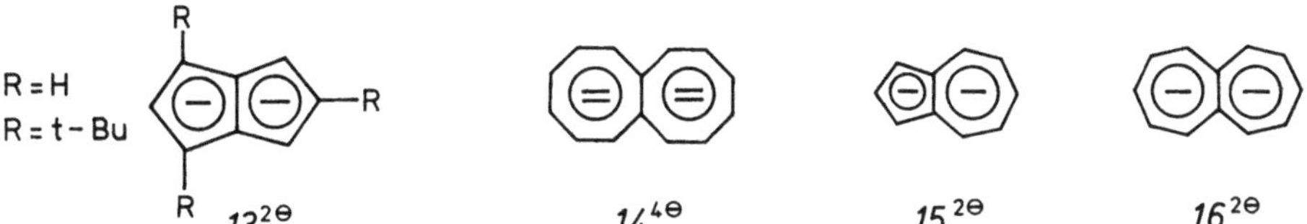

$13^{2\ominus}$ 10π $14^{4\ominus}$ 18π $15^{2\ominus}$ 12π $16^{2\ominus}$ 14π

(*15*) [92]. From the point of view of the character of each component, it seems that an aromatic and an antiaromatic [93] contribution should be taken into account. (c) The fusion of two antiaromatic rings $(4n_1\pi + 4n_2\pi)$ yields an aromatic $(4n + 2)\pi$-conjugated periphery while the two components are antiaromatic. Heptalene dianion (*16*) [94–96] with 14π-electrons in the conjugated perimeter represents this type of fusion. From the point of view of the perimeter model these systems are referred to as "de facto annulenes" with a perturbation of a σ-bond [8]. These systems are also often described by the term "annuleno-annulenes" [89]. The octalene tetraanion (14^{4-}) is a fourteen carbon system which contains four charges or one unit of charge per 3.5 carbon atoms, pentalene dianion (13^{2-}), has one charge per four carbon atoms. These systems seem to overcome the mutual charge repulsion which may prevail in doubly and quadruply charged hydrocarbons, due to aromatic stabilization.

5.1.1 Pentalene Dianion (13^{2-}) and Octalene Tetraanion (14^{4-})

The pentalene dianion (13^{2-}) was prepared by deprotonation of dihydropentalene (*17*) with n-butyllithium [91]. The pale yellow solution showed ^{1}H NMR signals at 4.98 and 5.73 ppm. The highfield position of these protons reflects the interplay between the diatropicity (low field shift) of the $(4n + 2)\pi$ system and the high field shift due to the effect of the two negative charges (s. Sect. 3.1.1). This bicyclic dianion is related to the moncyclic cyclooctatetraene dianion (*18*), the latter being perturbed by a σ bond, in the same sense as cyclooctatetraene (*18*) is related to pentalene (*17*). Charged tri-t-butyl pentalene was studied by Gerson and Hafner [61]. This pentalene ((*13*, R = t-Bu) is much more stable than the parent pentalene (*17*) and it can therefore be submitted to an electron transfer process to form the radical-anion [61].

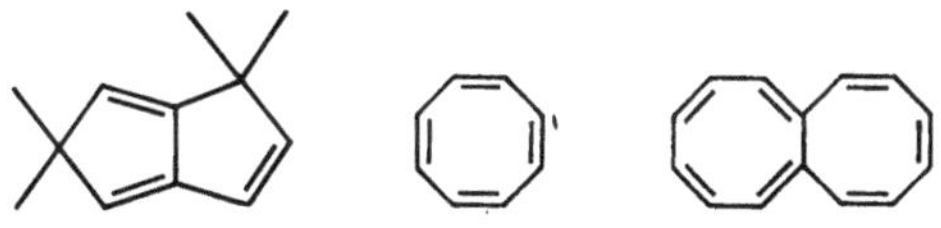

17 18 14

Octalene (*14*) is not planar and does not possess any element of symmetry [90]. Temperature dependent ^{1}H NMR measurements allow the detection of a rapid dynamic processes. These processes are interpreted in terms of inversion of the rings and a π-bond shift.

This system is a 14π-system but it does not exhibit a diamagnetic aromatic nature. It seems that the delocalization energy is not sufficient to overcome the energy that has to be "invested" in the flattening of the system. The radical-anion *14*$^{\cdot -}$, dianion *14*$^{2-}$, radical-trianion *14*$^{3\cdot -}$ and tetraanion *14*$^{4-}$ derived from *14* were prepared and characterized [90]. The dianion *14*$^{2-}$ has a D_{2h} symmetry as deduced from its ^{13}C NMR

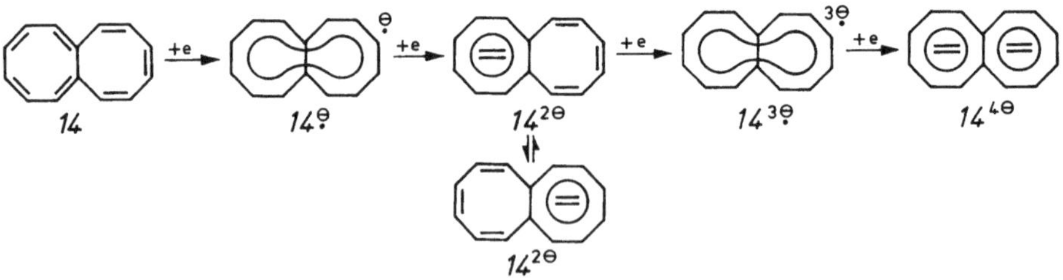

spectrum. This dianion is expected to be paratropic, however the ^{1}H NMR shifts do not show the expected high field shift. This behaviour is rationalized in terms of a rapid π-bond shift. This process is rapid on the ^{1}H NMR time scale even at temperatures as low as −150 °C [90].

5.1.2 Azulene (*15*$^{2-}$) and Heptalene (*16*$^{2-}$) Dianions

The azulene anion *15*$^{2-}$ was reinvestigated by Edlund [92] by the lithium reduction of the neutral azulene (*15*) in THF-d_8. This research followed an earlier study in which a dimeric dianion *19*$^{2-}$ was mistaken for the dianion *15*$^{2-}$ [93]. This anion is a 4nπ-

Table 1. Proton and carbon-13 chemical shifts for compounds 15, 19^{2-} and 15^{2-} [a]. Coupling constants (J/Hz) are given in parentheses.

	Structure	Solvent	Position 1, 3	2	3a, 8a	4, 8	5, 7	6	δ_{av}
^{1}H	15	cyclohexane	7.3	7.9		8.2	7.0	7.4	7.5
	19^{2-}	[$^{2}H_8$]THF	5.78	5.78		6.58	5.03	2.54	5.4
	15^{2-}	[$^{2}H_8$]THF	4.27 d	4.84 t		2.66 d	3.47 q	0.77	3.3
			(2)	(2)		(10)	(8, 10)	(7)	
^{13}C	15	$CDCl_3$	118.1	136.9	140.2	136.4	122.6	136.9	130.8
			(168)[b]	(163)		(152)	(156)	(152)	
		THF	104.6	105.3	121.4	127.8	117.6	43.8	
			(159)	(157)		(146)	(154)	(123)	
	19^{2-}	THF	105.0	106.9	122.2	128.1	114.8	44.0	
	15^{2-}	THF	110.6	100.5	123.5	99.2	133.7	67.9	110.2
			(155)	(162)		(155)	(140)	(151)	
q_{π}^{c}	15		1.10	0.98	1.01	0.88	1.05	0.95	
	15^{2-}		1.18	1.24	1.20	1.30	1.01	1.39	

[a] The n.m.r. spectra were obtained by using a Bruker WM-250 multinuclei Fourier transform spectrometer. Typical anion solutions were approximately 0.2 M, and the probe temperature was 25 °C. The chemical shifts are with reference to tetramethylsilane. For the carbon-13 measurements a small amount of cyclohexane was added to the carbanion solution as an internal reference using $\delta(SiMe_4) = \delta(\text{cyclohexane}) + 27.7$ p.p.m. [b] One bond ^{13}C—^{1}H coupling constants. [c] Hückel π-charges

electron system. However, the earlier reported data of the proton spectra show a slight upfield shift relative to the signals of azulene. These shift changes are of a magnitude that could be predicted from charge considerations alone and are not necessarily characteristic of a diatropic anionic system. In the case of *15*$^{2-}$ a 4nπ perimeter is expected from the fusion of two components [(4n + 2) π + 4nπ → 4nπ periphery]. Edlund [92] reports that the dianion formed first is actually the previously reported dimer (*19*$^{2-}$) which slowly undergoes a transformation to the azulene dianion (*15*$^{2-}$), which shows a significant degree of paratropicity (Table) due to its antiaromatic nature. These conclusions were also assisted by titration experiments and ^{7}Li NMR studies. The antiaromatic azulene dianion is a remarkably stable paratropic species.

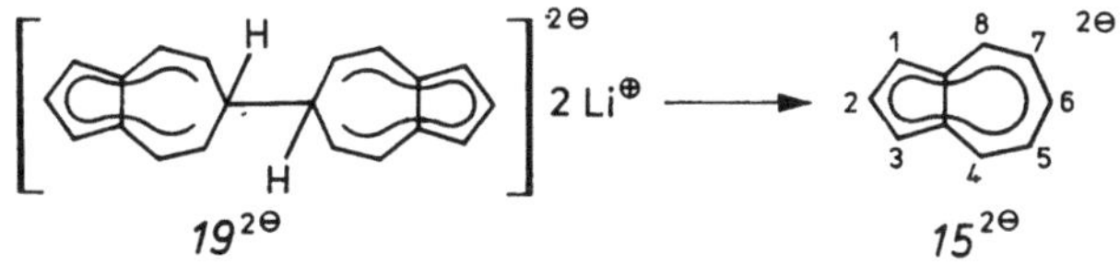

Heptalene, (*16*) a 12π system, was reduced to the dianion by Müllen and Vogel [94]. The resulting (4n + 2) π-system, i.e., *16*$^{2-}$ is obtained from *16* by lithium metal reduction. The radical anion of heptalene appears from its ESR spectroscopic data, to have a π-bond-delocalized ground state [95]. The number of experimental ESR coupling constants fits a C_{2v} or a D_{2h} symmetry [95].

Recently, the bicyclic system [14]annuleno[14]annulene (*20*) was charged [97]. This system forms a dianion, viz. *20*$^{2-}$ and a tetraanion, viz. *20*$^{4-}$ studied by Müllen [97]. This annuleno annulene is a (4n + 2)π + (4n + 2)π system and therefore diatropic. The di- and tetra-anions of *20* can be related to the octalene system (*14*) [90]. The annulenoannulene *20* forms upon alkali metal (Li, K) reduction all intermediate

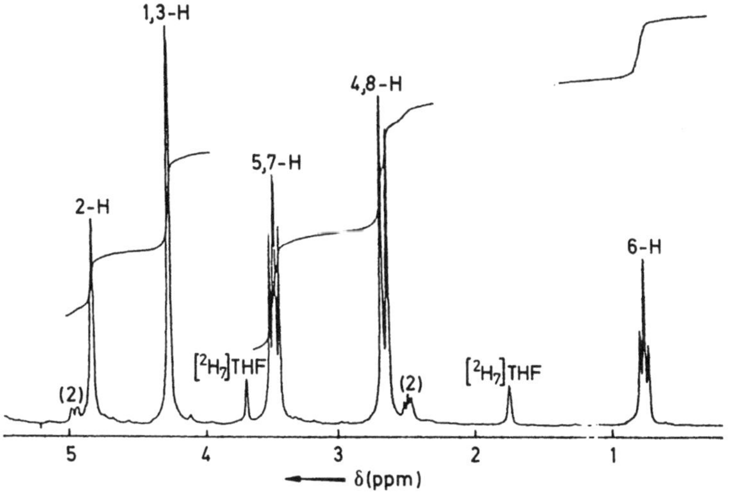

Fig. 5. 250.13 MHz ^{1}H NMR spectrum of the lithium dianion of azulene (*15*$^{2-}$) in [$^2H_8$]tetrahydrofuran. Residual signals are due to the doubly charged dimer *19*$^{2-}$

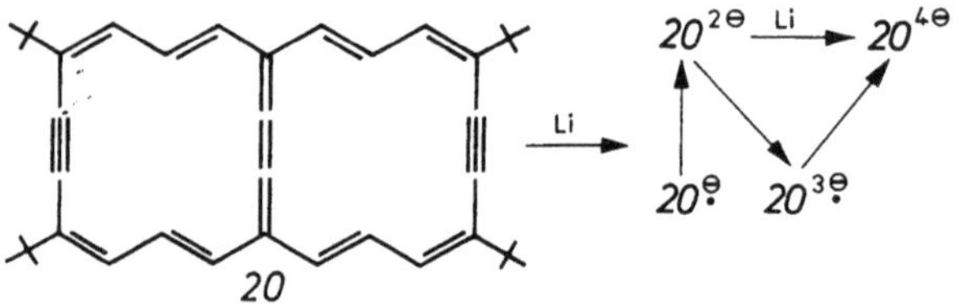

20

species, namely radical-anion, dianion, trianion-radical, and tetraanion. The paramagnetic species were characterized by ESR and ENDOR and the diamagnetic species by ^1H and ^{13}C NMR spectroscopies. In view of the spectroscopic data, the ionic species is described in terms of a charged macrocycle with peripheral delocalization [97]. The dianion 20^{2-} is a $4n\pi$ system (24 π-electrons) and therefore paratropic and the tetraanion 20^{4-} is a diatropic $(4n + 2)\pi$ system. They show extreme shift difference of inner and outer ring protons as has already been observed in annulene anions [97]. The charge distribution was derived from the proportionality between the charge-induced shifts and the ^{13}C NMR signals and were compared with HMO calculations. It is interesting that the anion behaves as a peripheral system while the neutral system does not [98,99]. These conclusions are in agreement with the nodal properties of the frontier orbitals involved in these anionic species.

5.2 Tri- and Tetracyclic Polyanions

5.2.1 Tricyclic Anions

Acepentalene dianion (21^{2-}) was reported recently by deMeijere [100]. This most interesting dianion as well as acepentalene itself (*21*) was of interest to theoretical chemists and spectroscopists. The dianion is prepared by reacting triquinacene (*22*) with n-butyl lithium and potassium-t-amyloxide in hexane (Lochmann-Schlosser reagent) [24]. This dianion showed a singlet in the ^1H NMR spectrum at 6.16 ppm. Quenching experiments with TMSCl ($ClSiMe_3$) gave products of the electrophilic attack at the central carbon atom (C-10) and at the methine carbons (C-1, C-4, C-7).

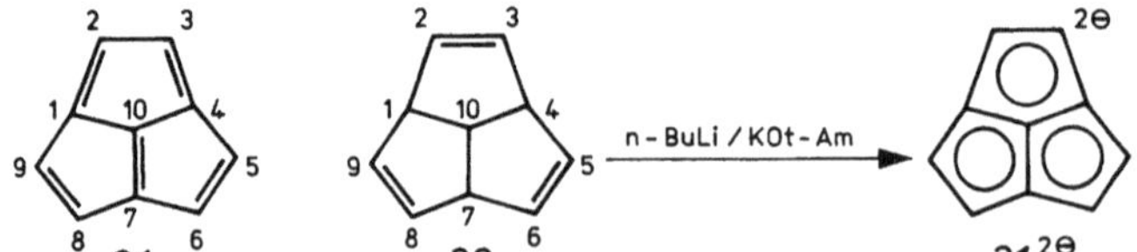

21 22 $21^{2\ominus}$

The dianion of the higher homologue, i.e. aceheptylene dianion (23^{2-}) was prepared by the metal reduction of aceheptylene (*23*). The hydrocarbon was prepared by Hafner [101] and was charged by reacting the hydrocarbon with lithium and sodium metals [84b]. The aceheptylene dianion shows a pattern of charge delocalization which may seem at first sight surprising and raises questions concerning the basic concepts of aromaticity in systems with a central carbon bridge and will be discussed in detail.

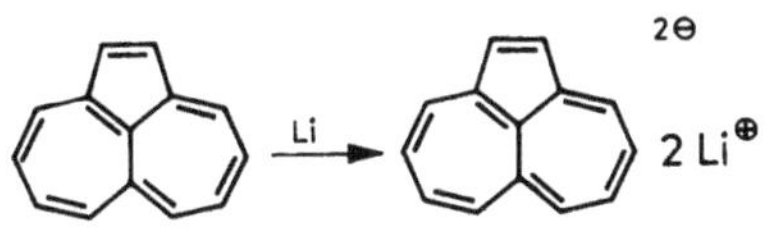

23 $23^{2\ominus}$

5.2.2 Priority of Paths of Delocalization in Dianions

The most comprehensive definition of the numerous definitions and criteria suggested to define aromaticity is the one based on the energy content of aromatic systems. Dewar [102] has defined aromatic molecules as cyclic species with a large resonance energy in which all the atoms in the ring take part in a single conjugated array. In other words, aromatic systems sustain a cyclic π-electron delocalization which reduces the energy content of the systems relative to that of corresponding model compounds without cyclic delocalization. Similarly, antiaromatic systems reveal a cyclic π-electron delocalization which leads to a strong destabilization and therefore to a high energy content in respect to analogous acyclic compounds. As a criterion of aromaticity or antiaromaticity, it is difficult to apply this energetically based definition in practice. The more practical definitions can be classified into two general groups: criteria based on purely theoretical concepts and those which refer to experimentally observable phenomena. The peripheral criterion suggested by Platt [87] exemplifies the first group of definitions by assigning aromatic character to a cyclic or polycyclic system with $(4n + 2)\pi$-electrons in its periphery. Systems with $4n\pi$-electrons in the path of conjugation would be inclined to reveal antiaromatic properties, while those with $(4n + 1)$ or $(4n + 3)$ peripheral conjugated π-electrons would be estimated as nonaromatic. This structural concept is based on the free electron theory and treats crosslinks and inner sp^2 carbons as small perturbations. The most useful definition among those which relate to experimentally observable phenomena is based on magnetic anisotropy [103,104]. Another experimentally based criterion for aromaticity refers to bond length alternation [102]. An aromatic compound is supposed to reveal a low degree of bond length alternation around the characteristic aromatic bond length (1.4 Å) in contrast to nonaromatic species in which a substantial bond alternation occurs.

The described definitions have been widely used to characterize aromaticity. However, the correlation between these criteria and aromatic (or antiaromatic) nature is by no means simple. Even more so, the aromatic character is, by definition, a relative property, namely, one can assign aromaticity only by referring to model compounds. Thus, even when the number of π-electrons calls for the existence of aromatic character, its extent cannot be predicted. Similarly the relationship between magnetic susceptibility of a polycycle and its conjugative stabilization is highly complex [105,106].

The aceheptylene *23* is a system of 14 carbon atoms of which 13 carbons compose its conjugated periphery. Accordingly, *23* can be considered as a perturbed [13] annulene or as a $(4n + 1)\pi$ conjugated system ($n = 3$) with an inner carbon atom. This system seems most suitable to examine the basic problem of patterns of delocalization in polycyclic anions. In view of the mentioned Platt's peripheral definition [87] such a compound is expected to be a nonaromatic polyvinylic system, not exhibiting any aromatic or antiaromatic properties. This expectation is unambiguously confirmed.

The ^{1}H NMR spectrum of *23* reveals a typical vinylic range with chemical shifts between 5.17 and 6.98 ppm. The nonaromatic nature of aceheptylene is clearly indicated by its high field chemical shift range (Table 2, Table 3) [84b)]. A comparison between the chemical shifts of *23* and those exhibited by the heptalene system [94)] reflects the nonantiaromatic nature of the former. Although antiaromaticity, imposed on heptalene by its [12]π-electron perimeter, is quenched, at least partially, by nonplanarity, its ^{1}H chemical shift center of gravity reveals a paratropic displacement relative to the corresponding value exhibited by 23 (5.48 vs. 6.00 ppm, respectively).

In the course of a metal reduction process of aceheptylene [84b)], two electrons are inserted into its π-framework to yield a system of 14 sp^2 carbon atoms and 16 π-electrons. In principle these π-electrons can accommodate three modes of different charge distributions over the system: In *23 a*$^{2-}$, the two extra π-electrons may be introduced into the heptalene moiety. The resultant species may be visualized as composed of a heptalene dianion coupled with a double bond. Such a mode of electron distribution will result in an aromatic character due to the aromaticity of the heptalene dianion moiety [94)]. The two extra π-electrons might be introduced into the periphery of the system. In this case a [13 C-15π] peripheral system is obtained (viz., *23 b*$^{2-}$) with a (4n + 3; n = 3) perimeter. According to the mentioned Platt's peripheral model, the dianion will reveal a nonaromatic character. The third possibility of π-electron distribution over the aceheptylene dianion system considers one of the two inserted electrons to be located at the inner carbon atom (C_{13}) and the other to be delocalized over the entire periphery. This mode would result in a [13 C-14π] periphery coupled with an inner negatively charged carbon (viz., *23 c*$^{2-}$). A Hückeloid periphery of 14π-electrons is therefore obtained and aromatic character is expected.

23 →

23 a *23 b* *23 c*

In view of the driving force involved in aromaticity (vide supra) the second possibility, leading to nonaromatic character, is ruled out. Kinetic as well as thermodynamic

Table 2. ^{1}H NMR Patterns[a] of 23 and 23^{2-} [84b)].

		$H_{1,2}$	$H_{3,10}$	$H_{4,9}$	$H_{5,8}$	$H_{6,7}$	^{1}H center of gravity of the heptalene moiety	overall ^{1}H center of gravity
23	1	6.98 (s)	6.71 (d, J = 9.1 Hz)	5.19 (dd, J = 10.6, 8.9 Hz)	5.79 (dd, J = 11.8, 8.6 Hz)	5.36 (d, J = 12.0 Hz)	5.77	6.00
23^{2-}	2	7.08 (s)	6.54 (d, J = 9.3 Hz)	6.75 (t, J = 9.0 Hz)	5.76 (t, J = 9.1 Hz)	7.61 (d, J = 9.1 Hz)	6.66	6.75

[a] ppm, referenced to Me_4Si

Table 3. ^{13}C NMR Patterns[a], Charge Densities, and Relevant Orbital Coefficients in *23* and 23^{2-} [84b)]

		$C_{1,2}$	$C_{3,10}$	$C_{4,9}$	$C'_{5,8}$	$C_{6,7}$	$C_{11,12}$	C_{13}	C_{14}
23	^{13}C	123.3	142.6	120.9	140.4	132.5	135.1	153.1	158.9
	charge density[b]	—0.0867	0.0450	—0.0136	0.0539	0.0013	—0.0674	0.0422	0.0930
	LUMO coefficients[b, c]	0.1160	0.3356	0.0005	0.3897	0.1406	0.1387	0.4415	0.4137
	HOMO coefficients[b, d]	0.2829	0.1578	0.3626	0.0758	0.3235	0.3913	0.0	0.0
23^{2-}	^{13}C	112.4	97.1	121.1	91.4	119.7	123.9	85.6	114.7
	charge density[b]	—0.1422	—0.1432	—0.0865	—0.1870	—0.1021	—0.1525	—0.2178	—0.1555

[a] ppm, referenced to Me_4Si; for numbering, see the schemes. [b] As obtained from ωβ calculations. [c] The orbital into which two electrons are added in the reduction process

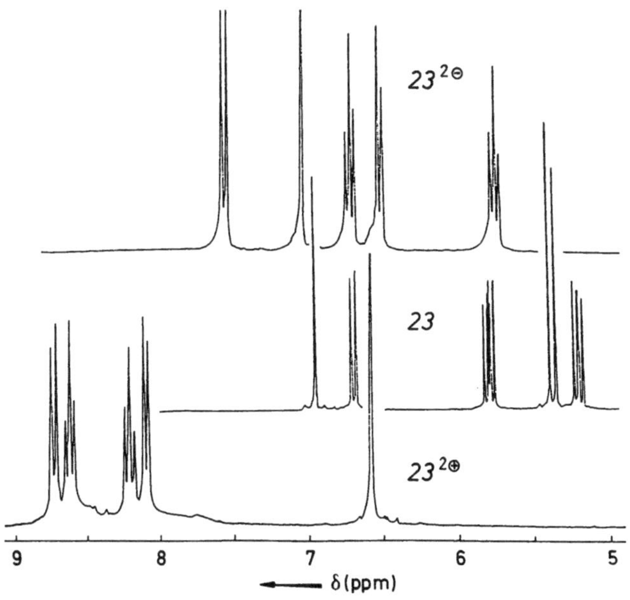

Fig. 6. ^1H NMR spectra of aceheptylene (*23*) and aceheptylene dianion (23^{2-}) and aceheptylene dication [84b)]

considerations lend strength to the third possibility. Theoretical calculations performed on aceheptylene (*23*) [84b)] point toward a very large atomic coefficient of the atomic orbital which belongs to the inner carbon C_{13} in the lowest vacant molecular orbital of the neutral compound. (The atomic coefficients of the LUMO of *23* — the orbital into which two electrons are added — are reported in Table 3). This indicates that the carbon is prone to acquire a substantial charge density.

Even more so, the aceheptylene, as starting material for the reduction process, was shown to be best described as a peripheral system coupled with an inner perturbation. Consequently, the distribution of electrons in the dianion as proposed in *23c* would minimize the reorganization of electrons in the course of the reduction process.

According to Hine's rule [107)], reactions will be favored when involving the least change in atomic positions and electron configurations, due to relatively small activation energies. While substantial changes of electron distribution and large atomic relocations (e.g., a shortening of the C_1–C_2 bond and an increase of the C_1–C_{11} and C_2–C_{12} bond lengths) are expected according to the first proposed distribution mode, no such phenomena should occur if the reduction proceeds via the third mode. Thermodynamically, the repulsion between two negative charges distributed only on the heptalene moiety may be stronger than the repulsion occurring when one negative charge is located on an inner atom and the other charge distributed over all the periphery. Indeed, ^{1}H and ^{13}C NMR spectra, as well as theoretical calculations performed on *23* point unequivocally toward an aromatic dianion in which nearly one negative charge is located on the inner carbon atom, while the second is delocalized on the periphery [84b)]. The center of gravity of the ^{1}H NMR spectrum of *23*$^{2-}$ is diatropically shifted in respect to that of 23 by 0.75 ppm. This low field displacement is rationalized in terms of the diamagnetic ring current related to aromaticity which is sustained in 23. It should be noted that the diamagnetic ring current effect is in fact much larger, the aromatic low-field shift being partially quenched by the shielding of the two negative charges.

The ^{13}C NMR bands of *23*$^{2-}$ reveal a total high field shift of 366 ppm with respect to the neutral compound ^{13}C chemical shift, indicating the presence of doubly charged species [31)]. This high value also indicates that the dianion is more diatropic than the neutral hydrocarbon [41)]. The inner carbon atom C-13 undergoes an unusually large paratropic shift of 67.6 ppm, a value rationalized by a large negative charge density on this carbon, in accord with the suggested electron distribution mode. This mode is further supported by ωβ calculations which were performed on the dianion *23*$^{2-}$. The calculations estimate (i) a large negative charge density on the inner carbon C-13 (Table 3) and (ii) a relatively small degree of bond lengths alternation on the periphery vs. a very long bond between C-13 and C-14 (1.4891 Å).

The suggestion of Rabinovitz and Hafner that this dianion is actually a 13-carbon 14π perimeter with one charge residing in the periphery and one being localized at the central atom (C-13 bridge) [84b)] is supported by Müllen's reductive alkylation studies of this dianion [108)]. The reductive alkylation of aceheptylene dianion (*23*$^{2-}$) yields three products and in all of them the central carbon atom C-13 is alkylated. It can be concluded that the quenching reaction proceeds as a kinetically controlled attack of the electrophilic agent at the position of the highest π-charge. This mechanistic view describes only the regioselective addition of the first electrophile to the dianion. The second step is controlled by the charge distribution within the intermediate monoanion.

Two related systems, viz., s-indacene (*24*) and as-indacene (*25*) form dianions containing 14π in their periphery [109,110)]. These dianions are relatively stable as compared with the parent hydrocarbons, viz. *24* and *25*, respectively. A detailed study on the s-indacenyl dianion (*24*$^{2-}$) shows that this dianion is a delocalized 14π-electron dianion where most of the charge is located at C-1, C-3, C-4 (and at C-5, C-7, C-8) carbon atoms. This dianion was obtained by treatment of 1,5- and 1,7-dihydro-s-indacenes in THF-d_8 or HMPA-d_8-THF mixture with BuLi. This preparation affords an easy access to this anion [109c)], by avoiding the reduction of the very unstable hydrocarbon *24*. The degree of charging of *24*$^{2-}$ is obtained from a simple charge-chemical shift

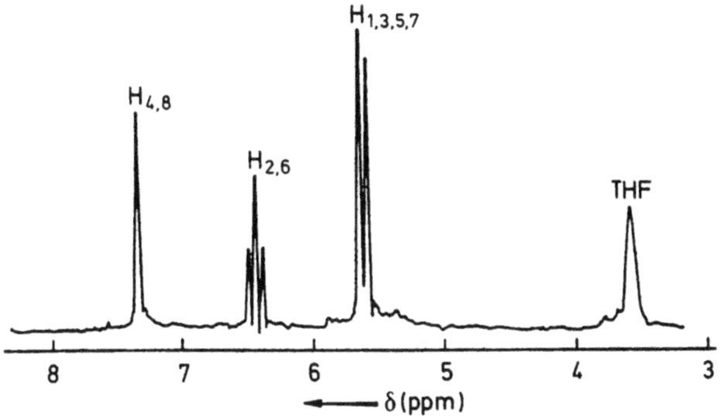

Fig. 7. ^{1}H NMR spectrum of s-Indacenyl Dianion [109]

relationship [26a, 31] of the ^{13}C NMR spectrum. The proton spectrum (Fig. 7) [109c] is consistent with a diatropic delocalized dianion. This spectrum shows the contradicting effects of charge and the ring-current effects, i.e. the diatropicity of the system on the one hand and the high field shift due to the shielding of the two negative charges on the other.

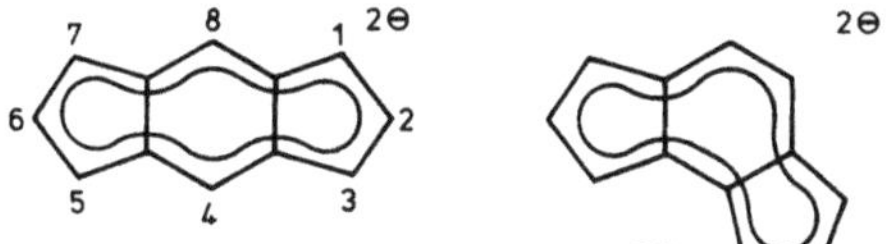

24$^{2\ominus}$ *25*$^{2\ominus}$

An interesting system is another 14π-electron dianion, viz. *27*$^{2-}$ which was prepared recently by Hafner [111] from *26*. This system also forms a monoanion, viz. *26*$^{1-}$ as outlined:

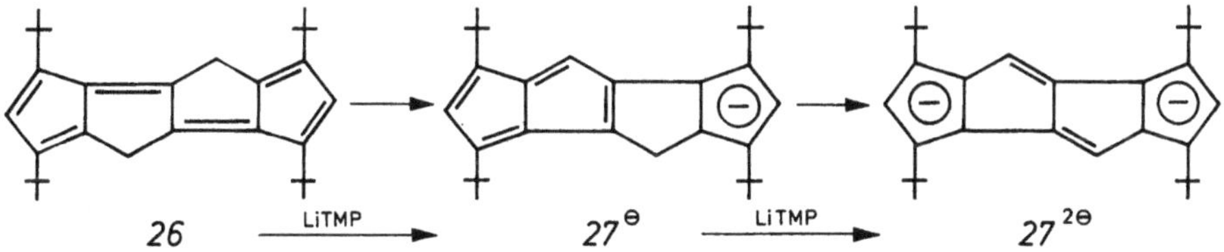

26 *27*$^{\ominus}$ *27*$^{2\ominus}$

A system which gave very interesting reduction products is the acepleiadylene cyclohept[f,g]acenaphthylene (*28*) which was synthesized in 1956 by Boekelheide [112]. This system can be viewed upon as a double bonded bridged annulene which fulfils Platt's [87] approach to π-conjugated systems as it contains 14π-electrons in its periphery. It can also be considered a polycyclic system with a central π bond as perturbation like the tetracyclic benzenoid pyrene (*29*) which also has 14π-electrons in its perimeter. Following the peripheral approach one can rationalize the alternate formation of paratropic and diatropic species, as well as intermediary paramagnetic species The systems with unpaired electrons, viz. *28*$^{\overline{\cdot}}$ and *28*$^{3\overline{\cdot}}$, were studied by ESR spectroscopy [113] and the paired-electron systems, viz., *28*$^{2-}$ and *28*$^{4-}$, were prepared and studied by Huber and Müllen applying NMR spectroscopy [114]. All charged

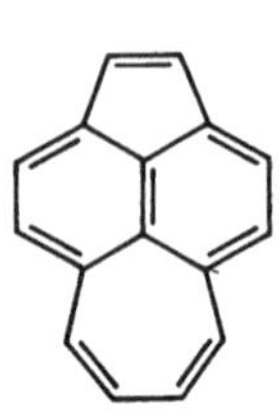 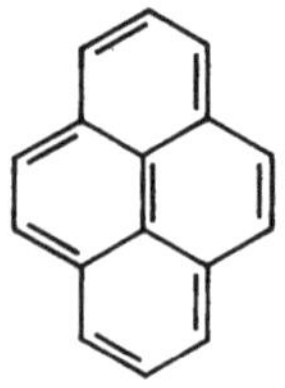

28 $14\pi + 2\pi$ | 29 $14\pi + 2\pi$

species were prepared by lithium metal reduction and studied spectroscopically and chemically.

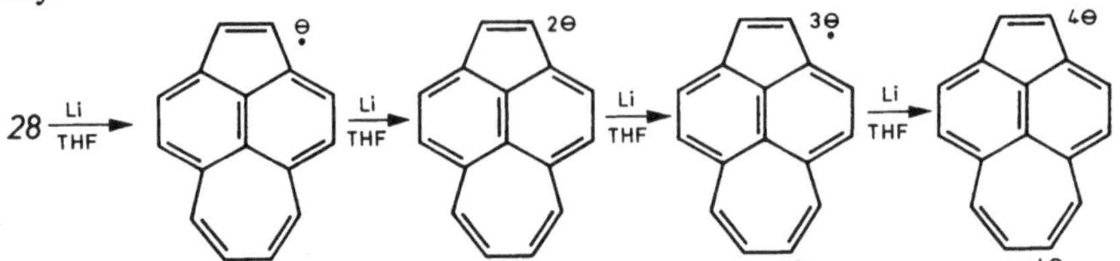

$28^{\dot{-}}$ ESR | 28^{2-} $16\pi+2\pi$ NMR | $28^{3\dot{-}}$ ESR | 28^{4-} $18\pi+2\pi$ NMR

When the reduction is monitored by ^1H NMR spectroscopy [114] the original spectrum disappears and the new diamagnetic species is formed. It shows a highly resolved spectrum whose center of gravity appears at a very high field ($\delta_{av} = 0.0$ ppm). From its proton ^1H NMR spectrum it is clear that this species is paratropic. Its paratropicity may stem from a $4n\pi$-electron system. This dianion has $4n\pi$-electrons only if the perimeter π-electrons are counted. It is therefore a $16\pi + 2\pi$ system. Further reduction gave rise to a spectrum with ($\delta_{av} = 5.0$ ppm) which is in line with a diatropic $(4n + 2)\pi$-electron system. The average chemical shift (δ_{av}) results from an interplay between the ring current effect which would shift the absorption to low field [42, 44, 45] and the effect of the shielding of four negative charges [29, 30]. Reaction of *28*$^{2-}$ with an electrophile, e.g. dimethylsulphate [108b] (Me_2SO_4) yielded a dimethyl derivative in which the methyl groups are located on the periphery. This behaviour is in sharp contrast to that of *23*$^{2-}$, and points to a peripheral delocalization of the negative charges.

5.3 Benzannelated Nonbenzenoid Anions

Benzannelated π-conjugated systems enable the assessment of the contributing components to the "aromaticity" or "antiaromaticity" of the system. On the one hand benzo rings help to keep the system planar and on the other the benzo group being "ultimate aromatic" systems, have to give up some of their aromatic nature so that their π-electrons will participate in the entire system.

5.3.1 Benzannelated Cyclobutadiene Cyclopentadiene and Pentalene

The synthesis of phenyl substituted cyclobutadiene dianions, *29a*$^{2-}$, *29b*$^{2-}$ was reported [115, 116]. These phenyl substituted dianions were shown by NMR data to delocalize the charge over the phenyl substituents. Until very recently the data concerning dibenzocyclobutadiene (biphenylene) dianion (30^{2-}) were not clear. Günther and Edlund [117] could prepare 30^{2-} by lithium metal reduction of *30* below —30 °C, using

$29b^{2\ominus}$

R = H, Phenyl

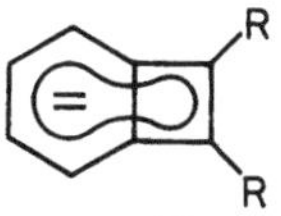

$29a^{2\ominus}$

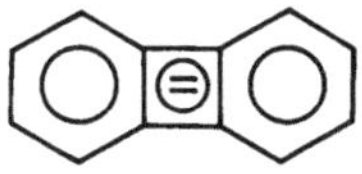

$30^{2\ominus}$

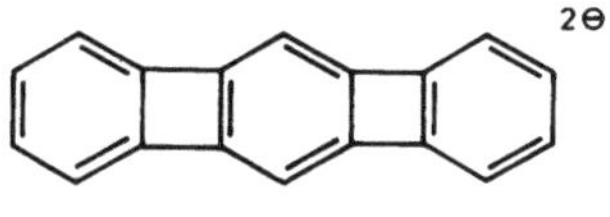

$31^{2\ominus}$

an ultrasonic bath. The dianion was characterized by its ^1H and ^{13}C NMR spectra. A strong effect on the spectrum is attributed to ion pairing. From calculated Günther's Q values [118], the system 30^{2-} exhibits a doubly charged central four-membered ring of $(4n + 2)\pi$-electrons. Günther's Q values are obtained from NMR coupling constants and are very useful for the estimation of the aromaticity of systems [118]. Comparison of the Q value of 30^{2-} (Q = 1.45) with that of $29b^{2-}$ [116] ($R_1 = R_2 = H\cdot$, $R_3 = R_4$ = Phenyl, Q = 1.05) demonstrates the high $(4n + 2)\pi$ character of the central ring of 30^{2-} (s. Sect. 5.3). (The irontricarbonyl complex of *29a* led to a Q value of 1.42). The ^{13}C NMR data support the formation of a doubly charged species. An overall carbon shift of 360 ppm ($\Sigma\Delta\delta\ ^{13}C$) as compared to the parent hydrocarbon shows the presence of two charges according to well accepted proportionality constants K_C in the Spiesecke and Schneider relation [30] (K_C = 160 ppm/e). The polyphenylenes synthesized by Vollhardt [119] also form dianions. The [3]phenylene *31* forms a dianion by reduction in the NMR tube by potassium metal in THF-d_8. The central hydrogens move downfield in contrast to expectations and the remaining hydrogens shift upfield less than anticipated. The dianion is a 20π-electron system and the strange NMR trends may be accounted for by an uneven distribution of charge which avoids antiaromaticity [119]. However, both systems 30^{2-} and 31^{2-} show similar proton spectra. Quenching of 31^{2-} by oxygen gave *31* showing that the overall topology of the system has not been disrupted.

Dibenzo[b,f]pentalene (indeno[2,1a]indene dianion) (3^{2-}) has attracted much interest [82]. Its preparation via lithium metal reduction of *3* and from *32* by deprotonation with BuLi is very easy. NMR studies (^1H, ^{13}C and ^7Li) of this system yield information on the electron distribution and the ion pair structure [60,81,82]. The intermediate step, i.e. the radical anion $3^{\overline{\cdot}}$ affords the estimation of charge densities. The starting material, viz. *3*, is a 4nπ system and is an excellent example of the Randić approach of polycyclic system [88]. The ^1H NMR spectrum of 3^{2-} shows the balance

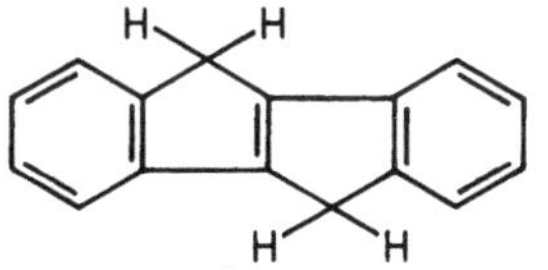

32

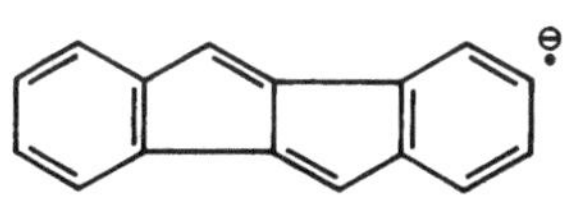

$3^{\ominus\cdot}$

between the two contradicting effects namely, the lowfield shift due to the ring current effect of the $(4n + 2)\pi$-electron system on the one hand and the charge shielding effect on the other hand. The average proton chemical shift of *3* and 3^{2-} is roughly the same. This system can be regarded as a peripheral aromatic [16]annulene dianion with delocalized 18π-electrons and cross-linked by three perturbing σ bonds.

Cyclic voltametry of **3** showed two one-electron reduction waves ($E^1_{pc} = -1.92$ V and $E^2_{pc} = -2.43$ V), the first step is reversible [82c)]. As already discussed, a careful ^{13}C NMR study of 3^{2-} reports the charge densities at the various carbon atoms and the ion-solvation equilibrium of its dilithium salt [60, 82c)], showing that the lithium salt exists mainly as a contact ion pair (c.i.p.).

Acenaphtylene (*8*$^{2-}$) dianion is obtained by alkali metal reduction (e.g. lithium or sodium) in an ether solvent. Early ^1H NMR studies were reported by Lawler and Ristagno [10, 11)]. This dianion was subject to detailed studies which concentrated on its mode of electron delocalization and the ion-pairing equilibrium. The mode of electron delocalization [84b)] is mainly deduced from ^1H and ^{13}C NMR chemical shifts in tetrahydrofuran (THF) (Fig. 8) while the ion-solvation equilibrium was deduced from ^1H, ^{13}C and ^7Li shifts in THF-d_8, 2-MeTHF, Et_2O and THF-HMPA-d_8 [83a)].

The dianion is composed of twelve carbons and 14π-electrons which can accommodate in principle two different modes of electron delocalization. In one mode, i.e. *8a*, the two electrons are delocalizede on the perimeter thus resulting in a (11 C-13π] system. In the second mode, i.e. *8b*, one of the two electrons is located on the inner carbon atom C_{11} and the other one is delocalized over the eleven carbon periphery analogous to the electron distribution in the aceheptylene dianion (*23*$^{2-}$). The resulting species would be a [11 C-12π] peripheral system which according to Platt's model, would be expected to reveal an antiaromatic character.

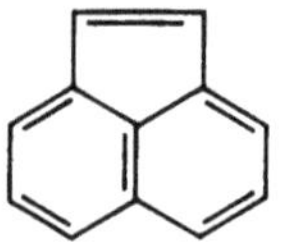

8

Calculations (ωβ) were applied as a means of assessing the portion of ^1H NMR chemical shift due solely to charge shielding. It was shown that in cyclic conjugated systems the proton chemical shifts are related linearly to the π-electron density [30, 31)].

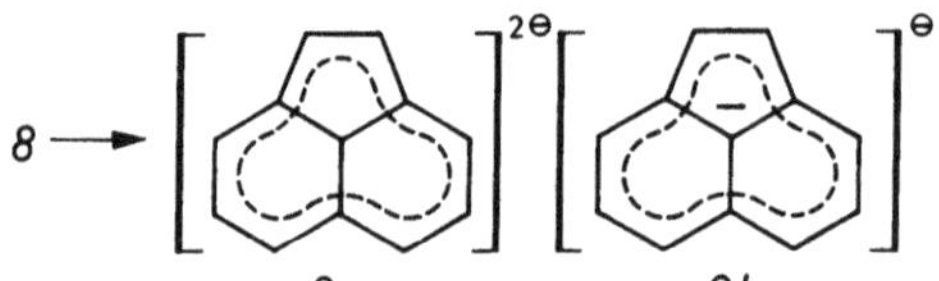

The theoretical charges on hydrogen-bearing carbon atoms of *8*$^{2-}$ were summed up and the sum was multiplied by 10.7 [30)]. The product was then divided by the number of protons and this average subtracted from the center of the ^1H NMR chemical shifts exhibited by the neutral acenaphthylene (*8*). The ^1H NMR center of gravity of *8*$^{2-}$ is found to be 7.56 ppm (Table 4); the theoretical high field displacement due to negative charge density is estimated as 1.99 ppm. Therefore, the center of gravity of *8*$^{2-}$ due only to negative shielding would be 5.57 ppm, while the exhibited center of gravity is 4.53 ppm. The difference between the two values (1.04 ppm) is relatively small (when compared with paratropic displacements due only to paramagnetic ring

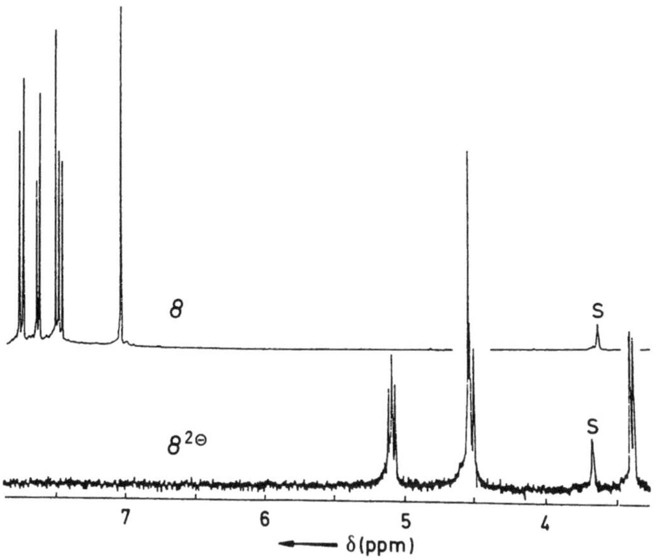

Fig. 8. ^{1}H NMR spectra of acenaphthylene (*8*) and acenpathylene dianion (8^{2-}) [84b)]

current in antiaromatic species such as anthracene dianion, chrysene dianion, or phenanthrene dianion with high field shifts of 2.90, 4.06 and 5.33 ppm, respectively, vide supra). On the basis of these observations and arguments, the total ^{1}H NMR high field chemical shift observed in the ion formation process is rationalized in terms of negative charge density shielding and the quench of the diatropic aromatic ring current which prevails in the neutral system. An antiaromatic character is not found to be generated in 8^{2-} as would be the case according to the second proposed possibility of electrons distribution mode in 8^{2-}, viz. $8b^{2-}$ in which one electron resides on the inner carbon and the second one resides on the periphery. A comparison between the ^{13}C NMR patterns of the neutral system and that exhibited by its corresponding dianion is particularly instructive. The total high field displacement of ^{13}C bands in the course of the reduction $8 \rightarrow 8^{2-}$ is considerably smaller than the total displacement observed in the process $23 \rightarrow 23^{2-}$ and is smaller than the shift expected [30)]. These results as well as other observations indicate that when diamagnetic, aromatic species are obtained in reduction or oxidation processes of neutral precursors, their total ^{13}C band displacements ($\Sigma \Delta\delta\, ^{13}$C) are in accord with the ^{13}C shift-charge density correlation [31)]. In contrast, when nonaromatic and, in particular, antiaromatic systems result, this correlation is not observed. The nonaromaticity of the acenaphthylene dianion (8^{2-}) vs. the aromatic character of the aceheptylene dianion (23^{2-}) is therefore emphasized. Another conspicuous difference between the ^{13}C patters of 8^{2-} and 23^{2-} is concerned with the absorptions exhibited by the inner carbon (C-11 and C-13, respectively). In the course of the reduction of *23* to 23^{2-} this band is paratropically shifted by 67.6 ppm, (Sect. 5.2.2) pointing toward a large negative charge on this carbon, whereas in the process $8 \rightarrow 8^{2-}$ the inner carbon NMR band is diatropically shifted by 9.47 ppm. This indicates that a very small portion of the negative charge resides on the inner carbon atom of *8*$^{2-}$ as suggested by the second possibility of electrons distribution modes in the dianion. Indeed, theoretical calculations [84b)] assign a very small negative charge density to C-11. Thus, the NMR results as well as theoretical calculations point unequivocally toward a dianion in which the two "extra" electrons are delocalized in the periphery to afford a nonaromatic [11 C-13π] system, as depicted in *8a*$^{2-}$. The clear difference in mode of electron distribution and in character between the dianion *8*$^{2-}$ and aceheptylene (*23*$^{2-}$) should be considered in view of the fact that the two systems are isoelectronic. These conclusions are supported by the X-ray structure [120)].

Table 4. ^{1}H NMR Patterns[a] of 8 and 8^{2-} [84b)]

	$H_{1,2}$	$H_{3,8}$	$H_{4,7}$	$H_{5,6}$	overall ^{1}H center of gravity
8	7.13 (s)	7.87 (d, J = 7.9 Hz)	7.59 (dd, J = 8.4, 8.2 Hz)	7.74 (d, J = 6.2 Hz)	7.56
8^{2-}	5.12 (s)	4.51 (d, J = 8.1 Hz)	5.10 (f, J = 6.7 Hz)	3.39 (d, J = 6.5 Hz)	4.53

[a] ppm, referenced to Me_4Si

Table 5. ^{13}C NMR Patterns[a], Charge Densities, and Relevant Orbital Coefficients in *8* and *8*$^{2-}$ [84b)]

		$C_{1,2}$	$C_{3,8}$	$C_{4,7}$	$C_{5,6}$	$C_{9,10}$	C_{11}	C_{12}
8	^{13}C	129.4	124.1	127.8	127.3	139.7	128.2	128.4
	charge densities[b]	−0.0313	0.0381	0.0043	0.0249	−0.0149	−0.0465	0.0043
	LUMO coefficients[b,c]	0.3175	0.3671	0.1519	0.3921	0.2961	0.0	0.0
8^{2-}	^{13}C	85.8	96.8	126.1	81.8	123.1	137.6	148.9
	charge densities[b]	−0.2272	−0.1772	−0.0882	−0.2438	−0.1697	−0.1218	−0.0661

[a] ppm, referenced to Me_4Si; for numbering, see the schemes. [b] As obtained from ωβ calculations.
[c] The orbital into which two electrons are added in the reduction process

The spectroscopic studies [83,84)] afforded chemical applications. The reductive alkylation of dianion *8*$^{2-}$ gave alkylated products only in the periphery [121)]. This observation confirms the structure *8a*$^{2-}$ suggested by Rabinovitz and Hafner [84b)]. Contrary to the reductive alkylation of *23*$^{2-}$ the quench of dianion *8*$^{2-}$ did not afford any alkylated product at the central atom C-11. The alkylated acenaphthylene served as a starting material for the preparation of acephenanthrylene *34*$^{2-}$ (vide supra) [121)].

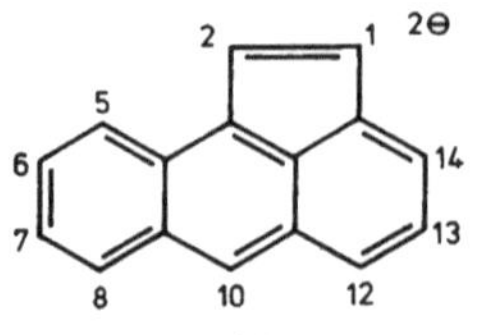

33$^{2\ominus}$

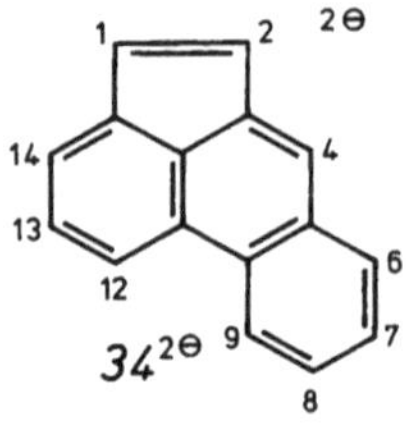

34$^{2\ominus}$

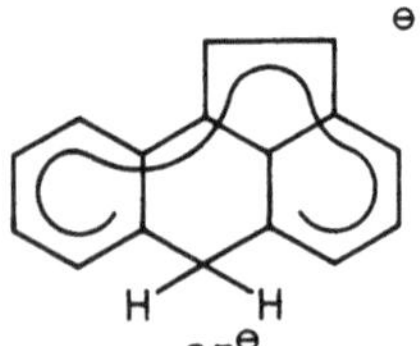

35$^{\ominus}$

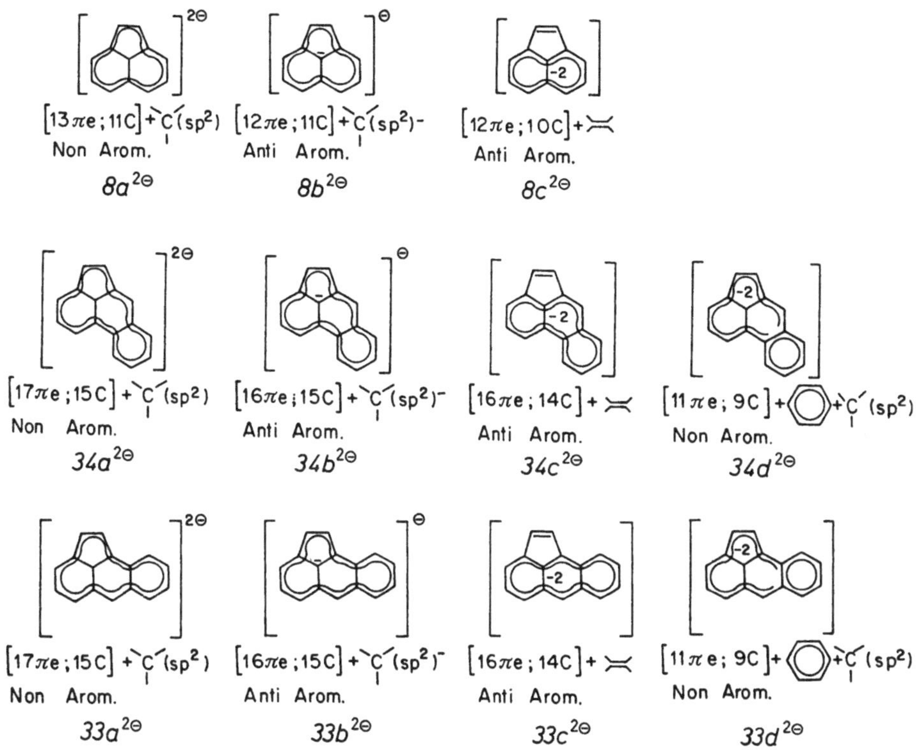

Fig. 9.

The higher homologs of acenaphthylene dianion 8^{2-} are the aceanthrylene dianion *33* and acephenanthrylene dianion *34* [122,123]. The convenient synthesis of the the hydrocarbon enabled a detailed investigation of their metal reduction and the exploration of their patterns of delocalization. Despite their being $(4n + 2)\pi$-electron systems, these anions are not diatropic as one may expect from counting their π-electrons.

The ^1H NMR spectrum shows bands in the region 4.0–6.0 ppm (See Table 6). Following the discussion of priority of paths of delocalization of aceheptylene dianion 23^{2-} and acenaphthylene dianion 8^{2-} also 33^{2-} and 34^{2-} show that specific paths of delocalization are favoured. While in the neutral structure *33* and *34* the "competition" is between aromatic and nonaromatic structures, in the respective dianions the "competition" is between nonaromatic and antiaromatic structures (Fig. 9). From the spectroscopic parameters, i.e., chemical shifts and coupling constants of the bridge protons it can be concluded that the neutral systems are best represented by structures with an aromatic skeleton connected to a virtually isolated double bond. In the charged systems, viz. 33^{2-} and 34^{2-} it seems that a nonaromatic path of conjugation is preferred to an antiaromatic path (Fig. 9). These considerations are also reflected in the carbon chemical shifts and in their HOMO-LUMO gap (ΔE) (vide infra) [122]. It can be concluded from all these observations that there is a tendency of aromatic systems to remain so and to avoid as much as possible paratropic antiaromatic contributions.

Table 6. ^1H NMR parameters of the charged system 33^{2-}, 34^{2-} and 35^{2-} in THF-d_8 [122].

System[a]	Temperature in °K	^1H NMR pattern[a,b]	neutral systems center of gravity[c]	charged systems center of gravity[c]
$34^{2-}/2\,Na^+$	293	6.02 (d, J = 7.6, H9); 5.69 (t, J = 7.5, H7); 5.47 (d, J = 8.5, H6); 5.40 (t, J = 8.4, H13) 5.29 (d, J = 8.1, H14); 4.88 (t, J = 7.3, H8); 4.78 (d, J = 2.1, H2); 4.72 (d, J = 2.0, H1) 4.42 (d, J = 6.6, H12); 4.27 (s, H4).	7.88	5.09
$34^{2-}/2\,Na^+$	213	6.04 (d, J = 7.4, H9); 5.68 (t, J = 7.1, H7); 5.46 (d, J = 8.0, H6); 5.41 (t, J = 6.8, H13); 5.28 (d, J = 8.1, H14); 4.88 (t, J = 6.8, H8); 4.74 (d, J = 2.4, H2); 4.68 (d, J = 2.2, H1); 4.45 (d, J = 6.7, H12); 4.31 (s, H4).	7.88	5.09
$34^{2-}/2\,Li^+$	293	6.32 (d, J = 7.9, H9); 5.88 (t, J = 7.3, H7); 5.70 (d, J = 8.5, H6); 5.62 (t, J = 7.5, H13); 5.43 (d, J = 8.2, H9); 5.17 (t, J = 6.6, H8); 4.78 (d, J = 6.7, H12); 4.63 (d, J = 1.8, H2); 4.56 (d, J = 1.8, H9); 4.31 (s, H4).	7.88	5.24
$34^{2-}/2\,Li^+$	213	6.36 (d, J = 7.4, H9); 5.86 (t, J = 6.4, H7); 5.66 (m, H6, H13); 5.42 (d, J = 7.6, H9); 5.10 (t, J = 6.6, H8); 4.81 (d, J = 6.6, H12); 4.58 (s, H2); 4.54 (s, H1); 4.44 (s, H4).	7.88	5.24
$33^{2-}/2\,Li^+$	293	6.32 (d, J = 6.9, H5); 5.96 (t, J = 8.2, H7); 5.72 (d, J = 8.7, H8); 5.55 (d, J = 2.3, H1); 5.52 (t, J = 6.9, H6); 5.50 (t, J = 8.4, H13); 5.09 (d, J = 8.1, H14); 5.02 (d, J = 3.0, H2); 4.14 (d, J = 6.5, H12); 3.85 (s, H10).	7.89	5.27
$35^{1-}/Li^+$[d]	293	7.10 (d, J = 7.3); 6.89 (d, J = 7.8); 6.79 (d, J = 7.2); 6.77 (t, J = 7.5); 6.66 (d, J = 3.2); 6.40 (t, J = 8.0); 6.32 (t, J = 7.5); 6.13 (d, J = 6.9); 5.78 (d, J = 3.2); 4.36 (s).		6.54[e]

[a] For number see Structures *33* and *34*. [b] Chemical shifts are given in ppm downfield from external TMS with $\delta = 3.67$ ppm for THF-d_8. The spin-spin couplings are given in Herz. [c] Center of gravity in THF-d_8. [d] No assignment is given for the monoanion 35^{1-}. [e] Center of gravity calculated excluding the singlet absorption at 4.36 ppm.

The enhanced stabilization gained by the peripheral delocalization of the perimeter (4n + 2)π-electrons (14π) is demonstrated by dibenzo[cd,gh]pentalene dianion (36^{2-}). The synthetic challenge of systems with a high strain energy like *36* was gratifying in the study of the stable dianion 36^{2-} [124,125]. It is necessary to bridge a distance of 3.31 Å by a single carbon atom, in this perturbed [12]annulenyl dianion. The un-

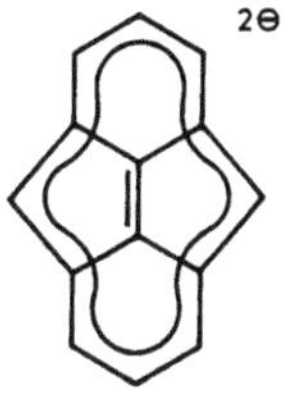

$36^{2\ominus}$

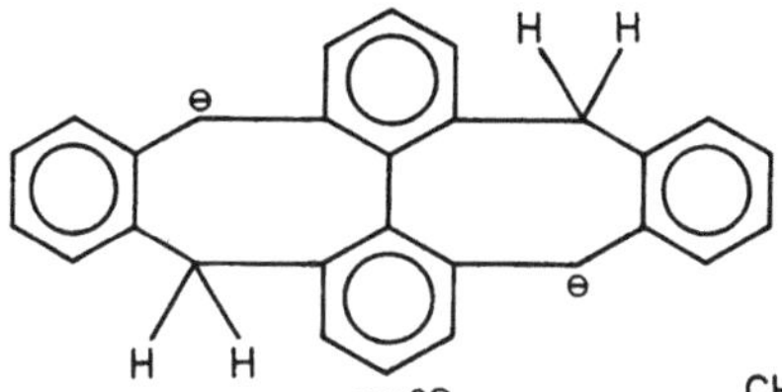

$37^{2\ominus}$

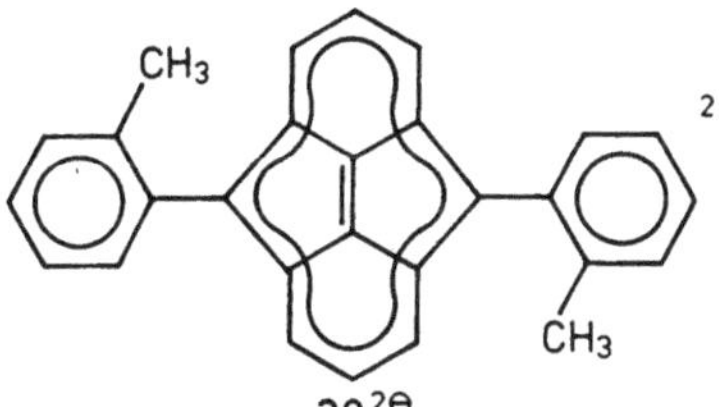

$38^{2\ominus}$

expected rearrangement of 37^{2-} to 38^{2-} reflects the energy gained by the formation of the (4n + 2)π dianion [126]. It should be noted, however, that the properties of 36^{2-} and 38^{2-} comply with the Randić model of conjugated circuits [88], as four such (4n + 2)π-electron circuits are present.

5.3.2 Benzannelated Cyclooctatetraene and Nonalene: Dianions and Tetraanions

Benzannelated cyclooctatetraene (COT) has attracted much interest due to the variability of isomers of the dibenzannelated COT and the possibility to prepare doubly and quadruply charged systems. Dibenzo[a,d]cyclooctatetraene dianion (39^{2-}) was prepared in 1965 [127]. The ^{1}H NMR spectrum of 39^{2-} was studied and compared with its isomer i.e. dibenzo[a,e]cyclooctatetraene dianion (40^{2-}). The two dianions were prepared by potassium metal reduction of *39* and *40*, respectively [128,129]. These

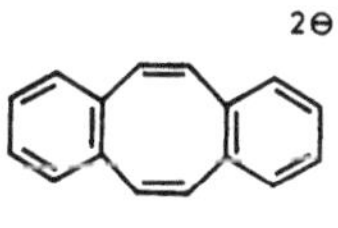

$39^{2\ominus}$

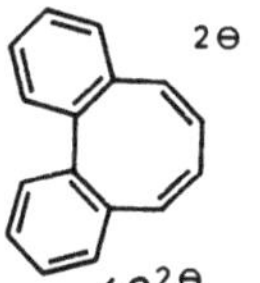

$40^{2\ominus}$

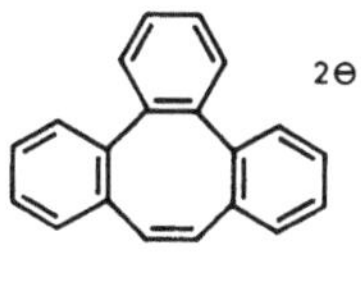

$41^{2\ominus}$

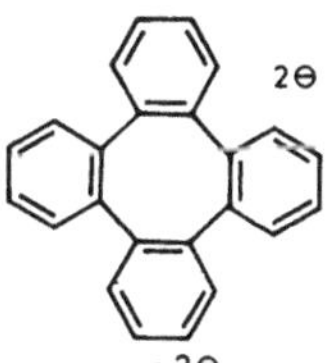

$42^{2\ominus}$

dianions are (4n + 2)π-electron species, planar, and diatropic [128]. The neutral systems are composed of aromatic and vinylic components due to the non planarity of the COT moiety [129]. The ^{1}H NMR spectra (Fig. 11) show that the diatropicity overcomes the charge shielding as shown by protons H-7 and H-8 of 39^{2-}. Protons H-2 and H-5 are shifted to as low as 8.0 ppm as a result of the planarization of this aromatic 18π dianion. Günther's Q value method for the classification of cyclic π-systems [118] was applied to these dibenzoannulene dianions. As the Q value estimation is based on ratios of vicinal coupling constants it requires a complete analysis of the ^1H NMR spectra [128]. It should be noted that this Q value method affords a very good parameter for assessing the type of π-bonding of a benzannelated annulene system (neutral or ionic) as mentioned in Section 5.3.1. This parameter is obtained from the ratio of the vicinal coupling constants is $^3J_{2,3}/^3J_{3,4} = Q$ of the attached benzene ring [118]. Neutral systems have Q values close to unit which is characteristic of vinylic systems without a π-bond delocalization [128]. The dianions *39*$^{2-}$ and *40*$^{2-}$ show larger Q values ca. 1.5 and point toward the formation of a completely delocalized π-system.

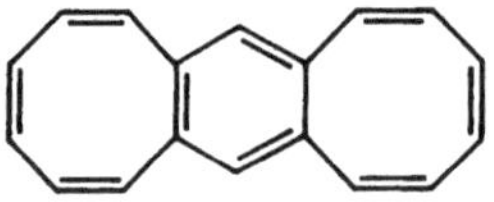

43

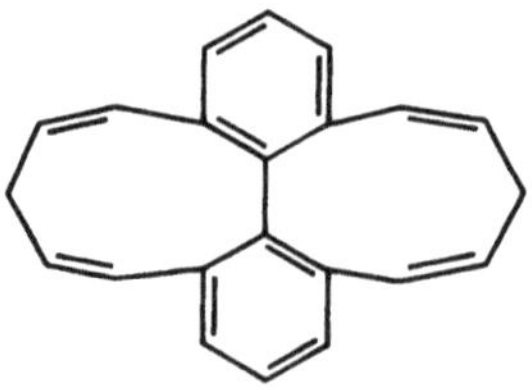

44

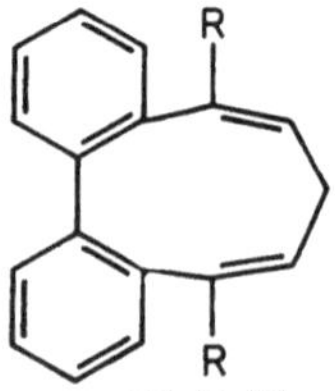

45 R=Ph

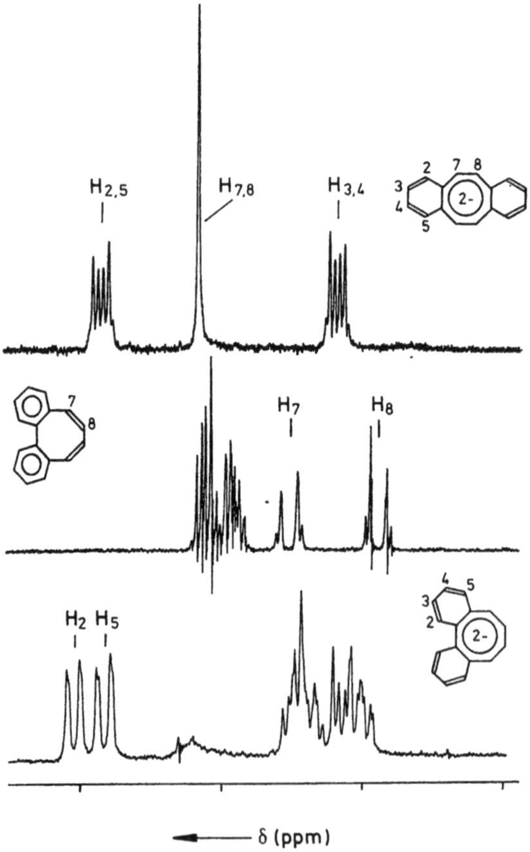

Fig. 10. ^1H NMR spectra of *39*$^{2-}$, *40* and *40*$^{2-}$ (90 MHz) [128]

Tribenzocyclooctatetraene dianion *41*$^{2-}$ and tetrabenzocyclooctatetraene dianion *42*$^{2-}$ were prepared by potassium and lithium metal reduction of *41*$^{2-}$ and *42*$^{2-}$ respectively [130]. The tetrabenzo derivative *42*$^{2-}$ is in fact a tetraphenylene dianion. This dianion shows a C_2 symmetry and therefore deviates from the D_{2d} symmetry of the tub shaped neutral *42* [131]. The radical-anion *42*$^{\overset{\cdot}{-}}$ possesses D_{2d} symmetry and is similar to the neutral compound. The deformation of *42*$^{2-}$ is a specific process of the (4n + 2)π-electron dianion. Under nonequilibrium conditions the radical-anion possesses the same distorted structure as that of the dianion. These observations of Huber and Müllen enable the monitoring of structural and energetic effects of electron transfer reactions [131].

The dicyclooctatetraeno[1,2:4,5]benzene (*43*) prepared by Paquette [132] has been reduced, by electrochemical methods and by treatment of Na/K alloy in THF-d_8, to a dianion. This system has the potential to form a tetraanion viz. *43*$^{4-}$. The dianion *43*$^{2-}$ can be rationalized by a dynamic process, viz. *43a* ⇄ *43b* or as suggested later by Staley that both eight membered rings are bent simultaneously (Structure *43c*) [133]. The tetraanion salt was too insoluble to be studied [132].

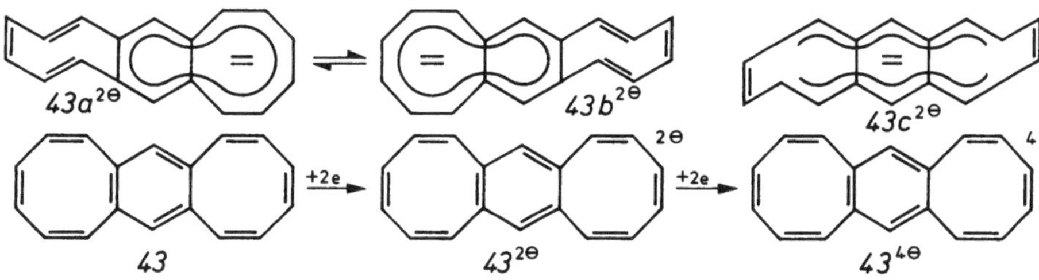

Dibenzo[gh,op]nonalene (*44*) forms a stable dianion *44*$^{2-}$ by the double deprotonation of *44*. This is a dibenzannelated dianion of the illusive annulenoannulene viz. nonalene. The dianion shows a diatropic ^{1}H NMR spectrum [134] which is closely related to monoanion *45*$^{1-}$ [135]. The structure was elucidated from its ^{1}H NMR spectrum and quench experiments. There has been a comment of Staley [136] concerning the structure elucidation of *45*$^{1-}$. The anions derived from this class are now under investigation [136b], where the possibility of a ring closure is being studied.

The pericondensed system diindeno[cd,lm]perylene (*12*) [85] has already been mentioned (Sect. 4). The neutral molecule appears to comply with the conjugated circuits model, as it is a diatropic system despite having 28π-electrons in the periphery. The ^{1}H NMR spectrum is centered at 8.00 ppm and points at its diatropicity. Calculations (ωβ) also predict a bond lengthening of the bonds which connect the fluoranthene moieites [82]. Treatment of *12* with sodium metal in THF-d_8 yields a two-electron reduction product. The formed dianion is also diatropic (7.75 ppm, center of gravity of the spectrum) despite the addition of only two electrons to a diatropic molecule [85]. In this case the dominance of a peripheral (4n + 2)π-system of 26π-electrons in the perimeter may account for the diatropicity of the dianion. Calculations (ωβ) show that *12*$^{2-}$ has a HOMO exhibiting a nodal plane through the central carbon atoms. Both *12* and *12*$^{2-}$ demonstrate that the π-electron distribution mode will be the one that results in diatropic aromatic nature, as was pointed out in detail also in the case of *8*$^{2-}$ and *23*$^{2-}$. It seems that the already mentioned tendency of conjugated systems to acquire aromaticity, or to remain aromatic is a general one. It suggests that the mode of π-electron distribution that prevails in a system and determines its character is the one that results in aromatic nature and reduces antiaromatic contributions. Thus, when the two theoretical criteria predict different results, the one that should be adopted is the criterion that assigns a prevailing aromatic contribution, or a reduced antiaromatic character.

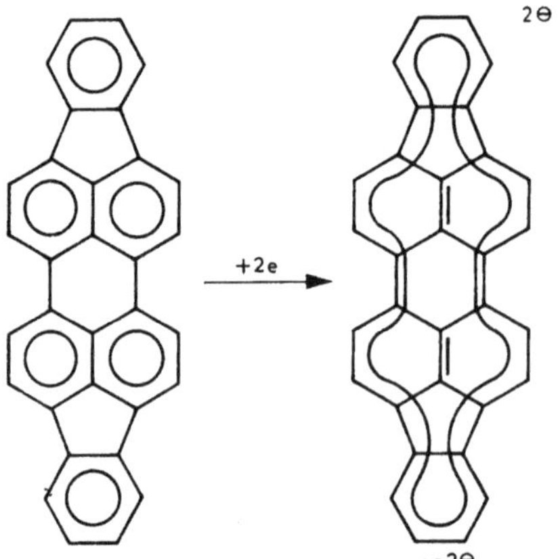

12 32π → *12*$^{2\ominus}$ $30\pi+4\pi$

6 Polybenzenoid Dianions

The polycyclic anions were first prepared by metal reduction in 1914 by Schlenk et al. [5a] and studied later by Schlenk and Bergmann [5b]. This class of conjugated anions opened a new era in carbanion chemistry by pointing out the electron transfer process as a source for charged species. The mechanism of the metal reduction of polycyclic hydrocarbons has been investigated and is well established [1,2,5,18,68]. The addition of two electrons to the fully conjugated $(4n + 2)\pi$-molecules yields $4n\pi$ paratropic systems [20,137–139]. The chemistry of this reaction is simple, with electrons initially on the alkali metal going to π-molecular orbitals associated with the aromatic hydrocarbon molecule (Eq. 13).

$$Ar + nM^0 \rightarrow (Ar^{n-})(M^+)_n \qquad (13)$$

It is assumed that there are at least three ways by which one may gauge the proclivity of a given aromatic hydrocarbon molecule to be reduced:

(a) Gas phase electron affinity as a predictor of aromatic reduction chemistry;
(b) Measurement of the electrochemical reduction potential;
(c) MO calculations as taking the coefficient m_{m+1} derived from the energy of the LUMO $\alpha_0 + m_{m+1}\beta_0$ from HMO.

Table 7 shows a comparison of these approaches [140] as depicted from the literature by Ebert.

Table 7. Three scales of the tendency of aromatic molecules to undergo one-electron reduction [140].

Molecule		EA (eV)[a]	Er (V vs. SCE)[b]	$-m_{m+1}$[c]
Naphthalene	(55)	0.074	−2.50	0.618
Triphenylene	(46)	0.251	−2.48	0.684
Phenanthrene	(53)	0.273	−2.49	0.605
Chrysene	(54)	0.516	−2.27	0.520
1,2-Benzanthracene	(49)	0.640	−2.02	0.452
Anthracene	(2)	0.653	−1.97	0.414
Pyrene	(29)	0.664	−2.04	0.445
Perylene	(60)	0.956	−1.66	0.347

[a] gas phase electron affinity, from Ref. 141; [b] voltammetric reduction peak potential +30 mV in acetonitrile solvent from Ref. 141; [c] $\alpha_0 + m_{m+1}\beta°$ is the energy of the lowest vacant orbital in units of the standard β in the HMO approximation with all α's and β's equal, from Ref. 20.

From the Table it can be seen that aromatic hydrocarbons tend to be more easily reduced as the molecular size increases. However, even for simple aromatic molecules both molecular size and molecular symmetry must be analysed for a quantitative understanding [37]. Detailed ^1H and ^{13}C NMR studies of reduced hydrocarbons enabled to correlate between the topology of each system, its magnetic properties, and anti-aromaticity. These studies will be discussed in the following sections.

HMO considerations differentiate between two classes of (4n + 2)π-conjugated polycyclic species: Systems endowed with C_3 or higher axial symmetry for which the highest occupied and lowest unoccupied orbitals appear in pairs vs. systems with lower axial symmetry in which no such orbital degenercies exist [54, 142]. The difference between the two classes becomes crucial when the polycycles are reduced to the corresponding 4nπ-conjugated dianions. In the first group the two additional electrons populate two different degenerate orbitals, which may lead to a triplet ground state [143], for example the triphenylene dianion (*46*$^{2-}$). However, configuration interaction could stabilize the singlet more than the corresponding triplet state [38, 39, 142]. Systems

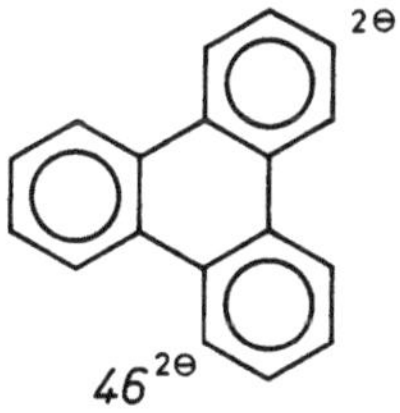

$46^{2\ominus}$

endowed with lower than threefold axial symmetry such as anthracene (2), phenanthrene (*53*) and chrysene (*54*) form upon reduction dianions in which the singlet state is lower in energy than the triplet state [20]. The rigidity of these systems requires planar or near planar geometry and even more important reduces substantially the extent of bond-length alternation. These consequences stand in obvious contrast with 4nπ monocyclic conjugated species [47]. As a result the LUMO-HOMO energy gap (ΔE) of doubly charged benzenoid polycycles with 4nπ-electrons in the path of conjugation is much smaller than in the monocyclic 4nπ-systems and even more so than that of the neutral (4n + 2)π-electron aromatic hydrocarbons. This is a key point for the understanding of the properties of 4nπ-conjugated dianions [20,42a,139]. It follows that most of the systems which are diamagnetic can be studied by NMR techniques. Such dianions were indeed studied by ^{1}H NMR and shown to be paratropic [10,11]. There are two NMR parameters which vary from system to system in the polybenenoid dianions, i.e. the chemical shift and the line shape. However, unlike the (4n + 2)π systems which show a well defined chemical shift in the range of ±1 ppm the 4nπ systems show chemical shifts in a much wider range — ca. 5 ppm. Apart from the wide range of chemical shifts they also vary in the NMR line-shape of their proton and carbon spectra. The following discussion will concentrate on these phenomena.

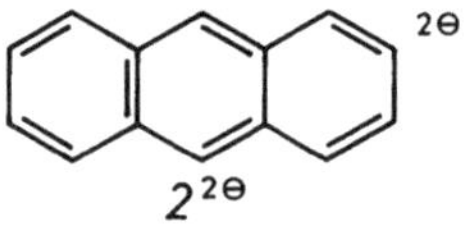
$2^{2\ominus}$

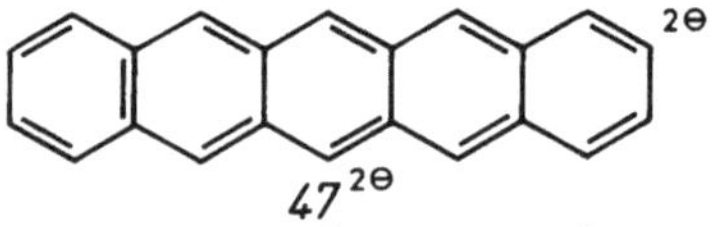
$47^{2\ominus}$

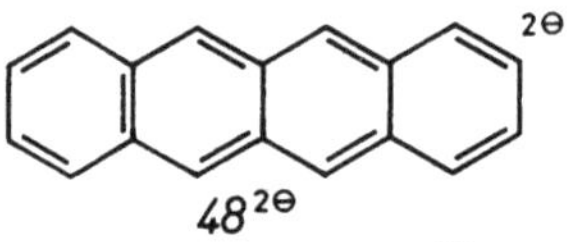
$48^{2\ominus}$

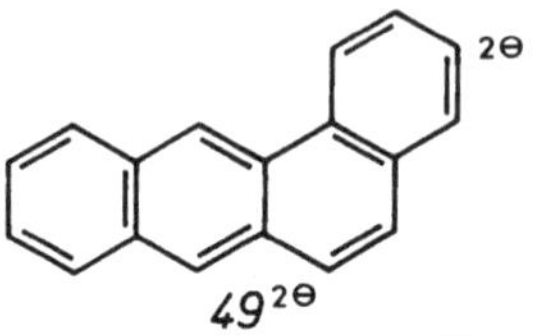
$49^{2\ominus}$

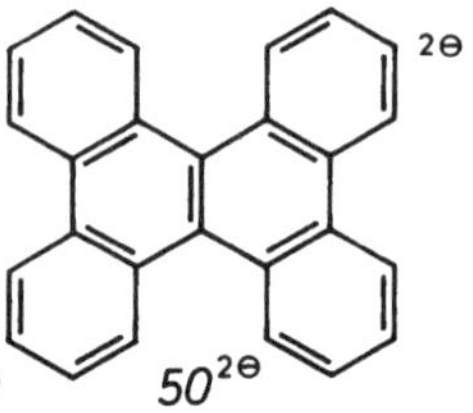
$50^{2\ominus}$

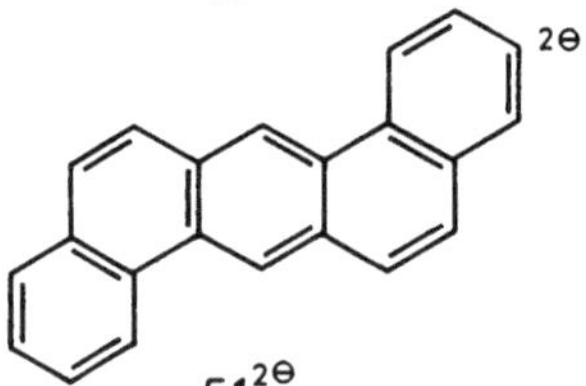
$51^{2\ominus}$

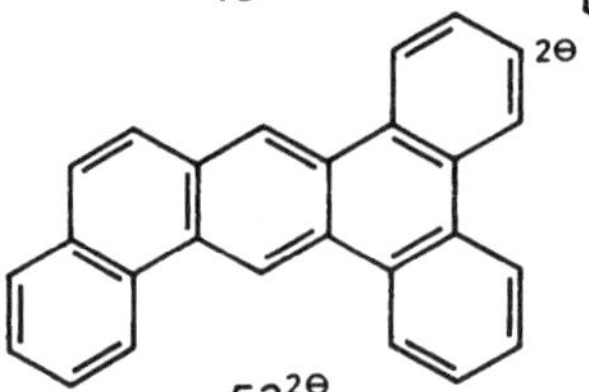
$52^{2\ominus}$

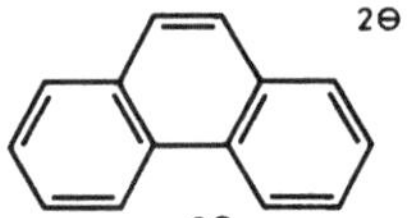
$53^{2\ominus}$

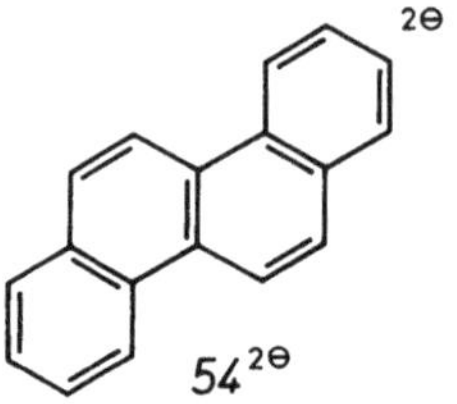
$54^{2\ominus}$

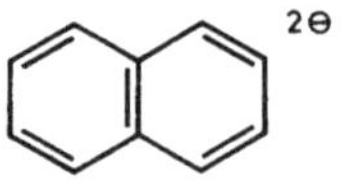
$55^{2\ominus}$

6.1 Paratropicity and Antiaromaticity in Polycyclic Dianions

An outstanding achievement of molecular orbital theory was Hückel's prediction [137] that conjugated monocyclic systems containing (4n + 2)π-electrons should be aromatic, i.e., conjugatively stabilized [138]. This proposition was modified by Platt to encompass neutral as well as charged polycyclic systems with (4n + 2)π-electrons in the path of conjugation [87]. Cyclic polyenes which did not fit the (4n + 2) rule, and indeed did not reveal a conjugative stabilization, were generally classified as "pseudoaromatic" until 1965, when Breslow coined the term "antiaromaticity" [38,139]. This term was meant to connote and emphasize a significant destabilization characteristic of certain 4nπ conjugated monocycles such as cyclobutadiene and cyclopropyl anion [38]. In agreement with Breslow's observation, theoretical calculations led to the estimation of negative conjugation energies [144,145].

Hückel's theory initiated a wealth of experimental work, all of which fitted with the prediction of enhanced stabilization in $(4n + 2)\pi$-systems. The measure of variegated physical and chemical quantities (such as heats of combustion and hydrogenation, magnetic susceptibilities, electronic spectra and the proclivity to react in Diels-Alder reactions) was found to correlate with the extent of aromatic character, as estimated by the theoretical methods [86].

The experimental support for antiaromaticity is by far less satisfactory. The inherent instability of antiaromatic systems, rationalized by their high energy content, results in a complete failure to isolate these compounds. Consequently, the possibility of corroborating theoretical predictions for systems in this class through confrontation with experiment, is disturbingly limited. Even more so, any attempt to theoretically estimate the relative magnitude of antiaromatic character in $4n\pi$ systems and then correlate this estimation with an experimental parameter, is bound to suffer from lack of support by such experimental data. The polybenzenoid $4n\pi$ dianions seemed to be the system of choice to examine the magnitude of the anisotropic paramagnetic susceptibilities, known to characterize $4n\pi$ cyclic conjugated systems, by means of their effect upon ^1H NMR chemical shifts (s. Sect. 3.1.2). The magnitude of this effect was suggested to be examined as a function of the energy gap between the lowest vacant (LUMO) and the highest occupied (HOMO) molecular orbitals of the corresponding systems [37]. The rationale for such an inspection is the allegation that antiaromaticity is a particular aspect of antibonding [38]; if so, the proximity of excited, antibonding, states to the ground state, and consequently, their exerted effect upon it, is expected to be directly related to the antiaromatic character of the system. NMR measurements have an obvious advantage over other techniques, as the unstable antiaromatic species are prepared from suitable, stable precursors, and studied in the reaction vessel — i.e., in the NMR tube.

The doubly charged polybenzenoid molecules are model compounds for antiaromaticity. It has been pointed out that there is a direct, unequivocal correlation, between the extent of paratropicity experienced by the $4n\pi$-conjugated systems, and the corresponding LUMO-HOMO energy gap, as estimated by SCF-MO calculations [35, 36].

6.1.1 Chemical Shifts and the LUMO-HOMO Gap of $4n\pi$ Polycyclic Dianions

Aromaticity — as well antiaromaticity — represents primarily a state of energy. Therefore, a parameter which directly reflects the energy content of antiaromatic systems or more specifically, the extent of destabilization due to antiaromatic character, seems to be a most suitable choice. Yet, methods designed to estimate the extent of this destabilization are inherently problematic in the same sense that methods aimed at the evaluation of aromatic stabilization remain ambiguous [86, 139]. At the crux of the problem lies the observation that the notion of conjugative stabilization or destabilization is, by definition, relative, and therefore requires reference systems that are devoid of cyclic delocalization. When aromatic systems are considered, the various methods designed to estimate the measure of the conjugative energy stabilization [58, 88, 146] generally agree [149]. This is hardly the case when antiaromatic contributions are expected; as the estimated magnitudes of the resulting destabilization due to antiaromaticity is not only sensitive to the choice of reference compounds, but is also highly susceptible to the level of sophistication of the quantum chemical calculation [39,50,144].

The diminution of geometrical distortion in the polycycles, when compared with the monocyclic series, is emphasized in polybenzenoid systems. Unless a large steric perturbation exists, the highly rigid benzenoid skeleton is bound to enforce planarity and approximate constancy in bond lengths not only in the neutral, but also in the doubly charged ions derived from the fused benzenoid systems. Thus, such doubly charged species, viz. 2^{2-} and 47^{2-}–54^{2-}, seem to meet all the requirements imposed by the peripheral electronic model for antiaromaticity: they are $4n\pi$ systems, in which, due to the suitable geometry, the peripheral delocalization of π-electrons is quite unperturbed. Moreover, in contrast to the neutral systems where aromatic contributions of the benzenoid moieties are expected, the doubly charged polybenzenoid ions should reveal a "pure" antiaromatic character.

As already mentioned, the predicted increase of the antiaromatic character, in soubly charged $4n\pi$ polybenzenoids, is directly related to a marked decrease in geometric distortion, that is, deviations from planarity and severe bond length alternation, due to the rigidity of the system. This is clearly indicated by experimental ^{1}H NMR data (vide infra). The decrease in geometric distortion is related in turn, to a larger perturbance of a triplet configuration to the singlet ground state — as this excited triplet lies closer to the ground state configuration. Consequently, a connection can be pointed out between the extent of antiaromaticity and the increase in perturbance due to the triplet state. SCF-MO calculations performed on the polycyclic system pyracyclene (*56*) indicate that its lowest lying triplet configuration is approximately 2.05 eV above the singlet ground state [125]. The gap between the lowest triplet and the singlet states estimated for the dibenzopentalene *36*, is only 0.32 eV [125]. The dissimilarity of the two perturbed [12]annulenes, *56* and *36* is suggestive. System *56* might be depicted as an aromatic naphthalene core perturbed by two double bonds which are connected to the aromatic moiety by long, virtually single, bonds. The alternation of bond lengths result in a quench of the peripheral $4n\pi$ antiaromatic contribution. In obvious contrast, when *36* is considered, no structure with a Kekulée ring can be drawn, thus aromatic contribution is expected. Even more so, the two canonical forms *36a* ↔ *36b* imply a diminution in bond length alternation. Doubly charged $4n\pi$ benzenoid polycycles exhibit disturbances from a triplet ground state which are estimated to correspond to 0.8–1.8 eV. It has been demonstrated, by means of NMR and ESR methods [37], that such doubly charged antiaromatic compounds possess a low-lying, thermally accessible triplet state which enables the existence of an equilibrium process with the singlet ground state. NMR and ESR lineshapes may depend on the extent and direction of this singlet-triplet equilibrium which, in turn, was found to be determined by the width of the energy gap between the LUMO and HOMO (ΔE) (vide supra). It has been concluded that the extent of perturbance from a triplet ground state of a $4n\pi$ conjugated compound, cyclic or polycyclic, constitutes a plausible index for the assessment of antiaromaticity. As a theoretical approach, this is a quantitative index, since it is related to the energy splitting between the LUMO and HOMO of the antiaromatic system.

Among the experimental methods established for the estimation of aromatic, or antiaromatic character, the most frequently employed are measures of properties that are related to sustained magnetic phenomena [39,42,139,148]. These phenomena

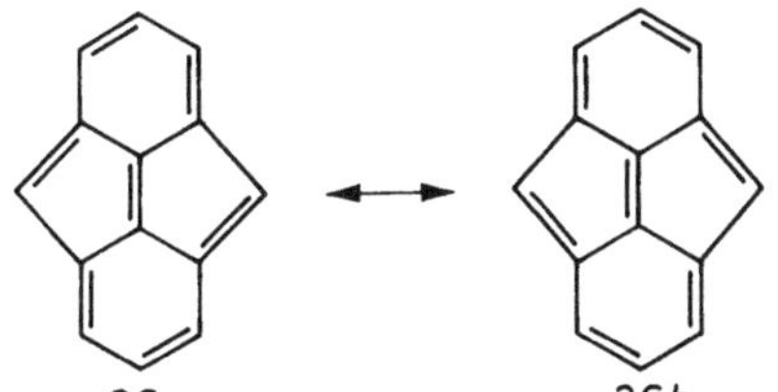

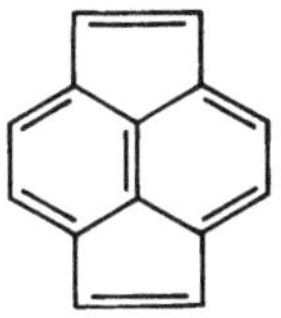

36a *36b* *56*

represent the outcome of two different contributions of the σ^1 term of the Ramsey equation (Eq. (10)): the diamagnetic contribution due to Larmor precession of π-electrons around the ring which depends only on electron density in bonding molecular orbitals — an exclusive ground state property. In contrast, the second, paramagnetic, contribution depends on electronically excited states, since it reflects magnetic dipole transitions between occupied and vacant molecular orbitals [148,149]. Obviously, such transitions would be particularly favored when the system possesses a low-lying excited state, or, in other words, when the energy separation (ΔE) between the LUMO and HOMO is small [104]. It seems that both paramagnetism and antiaromaticity depend on the LUMO-HOMO gap, both becoming more pronounced as this gap narrows. The extent of the paramagnetic contribution depends on the extent of mixing — in the magnetic field — between ground and excited states. It is, however, noted that the dependence of antiaromaticity and the anisotropy term upon the energy gap is not a mere correlation; both properties originate directly from a narrow LUMO-HOMO energy split.

A readily observed result of the secondary paramagnetic field, is a paratropic, high-field shift which characterizes the ^1H NMR absorptions of antiaromatic species. However, the information obtained from these high-field shifts was strictly qualitative and in most cases open to controversy and ambiguity [39]. This, again, is due to the scarcity of systems that reveal pure or even substantial antiaromatic contributions, and, even more so, to the lack of antiaromatic homologous series. It has been stated recently that "*doubly charged benzenoid polycycles seem to provide a good example of relatively large 4nπ series in which the various compounds differ either in topology or in the number of π-electrons* (i.e. the value of n), and none of them presents a difficulty due to a perturbed delocalization" [36]. The reduction process of condensed polycyclic hydrocarbons was carried out by sodium and lithium metals in THF [8,10]. The ^{13}C spectra of anions 2^{2-}, 47^{2-}–54^{2-} showed the highfield shifts due to the shielding of two negative charges [29–31]. The ^1H NMR spectra are highfield shifted due to the charge effect and the paratropic effect (see Table 8). Calculated ^{13}C charge densities were also obtained and compared with the experimental values [35,36]. The extent of the paratropic shift beyond the charge effect can be taken as the "pure" contribution of the antiaromatic paramagnetism. The HOMO-LUMO energy gap (ΔE) is correlated with the paratropic shift which solely arises from the secondary field. The correlation is shown in the Fig. 11. It can be seen from Table 8 that a large LUMO-HOMO gap is related to small paratropic shifts and vice versa, and that there is, indeed, a correlation between the paratropicity of the system and its LUMO-HOMO gap [142]. The paratropic ^1H NMR chemical shift is an outcome of an anisotropic paramagnetic field which is assumed to be sustained in 4nπ antiaromatic cyclic or polycyclic conjugat-

Table 8. ^{1}H NMR Paratropic Shifts of Polybenzenoid Dianions vs. LUMO-HOMO Energy Gaps [36].

System[a]	Neutral systems; *exptl.* ^{1}H NMR center of gravity[b]	Overall *paratropic* shift[b,c]	Charged systems; *calc.* ^{1}H NMR center of gravity[b,c]	Charged systems; *exptl.* ^{1}H NMR center of gravity[b,c]	$\Delta\delta$ *Calc.-exptl.*[b,c]	Calc. LUMO-HOMO gaps[c,d]
47	8.08	2.17	5.91	4.98	0.93	2.070
48	8.01	2.50	5.51	4.23	1.28	2.133
2	7.90	2.76	5.14	3.06	2.08	1.833
49	7.86	2.15	5.71	3.22	2.49	1.411
50	8.28	1.18	7.10	4.18	2.92	1.205
51	8.14	1.70	6.44	1.78	4.66	0.884
52	8.43	1.23	7.20[e]	3.51	3.69	0.990
53	7.80	1.90	5.90	2.42	3.48	1.333
54	7.67	2.32	5.35[e]	1.18	4.17	1.898

[a] For numbering — see text; [b] PPM down from $SiMe_4$; [c] For computation details — see Ref. 36, 160; [d] eV units, as furnished by SCF-MO method; [e] Unresolved, broad NMR absorptions.

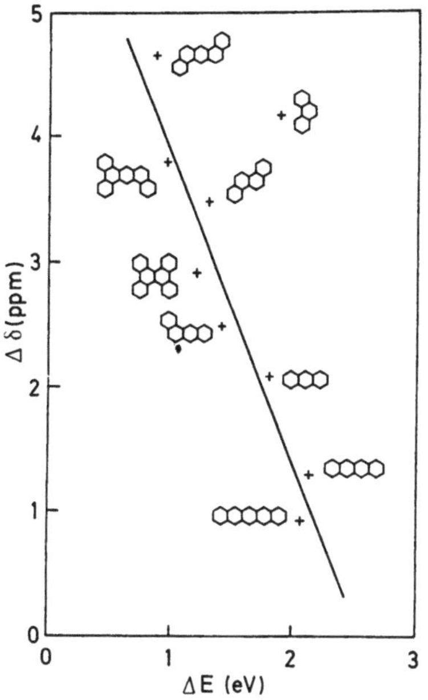

Fig. 11. ^{1}H NMR paratropic shifts of 4nπ polybenzenoid dianions vs. LUMO-HOMO energy gaps [36)]

ed systems. This field results, in turn, from a relatively large contribution of excited states to the ground state. Such a contribution was shown by Van Vleck to establish a positive, paramagnetic susceptibility [149)]. Efficient mixing process between the excited configuration and the ground state may exist. The efficiency of this process depends primarily upon the width of the splitting, between the lowest vacant and the highest occupied molecular orbitals — it increases as the gaps decrease. "Pure" antiaromatic species are expected to reveal triplet ground state, with one electron in each of two (highest occupied) orbitals. In polycyclic 4nπ systems such as the doubly charged benzenoid compounds, the HOMO's degeneracy is removed, thus resulting in a singlet configuration. However, the LUMO-HOMO energy splitting is still narrow enough to enable a significant contribution of the excited states and therefore to characterize antiaromatic systems [36)].

The dependence of NMR patterns upon states of solvation should not be overlooked. This aspect of polyanion chemistry has been discussed (Sect. 4). The influence of the counter cation upon the dianion though unaccounted for by simple calculations is supposed to vary from system to system. The different topologies of the anions and different modes of delocalization may be responsible for the deviations from a linear dependence of the correlation. From Table 9 one can arrive at conclusions on the "relative antiaromaticity" of the various anions. For example, benzanthracene dianion (*49*$^{2-}$) is "less antiaromatic" than chrysene dianion (*54*$^{2-}$), and dibenzoanthracene dianion (*51*$^{2-}$) is "more antiaromatic" than the pentacene dianion (*47*$^{2-}$). It is also proposed that ^{1}H NMR paratropic shifts may be used to "calibrate" the LUMO-HOMO energy gap.

6.1.2 ^{1}H NMR Line-Shape, Paratropicity and Antiaromaticity

Apart from the wide range of NMR chemical shifts of the 4nπ dianions discussed in the previous section there are also variations in the line shape of the condensed polycyclic dianions. An extreme case is the naphthalene dianion (*55*$^{2-}$) [18,150–158] which was very rigorously studied and characterized by UV-Vis-spectroscopy [1,2,150–158]

Table 9. LUMO-HOMO Energy Gaps Estimated (ωβ Method) for Dianions Derived from Benzenoid Polycycles[a].

	pentacene (47^{2-}) 0.488	tetracene (*48*$^{2-}$) 0.414	anthracene (2^{2-}) 0.310
phenanthrene (53^{2-}) 0.231	benzophenanthrene (57a^{2-}) 0.111	1,2-benzanthracene (49^{2-}) 0.226	

[a] LUMO-HOMO gaps, β units.

Table 10. Spectral Patterns and LUMO-HOMO Energy Gaps of the Three Subgroups of Doubly Charged Systems [37].

System	Doubly charged systems	LUMO-HOMO gap[a], β-units	^{1}H NMR patterns[b] 30 °C	−20 °C	−60 °C	ESR patterns, half-field line[c]
48^{2-}	tetracene-Li/THF	0.414	C	C	C	
	-Na/THF		C	C	C	N
62^{2-}	9,10-diphenyl-anthracene-Na/THF	0.420	C	C	C	N
2^{2-}	anthracene-Li/THF	0.310	C	C	C	
	-Na/THF		—	A	B	1590
	-Na/THF-DME		B	C	C	N
	-K/THF		—	—	A	
54^{2-}	chrysene-Li/THF	0.272	B	C	C	
	-Na/THF		—	A	B	1620
51^{2-}	1,2,5,6-dibenz-anthracene-Li/THF	0.251	B	C	C	
	-Na/THF		—	A	C	N
49^{2-}	1,2-benz-anthracene-Li/THF	0.226	A	C	C	
	-NaTHF		—	—	A	1600
53^{2-}	phen-anthrene-Li/THF	0.231	—	A	A	
	-Li/THF-DME		—	A	B	
	-Na/THF		—	—	—	1595
57^{2-}	1,2,3,4-dibenz-anthracene-Li/THF	0.157	—	—	—	
	-Na/THF		—	—	—	1605
57a^{2-}	3,4-benzophen-anthrene-Li/THF	0.111	—	—	—	
	-Na/THF		—	—	—	1625

[a] Obtained by ωβ calculations. [b] (—) No spectra obtained; (A) broad lines, unresolved structure; (B) sharp lines, unresolved structure, (C) sharp lines, fine resolved structure. [c] (N) No half-field signal. When half-field absorption is detected, its field is given in gauss.

The chemistry of naphthalene will be discussed later in detail. No ^{13}C NMR spectrum of this dianion is reported in the literature despite the extensive studies of its nature, its ^{1}H NMR chemical shift but no pattern was mentioned very briefly [11b]. Phenanthrene dianion (53^{2-}) shows only broad ^{1}H NMR and ^{13}C NMR lines while tetradibenzanthracene dianion 57^{2-} and 3,4-benzophenanthrene dianion $56b^{2-}$ includes reduction of polycyclic anions by lithium, sodium and potassium metal and their ^{1}H and ^{13}C NMR spectra showed that the line shape is dependent in some cases on the solvent, on the counter cation and on the temperature [37]. All these dianions belong to a lower symmetry group than C_3, viz. *47–54*, *2* and *56–61*, and therefore their singlet state is lower in energy than a triplet state. As already mentioned these are rigid planar $4n\pi$ systems with a narrow LUMO-HOMO energy gap. On the basis of the combined spectroscopic data (Table 10) the doubly charged benzenoids were differentiated into three subgroups. The first subgroup is represented by tetracene dianion (48^{2-}), and includes doubly charged species that reveal highly resolved ^{1}H NMR spectra and sharp ^{13}C NMR bands (subgroup C — Table 10). The proton line shape of 48^{2-} is independent of the reducing metal (Li, Na, or K), solvent or temperature (in the range between −70 °C and +40 °C). The second subgroup (A) — exemplified by 1,2,3,4-dibenzanthracene dianion 57^{2-} and 3,4 benzophenanthrene dianion $57a^{2-}$ includes species that do not produce NMR peaks irrespective of the experimental conditions. The ESR spectra of these dianions in frozen THF consist of a broad signal at 3200 to 3300 G along with a strong very sharp absorption at lowfield 1620 G (Table 10). The third subgroup (B) is represented by chrysene dianion (54^{2-}), (Figure 12) 1,2-benzanthracene dianion (49^{2-}) (Fig. 13), and anthracene dianion 2^{2-} (Fig. 14 and

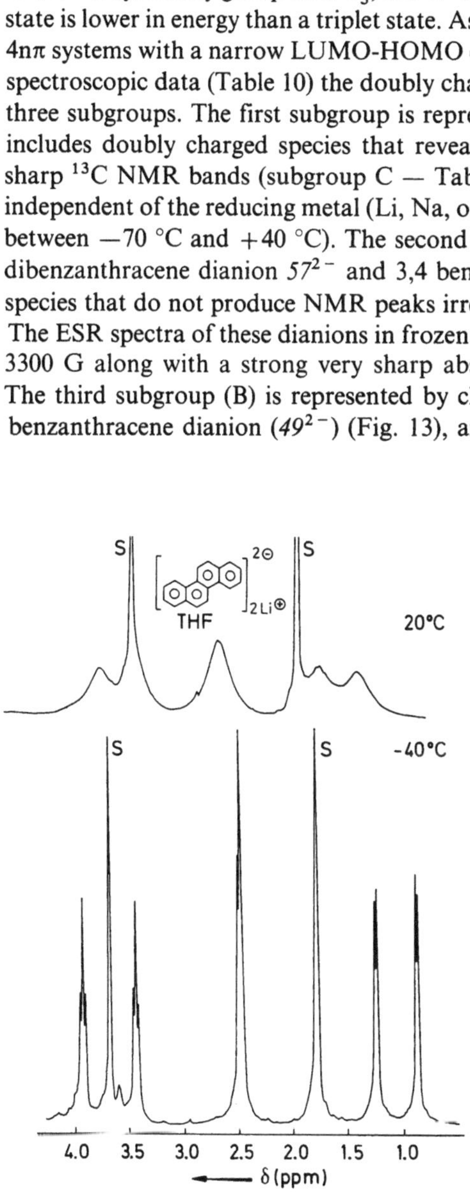

Fig. 12. ^{1}H NMR spectrum (300 MHz) of chrysene dianion (54^{2-}) as dilithium salt in THF [37]

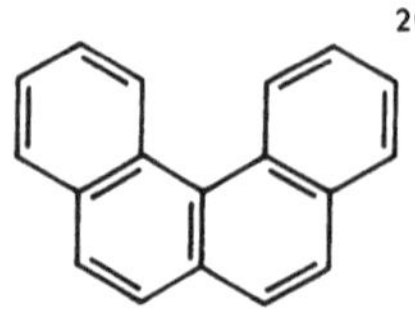

$57a^{2\ominus}$

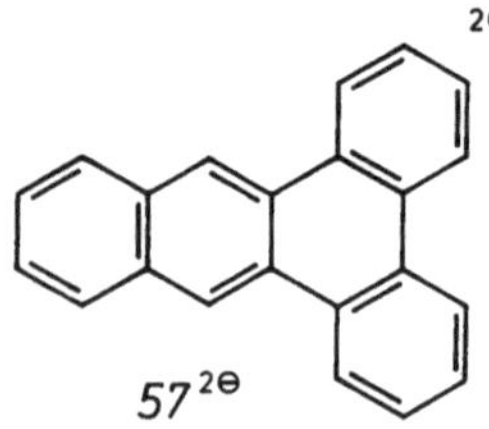

$57^{2\ominus}$

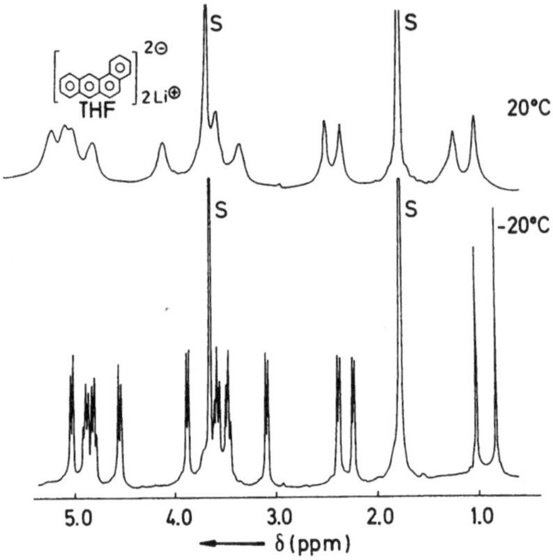

Fig. 13. ^1H NMR (300 MHz) spectrum of 1,2-benzanthracene-dianion (*49*$^{2-}$) as dilithium salt in THF [37]

15). The NMR spectra and color exhibited by the salts of subgroup B showed a strong dependence upon the countercation, the solvent and the temperature [37].

In an attempt to rationalize these phenomena, it was suggested [37] that the extent of the LUMO-HOMO energy separation obtained from Hückel MO calculations ($\omega\beta$) affects the line shape of these dianions from sharp lines in well resolved spectra via broad lines in others and up to a total disappearance of the spectrum (subgroups A–C). The suggestion of an equilibrium process between a singlet ground state and a low-lying thermally accessible triplet state has already been mentioned. The displacement toward the triplet is related to the energy difference between the LUMO and the HOMO of the dianion. Therefore, in the subgroup C the main contribution to the magnetic susceptibility comes from paired, filled orbitals. Consequently, the NMR properties of these species are determined by a diamagnetic character produced by the closed shell, and no line broadening is detected. The dianions which belong to subgroup A result in large population of the excited triplet with two unpaired electrons. Consequently, shortening of the NMR relaxation process up to the point where no NMR patterns can be detected will occur. In the subgroup B the line shape depends on the ion-solvation equilibrium which may affect the LUMO-HOMO gap. From HMO calculation the LUMO-HOMO gap for dianions of the subgroup C is estimated (β units) to be relatively large ($>0.4\beta$), e.g., tetracene dianion (*48*$^{2-}$) (Table 9). Systems of small energy gaps ($<0.2\beta$), e.g. *57a*$^{2-}$ (Tables 9, 10) do not show a spectrum

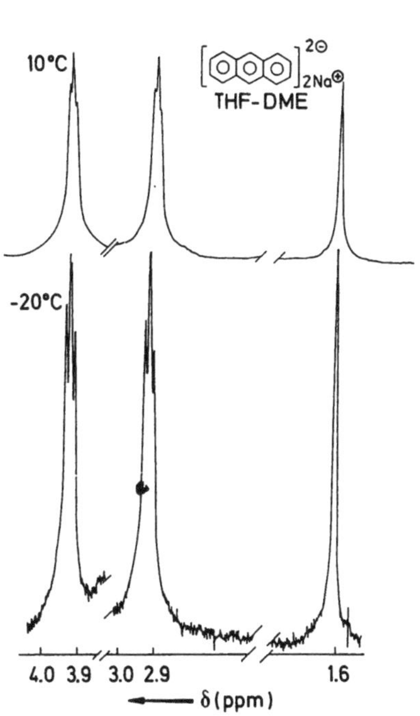

Fig. 14. ${}^{1}H$ NMR (300 MHz) spectrum of anthracene dianion (2^{2-}) THF-DME (95:5) [37]

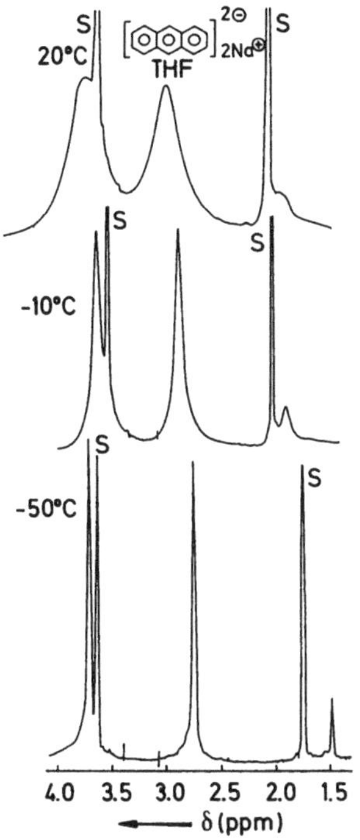

Fig. 15. ${}^{1}H$ NMR (3000 MHz) spectrum of anthracene dianion (2^{2-}) in THF at various temperatures [37]

and rationalize the ${}^{1}H$ NMR spectrum of naphthalene dianion (55^{2-}) (0.11β). It should be noted that azulene dianion 15^{2-} shows a relatively large HOMO-LUMO gap (0.427β). This value predicts no line broadening as already mentioned (Sect. 5.1.2) [37c]. This study pointed towards a relation between NMR line shapes and the LUMO-HOMO energy separation. One can now relate antiaromaticity with the paratropic shift and the line shape as obtained from NMR data.

6.2 Modes of Charge Distribution in Extended Polycyclic Dianions

The reduction of 1,2,3,4-dibenzotetracene [159] (*58*) to its corresponding dianion, viz. 58^{2-} was performed by lithium and sodium metals. The ${}^{1}H$ NMR shows a strange phenomenon [160]. The spectrum seems to be divided into two components. In the one

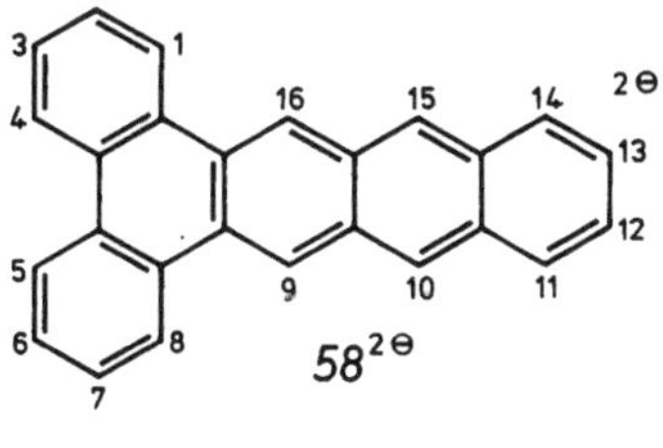

$58^{2\ominus}$

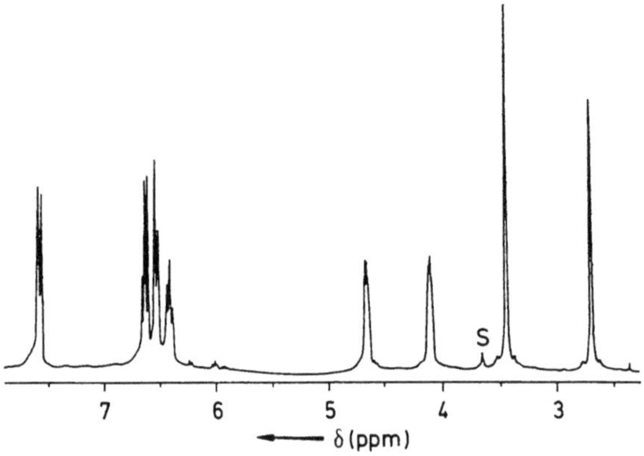

Fig. 16. ^{1}H NMR spectrum of *58*$^{2-}$ as disodium salt [160)]

component the proton bands appear in the region of 4–5 ppm and the other component appears in the region of 6.5–7.7 ppm (Fig. 16). The assignment of the spectrum is as follows: 7.64 ppm, H_1, H_8; 6.43–6.72 ppm, H_{2-4}, H_{5-7}; 2.70 ppm, $H_{10,15}$; 3.45 ppm, $H_{9,16}$; 4.12 ppm, $H_{11,14}$ and 4.68 ppm, $H_{12,13}$. This assignment (Table 12) of *58*$^{2-}$ suggests a rather unusual mode of charge density distribution over the dianion, i.e., a formal partitioning of the conjugated doubly charged system into two components: (A) a "phenanthrene" moiety composed of the carbon atoms C_{1-8}, $C_{4a,b}$, and $C_{8a,b}$, as well as $C_{16a,b}$ — in terms of the shielding parameters derived from the ^{1}H NMR spectrum [26)], this component seems to accomodate only a minor fraction of the overall negative charge density; (b) an "anthracene" moiety which includes all the other carbon atoms along with C_{8b} and C_{16a} which are shared by the two components — the ^{1}H NMR absorptions revealed by the protons attached to this moiety are paratropically shifted, indicating that most of the negative charge density resides over the carbon atoms which constitute the so-called "anthracene" component [37)].

This depiction of charge partitioning was strengthened by ^{13}C NMR as well as by theoretical calculations [160)] (SCF-MO method, Table 11). Both experimental and theoretical methods indicate a substantial localization of the negative charge density over those carbons which constitute the "anthracene" moiety. While such a mode of charge distribution is a characteristic feature of nonalternant systems (viz., azulene), it seems rather peculiar when pure benzenoid species are considered. It should be noted that the line shape of the ^{1}H NMR spectra of *58*$^{2-}$ as a sodium or lithium salt did not reveal any temperature dependence in the range of −50 to +40 °C. The ^{1}H and ^{13}C NMR as well as theoretical charge densities (Tables 11 and 12) [160)] led the authors to interpret the spectroscopic data in terms of a partitioning or segregation of the system as well as that of the MO's. The two negative charges are almost exclusively confined to the "less" antiaromatic "anthracene" moiety. Its carbons and protons reveal a larger highfield NMR shift due to the electronic shielding. This phenomenon would minimize antiaromatic contributions as the anthracene dianion moiety is "less antiaromatic" than the phenanthrene dianion moiety. As can be seen from Table 11 the existence of two rather than one molecular orbital system within *58*$^{2-}$ is plausible. The AO's coefficients of the HOMO and the LUMO of *58*$^{2-}$ are revealing. While in the highest occupied MO the atomic coefficients of those carbon atoms which

Table 11. ^{13}C NMR Chemical Shifts and Theoretical Parameters of *58*$^{2-}$ Dibenzotetracene Dianion (as Dilithium Salt)[161]

	C_1	C_2	C_3	C_4	C_{4a}	C_{8a}	C_{8b}	C_9	C_{10}	C_{11}	C_{12}	C_{9a}	C_{10a}
δ^a	152.4	126.5	123.7	146.8	119.5	122.3	116.7	89.0	86.9	112.8	117.6	130.3	126.0
ψ^b	0.982	1.113	1.113	1.027	1.056	1.007	0.980	1.324	1.365	1.161	1.120	0.844	0.906
HOMO^c coeff	0.041	0.141	0.127	0.057	0.184	0.195	0.155	0.362	0.367	0.237	0.173	0.024	0.099
LUMO^d coeff	0.300	0.351	0.020	0.273	0.278	0.184	0.168	0.032	0.147	0.155	0.081	0.131	0.033

^{a} Chemical shifts (ppm) refered to $SiMe_4$. For numbering see formula. ^{b} Charge densities as deduced from SCF-MO calculations. ^{c} Atomic orbital coefficients of the dianion HOMO. ^{d} Atomic orbital coefficients of the dianion LUMO

Table 12. ^1H NMR Chemical Shifts^a of *58*$^{2-}$ as Dilithium and Disodium Salts (THF-d_8)[161]

Anion	$H_{4,5}$	$H_{3,6}$	$H_{2,7}$	$H_{1,8}$	$H_{9,16}$	$H_{10,15}$	$H_{11,14}$	$H_{12,13}$
Na	7.63	6.45	6.66	6.57	3.45	2.70	4.12	4.68
Li	7.54	6.33	6.58	6.51	3.42	2.65	4.11	4.67

^{a} Chemical shifts (ppm) referenced to $SiMe_4$. For numbering see structure

constitute the "anthracene" are estimated as relatively large, they were found to be considerably smaller in the LUMO (Table 11). A reverse situation characterizes the AO coefficients of the "phenanthrene" moiety. Here, too, the minimization of anti-atomatic contributions is demonstrated in the extended dianions. However, unlike the situation with *8*$^{2-}$ and *23*$^{2-}$ here an alternant hydrocarbon forms the dianion. The same phenomenon was encountered also in the heterocyclic dianions (vide infra). Glidewell and Lloyd[162-164] studied *MNDO* bond orders in extended polycyclic hydrocarbons. Their conclusion is that the dominant fragment is always the 6π-system: only when this is inaccessible, do 10π fragments become important. If neither 6π nor 10π fragments are accessible, then strong bond fixation results. System *58*$^{2-}$ is considered to be composed of 6π, 10π and 14π components, the latter being the "phenanthrene" moiety[162].

6.3 Reductive Alkylation of Polycyclic Anions

The process now known as reductive alkylation of π-conjugated anions (quenching of anions) is as old as the preparation of the ions themselves[5]. The highly colored solutions obtained by the addition of alkali metals to solutions of aromatic hydrocarbons in ether were reacted with electrohpiles such as protons or alkyl halides (Scheme 2). The products of such a process are reduced hydrocarbons. The Birch reduction is one example of such a process, reaction of an anion with an alkyl halide leading to an alkylated reduced hydrocarbon is another example[165]. The complexity of the quenching experiments is demonstrated by the naphthalene radical anion[150-158]

$$A \xrightarrow{+e} A^{\overline{\cdot}} \xrightarrow{+e} A^{2-} \xrightarrow[-X^-]{RX} AR^- \xrightarrow[-X^-]{RX} AR_2 \qquad (a)$$

$$\downarrow +H^+ \qquad \downarrow +H^+$$

$$AH^{\cdot} \xrightarrow{e} AH^- \xrightarrow[-X^-]{RX} AHR \xrightarrow{B} \xrightarrow{RX} AR_2 \qquad (b)$$

Scheme 2. Possible routes for alkylation reactions in electron transfer reactions [170].

The chemistry of the naphthalene radical anion in quench reactions with water, carbon dioxide, or alky iodides involves several steps [165–167] (Eqs. 14–16):

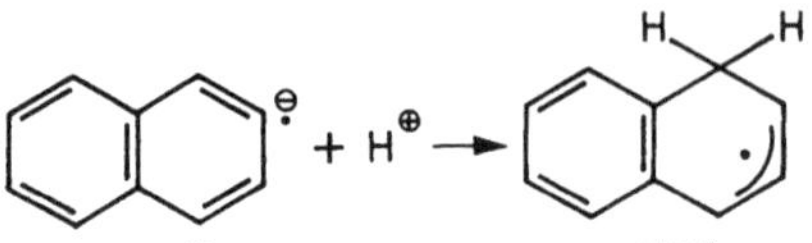

55⁻· 55·

55⁻· 55 H· 55 55 H⁻

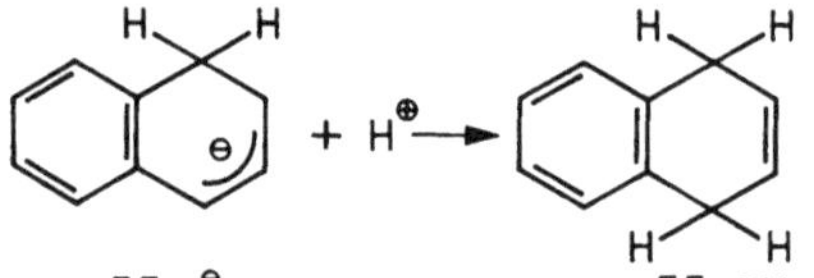

55 H⁻ 55-2H

In this mechanism two naphthalene radical anions will generate one molecule of 1,4-dihydronaphthalene and one molecule of naphthalene.

In general, however, one must be concerned with the possible dominance of chemistry by small amounts of dianions. Although not seen in electrochemistry, the naphthalene dianion has been reported in the literature [11, 150–159, 167] and could dictate the results of quench reactions. In the specific case of sodium naphthalene in tetrahydrofuran, kinetic analysis of a water quench directly implicites the radical anion as the chemically dominant species [150–158, 167]. In the case of the larger aromatic molecule, perylene, however, the dianion and not the radical anion is the species quenched [167a)].

Quenches involving alkyl halides are more complex than those involving water. One can obtain different products depending on quench sequence (radical anion added to halide or halide added to radical anion) [167b)] and one can also obtain products other than those predicted by Equations 17–19. The naphthalene radical anion can reduce alkyl halides (especially iodides) to form alkyl radicals and halide ions. The alkyl radicals, once formed, can both dimerize and disproportionate [167d–f)].

$$RCH_2CH_2I + C_{10}H_8^{\overline{\cdot}} \rightarrow RCH_2CH_2\cdot + C_{10}H_8 + I^- \qquad (17)$$

$$2\,RCH_2CH_2\cdot \xrightarrow{\text{dimerize}} R(CH_2)_4R \qquad (18)$$

$$2\,RCH_2CH_2\cdot \xrightarrow{\text{disproportionate}} RCH_2CH_3 + RCH{=}CH_2 \qquad (19)$$

In the reaction of benzyl chloride with sodium naphthalenide, the dimerization product is obtained in 80% yield, indicating that reactions 17–19 can completely overshadow reactions 14–16 in some cases [167f)]. This knowledge led Harvey [25, 168)] and Rabideau [169)] to selectively prepare the relevant anionic intermediates.

The reduction of a wide range of polycyclic systems with alkali metals in liquid ammonia was studied by Müllen [170a, b)]. It was noticed that polycyclic benzenoid and nonbenzenoid dianions are either persistent as dianions or undergo a single or multiple protonation by ammonia. This process shows a very high degree of regioselectivity. The alkylation reactions according to route (a) or (b) Scheme 2 [170)] are governed by the different π-charge distributions in the monoanion and the dianion (AH^- and A^{2-}, respectively). It has been shown that controlling the reaction mechanism has a significant synthetic utility [170)].

Two routes seem to prevail when polycycles are being reduced in ammonia:

(a) electron transfer generates a radical anion or dianion, and
(b) the species can be protonated by ammonia.

The different modes of behaviour are rationalized by comparing the π-electron energy of the dianion with that of the corresponding monoanion. From this different, i.e. $E_\pi(A^{2-})$—$E_\pi(A\text{—}H^{1-})$, the atom localization energies, Aμ, are obtained [14–17)]. This difference is a measure of the basicities of the monoanion and the dianion [14–17)]. The tendency toward protonation increases with a decreasing Aμ value. Each protonation site is calculated differently. This approach seemed to work very nicely, predicted on the basis of the relevant atom localization energies [170)]. This study has also shown that what has been eroneously assigned as pyrene tetraanion (*29*$^{4-}$) [171)] is, in fact, mono-protonated-monoanion of pyrene (*61*$^{1-}$) [172)]. Ebert [173)] has shown that also perylene (*60*) does not form a tetraanion.

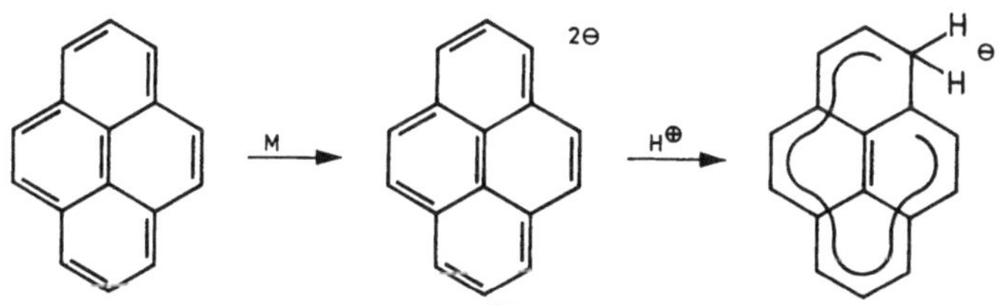

29 *59*$^{2\ominus}$ *61*$^{\ominus}$

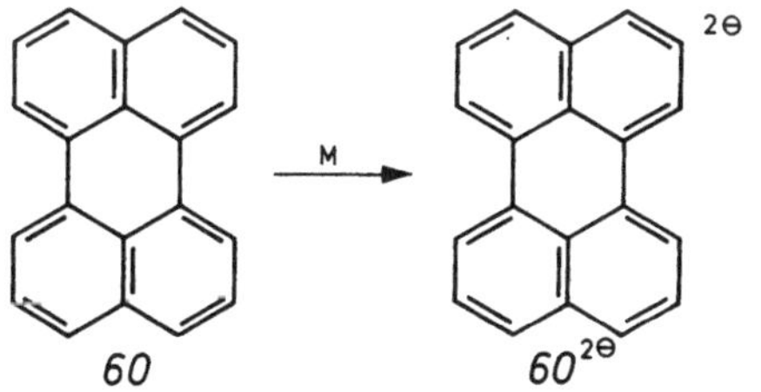

60 *60*$^{2\ominus}$

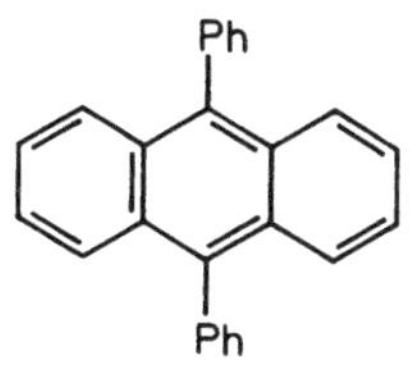

62

In the reaction of perylene (*60*) with potassium in tetrahydrofuran, both the potassium consumption values and products with methyl iodide were studied. These studies suggested the intermediacy of the dianion, rather than of the tetraanion of *60*. Quenches with protic sources form species which are unstable to air and/or the analysis conditions themselves. Studies on 60^{2-} quenched by methanol-d_4 suggest [173] that dideuterio dihydro perylene was not formed. These authors suggest a disproportionation of the dideuterio-dihydro product into tetradeuterio tetrahydro perylene and perylene itself. Quench with methyl iodide afforded dimethyl dihydroperylene as the most abudant product. The errors in assigning tetraanions [171] arise from relying mainly on spectroscopic methods. However, quenching experiments such as the experiment with methanol-d_4 are not as simple as they seem at first glance. Another difficulty to analyse highly charged species may originate from the instability and the insolubility of the salts. This situation may be responsible for the discrepancy between quenching experiments and spectroscopy. However, three-anion radicals were obtained recently by Gerson and Vögtle for 1,8-diphenylnaphthalene (*63*) and similar systems. It has been shown that a significant portion of the charge resides on the phenyl substituents of these substituted naphthalenes (s. Sect. 8.2).

Reductive alkylation of coal and graphite and other fossil fuels has been discussed in detail [140].

6.4 Thermodynamic Parameters for Polycyclic Dianions

In a series of publications, Stevenson et al. have demonstrated that the enthalpies of generation of polyacene dianions can be obtained by calorimetric measurements. Compared to their calculated instability in the gas-phase [70], THF solutions of polycyclic dianions are thermodynamically and kinetically stable as evidenced by their spontaneous formation and persistence. The solvation energy of the dianions plus that of two cations must overcome the repulsive interaction of the charges. This aspect has been demonstrated by Stevenson in his studies on the cyclooctatetraene dianion [70]. The dianion sodium salts were prepared in THF in thin-walled glass bulbs and the bulbs were crushed under water in a calorimeter system. The heat of aquation of the solvent is taken into account and thus, the net change in the heat content of the calorimeter vs. the millimoles of dianion salt is obtained. The plots are linear and the slopes represent the enthalpies of the following reaction (Eq. 20):

$$A^{2-}, 2\,Na^{+}(THF) + 2\,H_2O(l) \rightarrow AH_2(s) + 2\,NaOH(aq)\,. \qquad (20)$$

From this equation the heat of generation of the dianion from the metal and solvated polycyclic dianion can be derived. The heats of generation of the dianions was also obtained from the reaction of the dianion with iodine. From these data the following heats of generation of the anions were obtained (Table 13). It is interesting that the dianions studied have heats of generation very similar in magnitude [70a], and that these enthalpies are about the same as the heat of reaction of sodium metal with water. It can therefore be concluded that the solvation enthalpies of the dianions must be very negative.

Table 13. Enthalpies of Reaction (ΔH°_{rxn}) of THF Solvated Dianions with Water and Iodine and the Heats of Generation (ΔH°_{gen}) of the Dianions in THF (kcal/mol) [70a)].

Dianion	ΔH°_{rxn} with		ΔH°_{gen} with	
	H_2O	I_2	H_2O	I_2
anthracene (*2*$^{2-}$)	−64.9 ± 1.9	−99.5 ± 4.4	−40.3 ± 2.0	−41.6 ± 4.6
tetracene (*48*$^{2-}$)	−62.9 ± 2.1		−37.8 ± 2.2	
pentacene (*47*$^{2-}$)	−59.6 ± 2.1		−40.3 ± 2.2	
pyrene (*59*$^{2-}$)	−33.4 ± 1.6	−97.2 ± 2.7	−46.1 ± 1.7	−47.7 ± 2.8
perylene (*60*$^{2-}$)	−57.8 ± 1.5		−44.5 ± 1.7	

7 Polycyclic Dianions Containing Heteroatoms

Contrary to the comprehensive studies of charged carbocyclic systems, information concerning heterocyclic 4nπ conjugated dianions is limited. On the one hand there were reports that claimed that these dianions are extremely unstable [174)]. On the other, the chemistry of such systems has been reported [5,175–177)]. The introduction of a heteroatom into the path of conjugation of a polycyclic dianion raises several questions resulting from: (a) the topology of the system, (b) the influence of the heteroatom on the charge distribution pattern and on the HOMO-LUMO gap, and (c) the quench of paratropicity of the heterocyclic dianions as compared with the carbocyclic systems. To answer these questions a spectroscopic study, and more specifically a ^1H and ^{13}C NMR study of these anions was required. Such studies appeared in the literature since 1985 on nitrogen containing stable dianions [178–180a)] and later on sulfur and oxygen containing dianions [184,185)]. The reduction of the heterocyclic system is carried out by sodium metal and undergoes the same stages observed in the carbocyclic series. A radical anion is formed as the primary reduction process followed by the formation of the diamagnetic 4nπ paratropic dianion [178,179)]. Linear and angular systems show that their appearance in the NMR spectrum, namely line-shape and degree of paratropicity depend on their topology. The heterocyclic systems also exhibit an electrocyclic ring closure process [179,181)].

7.1 Nitrogen Containing Dianions

The 4nπ-electron systems are intriguing in view of their paratropicity and tendency to exist at least to some extent as ground-state triplets [151)]. There is some parallelism between conjugated 4nπ series derived from the polycyclic hydrocarbon dianions and the respective aza derivatives. The calculated HOMO-LUMO gap is narrow in both series, as expected for 4nπ conjugated systems. This narrow gap influences the line shape and the degree of the paratropicity of the dianions. The presence of a heteroatom influences the charge densities and, to some extent, the HOMO-LUMO gap, so that proton and carbon NMR parameters of the dianions containing even two fused rings can be studied. The paratropicity of the hydrocarbon dianions arises from an effective delocalization of 4nπ electrons. The introduction of electronegative nitrogen atoms into the path of conjugation may quench this electron distribution and hence the systems' paratropicity. The following systems, viz., *64–72* were prepared and some will be discussed in detail.

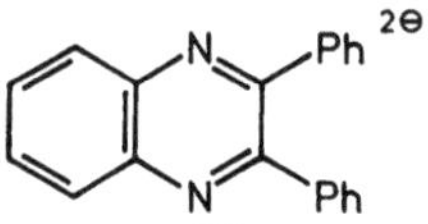

$64^{2\ominus}$

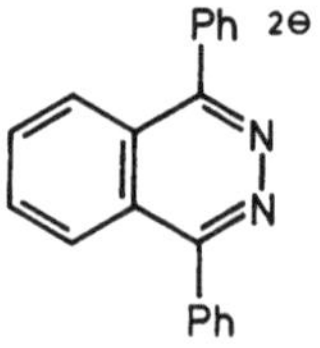

$65^{2\ominus}$

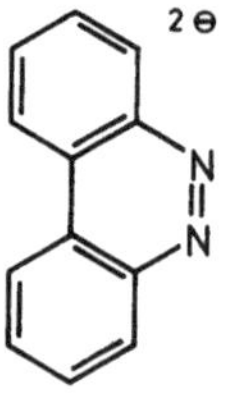

$66^{2\ominus}$

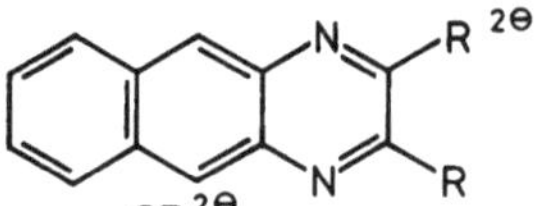

$67^{2\ominus}$

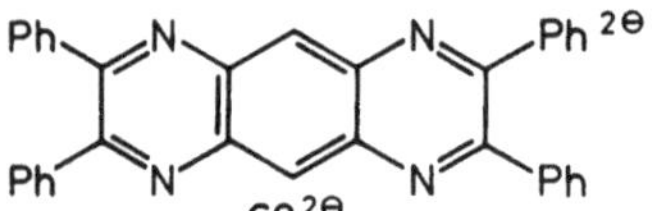

$68^{2\ominus}$

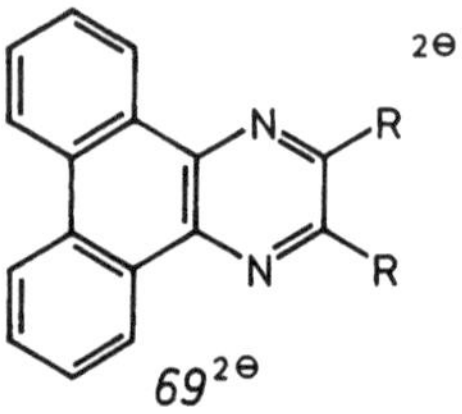

$69^{2\ominus}$

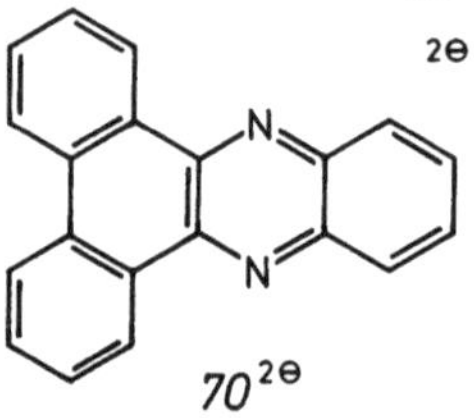

$70^{2\ominus}$

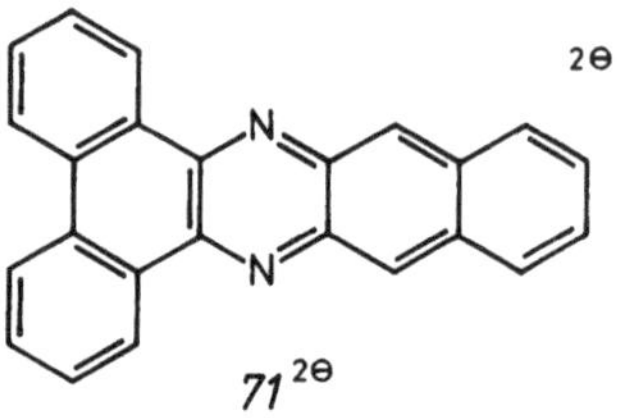

$71^{2\ominus}$

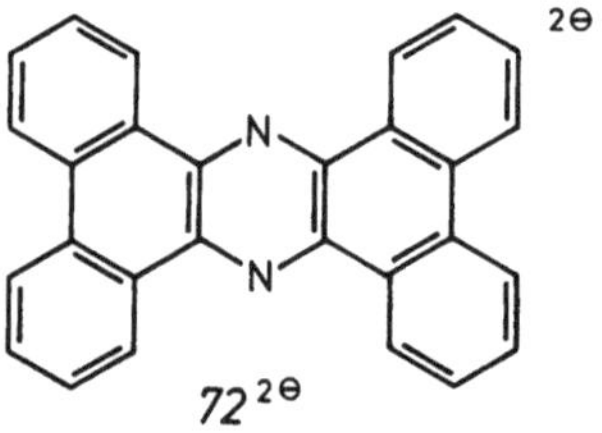

$72^{2\ominus}$

The ^{1}H and ^{13}C spectroscopic data of some of the dianions are given in Tables 14 and 15.

The fact that heterocyclic dianions of the naphthalene skeleton, viz. 64^{2-} and 65^{2-}, could be monitored and studied shows that a partial quenching of the paratropicity of the system has occurred. This effect could also originate from a partial deviation from coplanarity of the system. Although *64* and *65* are related compounds, the solution of 64^{2-} and the solution of 65^{2-} revealed entirely different ^{1}H NMR spectra (Fig. 17). The main differences are (i) the ^{1}H NMR spectrum of 64^{2-} shows line broadening while that of 65^{2-} does not; (ii) The spectrum of 65^{2-} is almost unaffected by temperature while 64^{2-} shows a strong temperature dependence; and (iii) in 2,3-diphenylquinoxaline dianion (64^{2-}) only a small portion of the negative charge is located on the phenyl moieties. This was deduced from calculations and from the proton absorption of the para hydrogen of the phenyl group. The band attributed to the para hydrogen appears at 6.58 ppm in 64^{2-} vs. 5.91 ppm in 65^{2-}.

It was construed that line-broadening and, in extreme cases, disappearance of the spectrum of doubly charged $4n\pi$-electrons carbocyclic dianions arise from paramagnetic dilution due to thermal equilibrium between the ground state and a low-lying triplet

Table 14. ^{1}H NMR Parameters of Doubly Charged Heterocyclic System of THF—d_8 [179].

System[a]	Temp., K	^{1}H NMR pattern[a,b]	Neutral system center of gravity	Charged system center of gravity	Calcd high field[e]
64^{2-}/2 Na^+	292	6.88 (d, J = 7.5, 4 H, $H_{2',6'}$), 6.76 (t, J = 7.6, 4 H, $H_{3',5'}$), 6.58 (t, J = 7.0, 2 H, $H_{4'}$), 4.67 (br s, 2 H, $H_{6,7}$), 4.13 (br s, 2 H, $H_{5,8}$)	7.64	6.09	2.03
64^{2-}/2 Na^+	203	6.75 (d, J = 7.4, 4 H, $H_{2',6'}$), 6.67 (t, J = 7.1, 4 H, $H_{3',5'}$), 6.52 (t, J = 6.8, 2 H, $H_{4'}$), 4.44 (br s, 2 H, $H_{6,7}$), 3.62 (br s, 2 H, $H_{5,8}$)	7.64	5.92	2.25
65^{2-}/2 Na^+	300	7.21 (d, J = 8.1, 4 H, $H_{2',6'}$), 6.71 (t, J = 7.1, 4 H, $H_{3',5'}$), 6.33, 5.84 (AA'BB', 4 H, $H_{6,7}$, $H_{5,8}$), 5.96 (t, J = 7.1, 2 H, $H_{4'}$)	7.86	6.57	1.69
65^{2-}/2 Na^+	213	7.20 (d, J = 8.1, 4 H, $H_{2',6'}$), 6.70 (t, J = 7.6, 4 H, $H_{3',5'}$), 6.31, 5.78 (AA'BB', 4 H, $H_{6,7,5,8}$), 5.91 (t, J = 7.1, 2 H, $H_{4'}$)	7.86	6.54	1.73
68^{2-}/2 Na^+	298	7.12 (d, J = 7.2, 8 H, $H_{2',6'}$), 6.95 (t, J = 7.6, 8 H, $H_{3',5'}$), 6.85 (t, J = 6.9, 4 H, $H_{4'}$), 4.48 (s, 2 H, $H_{5,10}$)	7.69	6.77	1.89
71^{2-}/2 Li^+	295	8.09 (d, J = 8.0, 2 H, $H_{4,5}$), 7.71 (d, J = 6.8, 2 H, $H_{1,8}$), 7.00 (t, J = 6.9, 2 H, $H_{2,7}$), 6.81 (t, J = 7.3, 2 H, $H_{3,6}$), 5.93 (br s, 4 H, $H_{11,12,13,14}$), 4.70 (s, 2 H, $H_{10,15}$)	8.43	6.60	2.39
71^{2-}/2 Na^+	293	8.04 (d, J = 7.9, 2 H, $H_{4,5}$), 7.76 (d, J = 8.2, 2 H, $H_{1,8}$), 7.01 (t, J = 7.0, 2 H, $H_{2,7}$), 6.75 (t, J = 7.5, 2 H, $H_{3,6}$), 5.71 (br s, 4 H, $H_{11,12,13,14}$), 4.49 (s, 2 H, $H_{10,15}$)	8.43	6.50	2.53
71^{2-}/2 Na^+	203	8.01 (d, J = 8.1, 2 H, $H_{4,5}$), 7.62 (d, J = 8.6, 2 H, $H_{1,8}$), 6.95 (t, J = 7.7, 2 H, $H_{2,7}$), 6.73 (t, J = 7.6, 2 H, $H_{3,6}$), 5.58 (AA'BB', 4 H, $H_{11,12,13,14}$), 4.27 (s, 2 H, $H_{10,15}$)	8.43	6.25	2.85
72^{2-}/2 Na^+	213	7.55 (d, J = 8.3, 4 H, $H_{4,5,13,14}$), 6.80 (d, J = 8.3, 4 H, $H_{1,8,10,17}$), d 6.71 (t, J = 7.3, 4 H, $H_{2,7,11,16}$), 6.39 (t, J = 6.7, 4 H, $H_{3,6,12,15}$)	d	6.86	d

[a] For numbering, see Chart I. [b] Chemical shifts are given in ppm referenced to Me_4Si (for numbering see the chart and schemes; coupling constants are given in Hz). [c] Abbreviations: s = singlet, d = doublet, t = triplet, br = broad bands. [d] No ^{1}H spectrum could be observed for *72* in THF. [e] In units of charge; calculated according to the chemical shifts charge density correlations for carbocyclic systems.

Table 15. ^{13}C NMR Parameters of the Doubly Charged Heterocyclic Systems in THF [179].

System[a]	Temp., K	NMR chemical shift, ppm[b]	Neutral system center of gravity	Charged system center of gravity	$\Delta\delta(^{13}C)$
64^{2-}/2 Na^+	213	163.6, 145.6, 139.2, 126.8, 126.3, 120.6, 116.4, 100.3	133.6	129.2	88
65^{2-}/2 Na^+	213	140.9, 138.7, 127.7, 121.2, 120.8, 115.2, 109.5, 107.9	132.6	122.5	202
65^{2-}/2 Na^+	273	140.6, 138.9, 127.9, 121.5, 120.5, 115.5, 110.3, 108.9	132.6	122.7	198
68^{2-}/2 Na^+	293	157.2, 144.7, 138.2, 128.8, 127.6, 124.3, 93.8	134.8	132.2	88.4
71^{2-}/2 Na^+	293	159.9, 139.7, 139.0, 128.2, 125.5, 123.9, 121.8, 120.0, 118.3, 118.1, 117.9, 96.8	129.3	125.8	84
72^{2-}/2 Na^+	213	151.1, 126.4, 124.2, 123.3, 121.7, 119.1, 116.2	c	126.0	c

[a] For numbering, see Chart **0**. [b] Chemical shifts are given in ppm referenced to Me_4Si. [c] These data could not be obtained because of the limited solubility of *72* in THF.

[37]. A correlation was demonstrated between the extent of line broadening and the HOMO-LUMO energy gap, as given by ωβ. The calculated gaps for *64*$^{2-}$ and *65*$^{2-}$ are 0.19β and 0.29β, respectively (Table 16). In *65*$^{2-}$, the gap is relatively high, predicting negligible concentration of the triplet and therefore no line broadening is expected. The high-field shift of the para hydrogen of the phenyls in *65*$^{2-}$, relative to those of *64*$^{2-}$, can be understood in terms of charge delocalization. In *65*$^{2-}$ the phenyl rings are less hindered sterically and can attain better coplanarity and orbital overlap with the polycyclic backbone. Even if the same extent of coplanarity is assumed in *64*$^{2-}$ and *65*$^{2-}$, as was done in calculating these two dianions, a higher charge density is predicted at the phenyl moiety of *65*$^{2-}$ than in *64*$^{2-}$: —0.226 vs. —0.092 electron units, respectively. The better overlapping between the aromatic moiety and the phenyl groups in the 1,4-diphenylphthalazine dianion *65*$^{2-}$ than that in *64*$^{2-}$ is also manifested by the variations in the bond lengths, as derived from the calculations. The bond length of C2—C1′ in the neutral molecule *64* is 1.495 Å and that of C1—C1′ in *65* are 1.484 Å for *64*$^{2-}$ and 1.462 Å for *65*$^{2-}$. The calculations show a higher bond density as a result of the reduction process of *65* relative to *64*.

On the basis of the ^{1}H NMR spectra of *64*$^{2-}$ and *65*$^{2-}$ it seems that the charge density on the diazanaphthalene nucleus in *64*$^{2-}$ is higher and shows a more pronounced paratropicity (line broadening and a highfield shift). This conclusion is supported by the differences in the chemical shifts of the various hydrogens, and carbons in the NMR spectra, which are by far more pronounced in the dianion derived from the quinoxaline system. The range of the ^{1}H and ^{13}C NMR bands of *64*$^{2-}$ is 3.1 ppm and 63.3 ppm, respectively, while that of *65*$^{2-}$ is much narrower, 1.4 for ^{1}H and 33.0 ppm for the ^{13}C spectrum. Such a nonuniform charge distribution, as in *64*$^{2-}$, is characteristic of paratropic 4nπ-electron antiaromatic system.

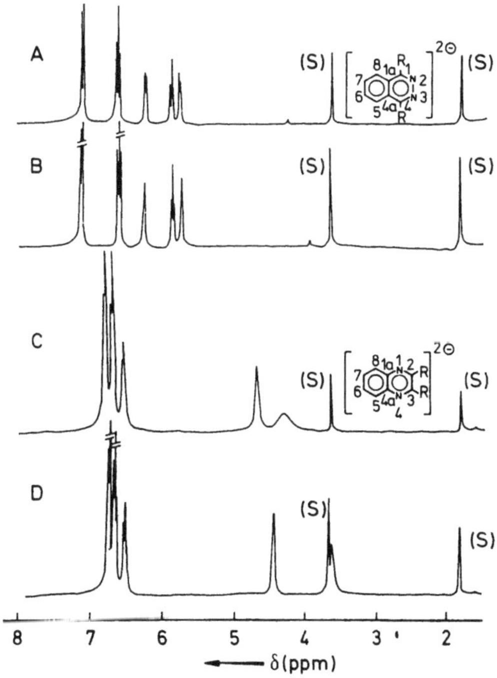

Fig. 17. ^{1}H NMR spectra of 1,4-diphenylphthalazine dianion (*65*$^{2-}$) and 2,3-diphenylquinoxaline dianion (*64*$^{2-}$) in THF-d_8. (A) *65*$^{2-}$/2 Na^+, 300 °K; (B) *65*$^{2-}$/2 Na^+, 213 °K; (C) *64*$^{2-}$/2 Na^+, 292 °K; (D) *64*$^{2-}$/2 Na^+, 203 °K [179]

Table 16. HOMO-LUMO Energy Gap, Line Broadening, and Charge Density of the Heterocyclic Dianions [179].

HOMO-LUMO energy gap[b,c]	68^{2-} [d]	71^{2-}	65^{2-} [d]	64^{2-} [d]	72^{2-} [e]	69^{2-} [d,e]
HOMO-LUMO energy gap[b,c]	0.38	0.31	0.29	0.19	0.11	0.11
detectable line broadening				+	+	*f*
net charge on nitrogen	−0.265	−0.321	−0.205	−0.418	−0.372	−0.364
maximum net charge on carbon	−0.219	−0.192	−0.241	−0.221	−0.079	−0.132

[a] For numbering, see Chart. [b] As obtained by ωβ calculation. [c] The HOMO-LUMO energy gaps are given in β units. [d] The calculations were carried out assuming β as 0.5 for the bond between the phenyl groups and the aromatic skeleton. [e] Calculations were run for these systems with ω = 1.0. [f] No spectrum could be recorded at all.

Apart from its spectroscopic properties, 2,3-diphenylqunoxaline dianion (*64*$^{2-}$) shows a very interesting ring closure to the dibenzo[a,c]phenazine dianion (*70*$^{2-}$) (Scheme 3) [179,180b]. The deep purple solution of *64*$^{2-}$ turned red, and changes in the ^{1}H and ^{13}C apectra were recorded. The new absorption bands of the ^{1}H and ^{13}C NMR which appeared after one week are identical with those of *70*$^{2-}$. The transformation of *64*$^{2-}$ to *70*$^{2-}$ requires a formal loss of H_2, a rather strange event under the reductive reaction conditions. It is well-known that in the photocyclization of stilbene and its derivatives an oxidant is employed in order to push the dehydrogenation to the final aromatized product. The closure seems to constitute a good synthetic route to polyheterocyclic systems as in this case, a photochemical route from neutral *64*$^{2-}$

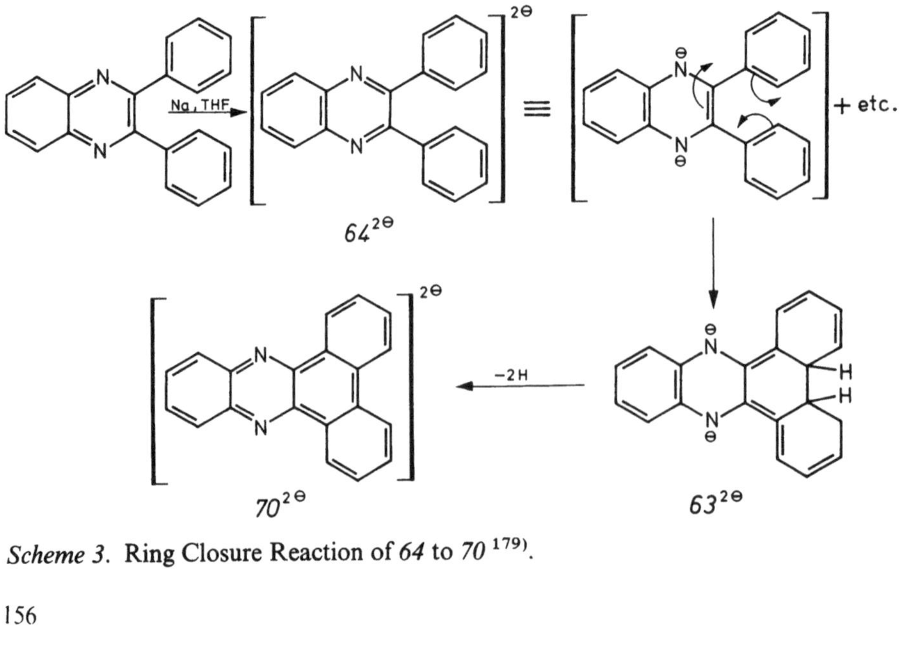

64$^{2\ominus}$ *70*$^{2\ominus}$ *63*$^{2\ominus}$

Scheme 3. Ring Closure Reaction of *64* to *70* [179].

to neutral 70^{2-} is unattainable [181]. The lengthy reaction time is overcome by employing C_8K as a reagent [181]. A possible mechanism involves an electrocyclic conversion of the dianion (Scheme 3). Yet, the only species which could be observed in the course of the reaction are the starting material 64^{2-} and the final product 70^{2-}.

Another interesting phenomenon is encountered in the reduction of *71*. This system resembles the dibenzotetracene *58* in which a partitioning of MO's was observed in the dianion [160, 161]. Reduction of *71* afforded a paratropic dianion 71^{2-} being a 4nπ-electron system. No line broadening was observed, in line with the relatively wide HOMO-LUMO gap (0.31β). A very interesting phenomenon ensues, as can be gathered from the ^{1}H NMR spectrum of 71^{2-} (Fig. 18). A formal segregation of the charged conjugated system into two components: (a) a "phenanthrene"-like moiety which accommodates only a minor fraction of the negative charge: (b) a benzo[g]-quinoxaline-like moiety which accomodates the major part of negative charge was suggested. This mode of delocalization is quite unusual, since the system is alternant while such a phenomenon is characteristic of nonalternant systems [179]. The partitioning is corroborated by ωβ calculations. The interpretation given for 58^{2-} seems to apply here as well. Tribenzo[a,c,i]phenazine (*71*) can be regarded as built up formally of benzo[g]quinoxaline and phenanthrene-like components. Phenanthrene dianion is rather antiaromatic, as shown by its very pronounced paratropicity as well as by the calculated HOMO-LUMO energy gap which is relatively low. Benzo[g]quinoxaline dianion, which is the heterocyclic analogue of anthracene dianion, is "much less antiaromatic" than phenanthrene dianion. This comes out both from the ^{1}H NMR spectrum of 2,3-di-*p*-tolylbenzo[g]quinoxaline dianion, and from the computed HOMO-LUMO gep, 0.34β. It seems that when *71* is charged to form 71^{2-} it would "lose" more if the major part of the negative charge were directed to the phenanthrene moiety. Therefore, the alternative pathway is preferred, resulting in a formal benzo[g]-qunoxaline dianion-like moiety and a virtually uncharged phenanthrene. In an alternant system it is reasonable, even though somewhat unexpected, that a molecule will choose a delocalization path that minimizes antiaromatic contributions. These data strengthend the viewpoint that phenanthrene dianion is a highly paratropic antiaromatic system [37]. It also reflects the tendency of an aromatic system to remain so [84b, 179].

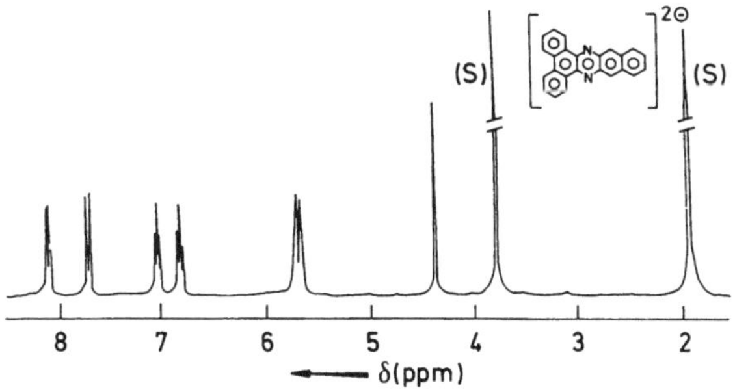

Fig. 18. ^{1}H NMR spectrum (300 MHz) of tribenzo[a,c,i]phenazine dianion 71^{2-} as the disodium salt in THF-d_8 at 203 K [179]

7.2 Sulfur and Oxygen Heterocyclic Dianions

Dianions derived from sulfur and oxygen containing polycyclic systems have not been reported until very recently. The weak carbon-sulfur bond could discourage attempts to prepare such dianions. Sulfur removal studies by alkali metals were successful due to the properties of the carbon-sulfur bond [182, 183]. Reductive alkylation studies on model compounds which are relevant to sulfur containing systems demonstrated the stability of hydrocarbon dianions [182]. A spectroscopic study performed under mild conditions afforded the characterization of sulfur and oxygen containing dianions [184–186]. Benzo[b]thiophene, (*73*), 1,3-diphenylbenzo[c]thiophene (*74*) and 1,3-diphenylbenzo[c]furan (*75*) form the respective dianions at low temperature.

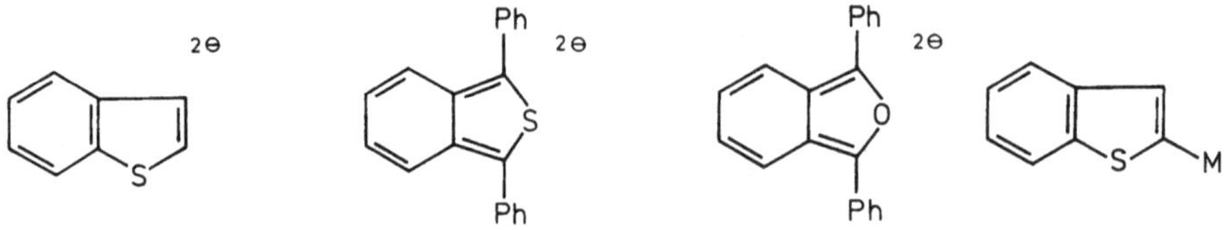

73$^{2\ominus}$ *74*$^{2\ominus}$ *75*$^{2\ominus}$ *76*

The formation of dianion *73*$^{2-}$ is accompanied by a metallation reaction at the α-position. This reaction was observed when both lithium and sodium were applied. When prepared at —78 °C with sodium metal, *73*$^{2-}$ afforded mainly the dianion. However, at —20 °C the main product was the α-metallated system *76*. The assignments of systems *73*$^{2-}$–*75*$^{2-}$ were derived from sophisticated NMR techniques and quench experiments. The various reaction paths of benzo[b]thiophene are shown in Scheme 4.

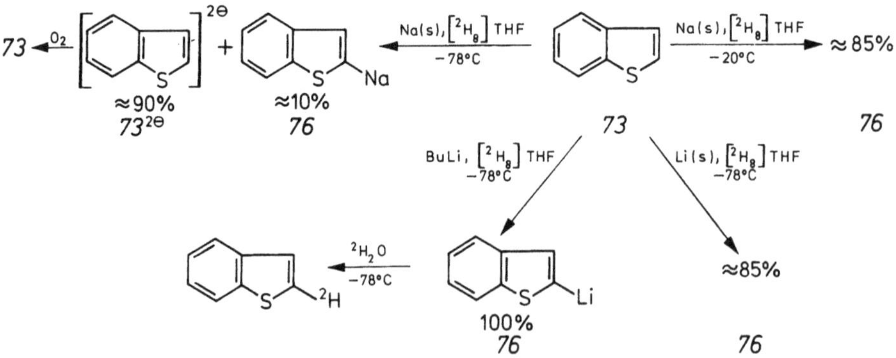

Scheme 4. The reaction of sodium and lithium with *73* in THF-d_8.

A much higher stability was shown by the dianion of benzo[c]thiophene, viz. *74*$^{2-}$. Here, the α-positions are blocked, a situation which may be responsible for its stability. The NMR spectroscopic assignment of both carbon-13 and protons showed interesting phenomena. The dianion *74*$^{2-}$ was prepared by sodium metal reduction and exhibits a high barrier to rotation about the carbon-carbon bond linking the phenyl rings to the heterocyclic system. A high degree of charge is localized in position α to the sulfur atom, thus pointing at the effect of charge polarization by sulfur in the dianion. In the ^{1}H NMR spectrum of the disodium salt the lines are very sharp and well resolved. This salt was also much more stable than the respective dipotassium and dilithium salts. The disodium salt can be stored for weeks at —40 °C without any detectable decomposition. Dianions were also prepared from 1,3-diphenylbenzo[c]-furan (*75*). The ^{1}H NMR spectrum of the benzo[c]furan dianion 75^{2-} is not really paratropic. This observation can be rationalized as follows. Benzo[c]furan has only a low resonance energy and a narrow HOMO-LUMO gap [187]. Thus, despite its $(4n + 2)\pi$-electrons, it cannot be classified as a classical aromatic system. In such a case one would expect that the related $4n\pi$ dianion will exhibit negligible paratropicity. This argument may also apply to 75^{2-}. However, there is still a probability that the carbon atoms which accommodate a high degree of charge density (α position to the heteroatom) are pyramidal [188]. Such a structure was suggested by Smith [189] in his studies on the chemistry of the system. From the NMR data it has been concluded that potassium forms contact ion pairs while the small lithium cation shifts the equilibrium toward solvent separated ion pairs [185]. The solvated lithium cation can cause a lesser polarization of the charge towards the benzylic positions (α positions of the furan moiety). This ion-solvation equilibrium situation explains the pronounced stereoselectivity reported by Smith and McCall [189] on the alkylation of $75^{2-}/2\,Na^+$ and $75^{2-}/2\,K^+$ versus the lower stereoselectivity observed in $75^{2-}/2\,Li^+$. As the metal cation is removed from the carbanionic center, the chance for planarization of this center enhances, thus rendering the reaction less stereoselective. The stereoselectivity of the sodium and the potassium salts of 75^{2-} is nearly the same and they also showed similar ^{1}H NMR spectra. The spectrum of the lithium salt is different as expected for a salt with a high degree of solvent separated ion pairs, which, in turn, results in a reduced influence of the anisotropic effects of the alkali metal.

8 Charged Cyclophanes: From Dianions to Tetraanions

Anions derived from cyclophane represent a new class of carbanions in which new modes of electron delocalization were encountered. Two main groups of cyclophane anions were studied. In the $[2_n]$ (n = 2) paracyclophane [190,191] several modes of electron delocalization may occur as they have an inner and outer periphery. Another group of interesting cyclophanes is the [2.2]paracyclphanes in which through space interactions may prevail.

8.1 [2.2]Paracyclophanes

An outstanding characteristic of [2.2]paracyclphanes is the through-space interaction of the two fully conjugated layers ad demonstrated by their electronic properties [192]. The two interpenetrating decks afford one overall π-system [193]. Very little is known about this interaction in dianionic [2.2]paracyclophanes. The extension of the delocalization could also be extended through transition-metal complexation. This approach was reported by Boekelheide with the aim to form polymers with interesting electrical properties [192g]. Fluorene cyclophanes were prepared by Haenel [194], and the ^{1}H

NMR spectra of *syn* and *anti* [2.2](2,7)fluorenophane dianions did not show a significant change as compared with fluorene anion itself [194]. Indacene cyclophane dianions were prepared by Hafner [195].

In view of the synthetic availability of *syn* and *anti* [2.2]indenophanes [196], a detailed spectroscopic study of *syn* and *anti* [2.2]indenophane dianions, viz. 77^{2-} and 78^{2-} could be carried out [191]. These dianions showed a through space interaction between the two charged $(4n + 2)\pi$ layers. The through-space magnetic and/or electronic interaction was also observed in the monoanion 79^{2-}. Interesting is the [2.2](1,4)-

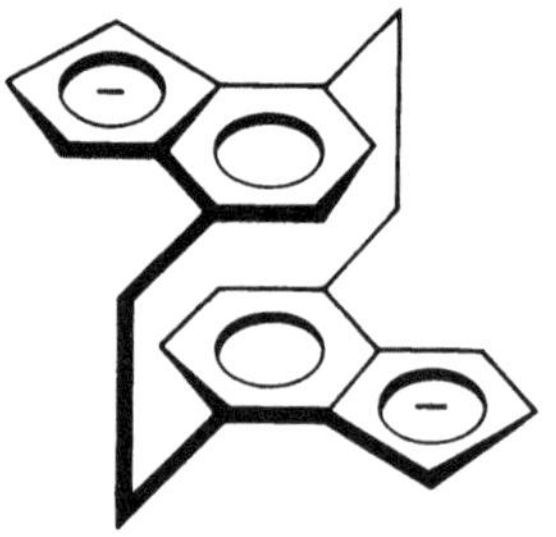

(anti) $77^{2\ominus}$

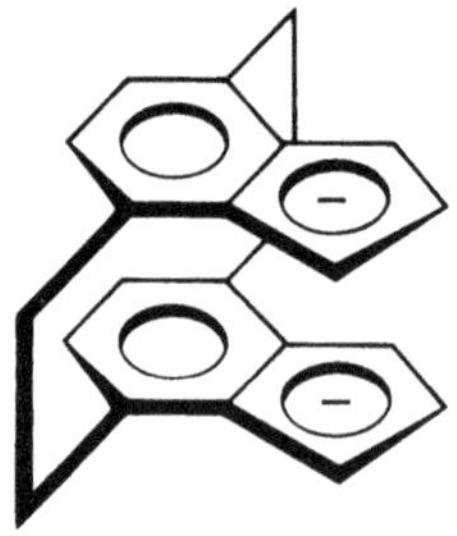

(syn) $78^{2\ominus}$

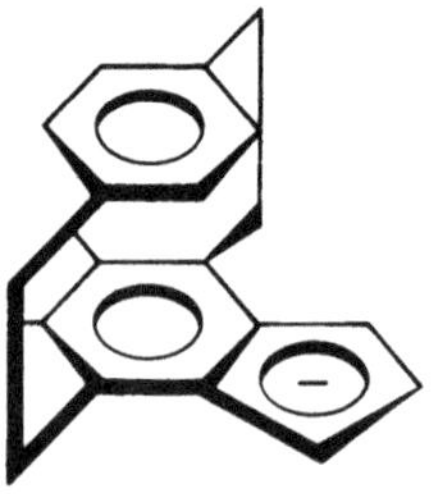

$79^{\ominus}$

benzo[g]chrysenoparacyclophane dianion (80^{2-}) which represents the family of paratropic layered dianions. It is so far the only member of the series and it shows a very significant shift of the protons of the neutral benzene layer [197]. While the chrysene dianion layer shows the expected paratropic spectrum for a $4n\pi$ dianion, the neutral

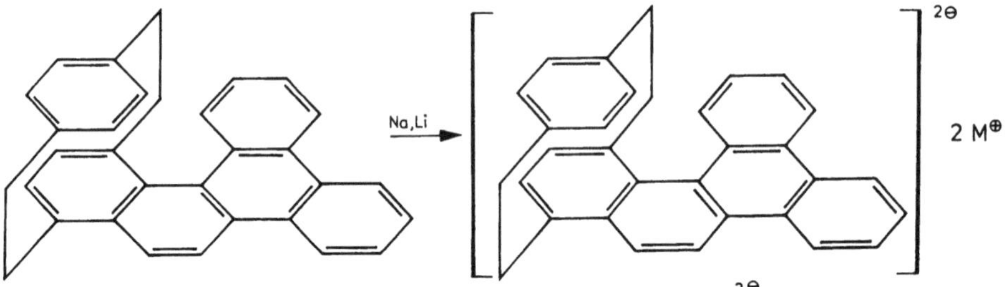

80 $\qquad$ $80^{2\ominus}$

benzene protons exhibit very significant lowfield shifts. These shifts are in the region of 10.31–6.46 ppm (four doublets) for $80^{2-}/2\ Li^+$ and 9.81–6.49 ppm (four doublets) for $80^{2-}/2\ Na^+$ [197]. The wide range of the chemical shifts of the neutral layer depends on its relative location in respect to the $4n\pi$ dianionic layer. This lowfield shift of the neutral benzene layer is attributed mainly to through-space anisotropy effects.

8.2 Macrocyclic Cyclophane Anions

There is a close relationship between polycyclic compounds such as coronene *81* [198] or Staab's kekulene *82* [199] and the macrocyclic cyclophanes such as *83* [200]. They can be described as perimeter structures in which an "outer" and "inner" perimeter can be considered.

81

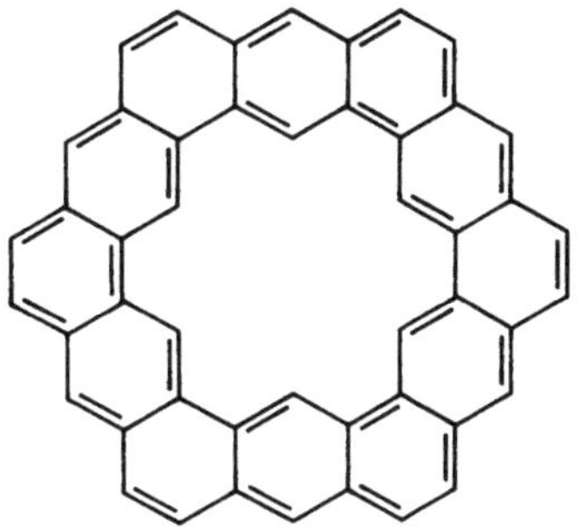

82

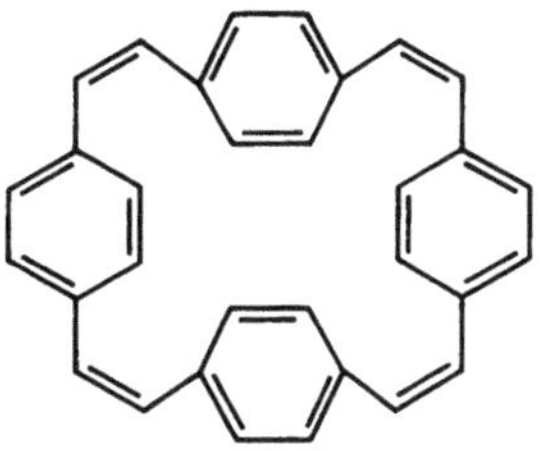

83

The elegant study of Müllen and Wennerström on the metal reduction of *83* and *84* afforded dianions and tetraanions, while the reduction of cyclophenes *85* and *86* afforded dianions [201, 203].

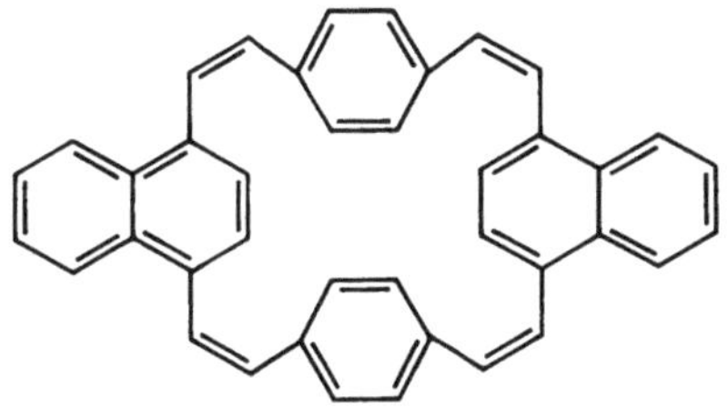

84

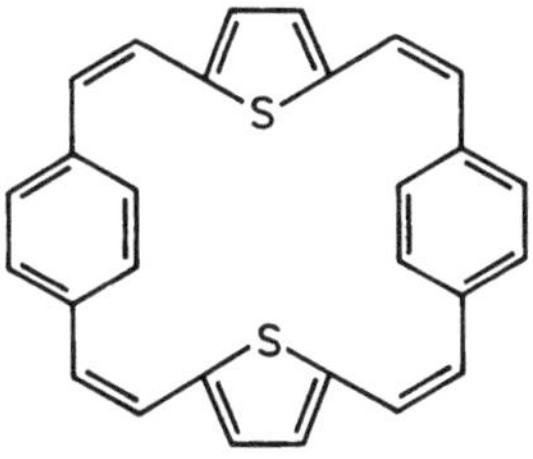

85

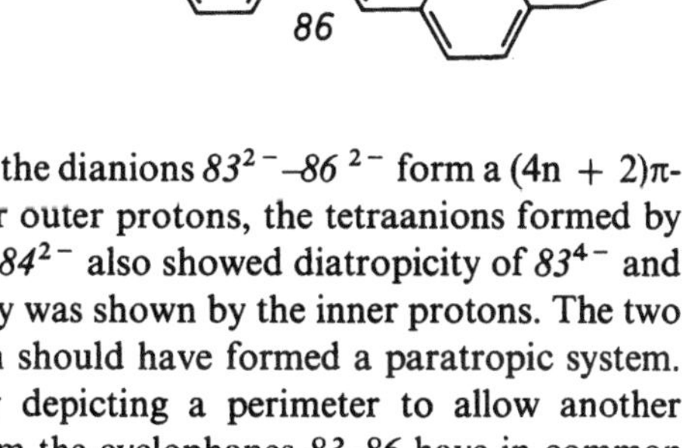

86

The interesting observation is that while the dianions 83^{2-}–86^{2-} form a (4n + 2)π-perimeter and showed diatropicity of their outer protons, the tetraanions formed by the addition of two electrons to 83^{2-} and 84^{2-} also showed diatropicity of 83^{4-} and 84^{4-}. However, in this case the diatropicity was shown by the inner protons. The two electron reduction of a (4n + 2)π-dianion should have formed a paratropic system. The system overcomes this difficulty by depicting a perimeter to allow another (4n + 2)π-periphery. All the dianions from the cyclophanes *83–86* have in common that inner (outer) protons are strongly shielded (deshielded). The large chemical shifts are typical for (4n + 2)π-annulenes and are caused by large diatropic ring currents. In contrast to the neutral compounds, the rotation of the paraphenylene moieties about the adjacent single bonds is slow in most of the dianions on the NMR time scale. The neutral cyclophanes which are 4nπ systems form by the addition of two electrons a new (4n + 2)π-dianion. The behaviour of the dianions as demonstrated by the large "ring current" affects the ^1H NMR shifts. The corresponding tetraanions have 4nπ-perimeters and show paratropic properties. Upon each stage of a two electron reduction the inner and outer protons are reversed. The studies on the preferred perimeter of cyclophane anions were made possible as the rotation of the phenyl rings is slow on the NMR time-scale. This is another demonstration of the principle of maximum aromatic contributions as shown in Sections 5.2.2 and 5.3.1. These results have also an important implication on Wennerström's studies on conjugated polymers consisting of chains of paraphenylene rings linked by vinylene bridges [203]. It should also be pointed out that in the polyheterocyclic dianions 73^{2-} and 74^{2-} the sulphur participates in the course of delocalization, while in 85^{2-} the path of delocalization does not include the sulphur atoms.

Another class of cyclophanes are derived from 1,8-diphenylnaphthalene with a bridge connecting the para positions of the phenyl groups. Such cyclophanes were prepared by Vögtle [204].

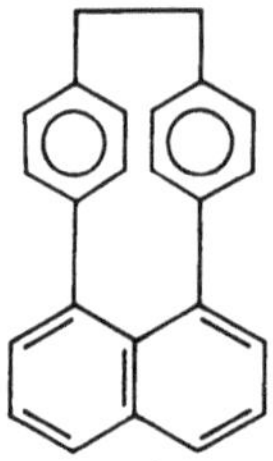

$87^{3\dot{-}}$

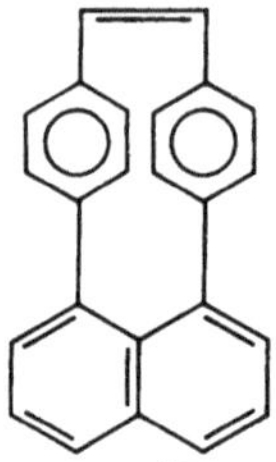

$88^{3\dot{-}}$

These cyclophanes [205] form upon charging trianion-radicals, viz. $87^{3\dot{-}}$ and $88^{3\dot{-}}$, respectively. It is surprising that such a perimeter can accommodate three negative charges [205]. However, in these anions the bulk of spin population resides on the two benzene rings so that these radical trianions can be regarded as the radical anions of an "open chain" cyclophane with a fused naphthalene π-system. The authors conclude that the naphthalene ring bears almost two charges. These unexpected charge delocalization patterns demonstrate that a complete understanding of charge delocalization pattern is far from being achieved. Such a goal will be the finding of a set of simple rules which will enable an immediate recognition of a preferred mode of delocalization.

9 Concluding Remarks

Polycyclic dianions and tetraanions show conspicuous features which are unprecedented in the neutral series. These features shed light on the general problem of electron structure of π-conjugated systems.

A) A general tendency to acquire aromatic character or to avoid antiaromaticity is exhibited by polycyclic anions. This tendency is fulfilled by sustaining modes of electron delocalization and charge distributions which would result in aromatic character or reduced antiaromatic contributions. A (4n + 2)π-electron path of conjugation offers the largest contribution to the character of the system. Applying this approach coincides with theoretical models and the magnetic experimental criterion for aromaticity. The preferred paths of delocalization of pericondensed anions, large ring cyclophane anions and the partitioning of charge and MO's in polycyclic carbocyclic and heterocyclic anions manifest this approach. Two main models were suggested, i.e. the peripheral model (Platt) and the conjugate circuits (Randić). It could be demonstrated by anions such as 12^{2-}, 20^{2-}, 23^{2-}, 33^{2-} and 34^{2-}, that the preferred path of delocalization will be the one which will afford a $(4n + 2)\pi$-array or arrays of electrons.

B) Polycyclic dianions in which an efficient unquenched delocalization of $4n\pi$-electron exists, reveal paratropic ^1H and ^{13}C chemical shifts. There is an unequivocal correlation between the magnitude of these paratropic shifts and the estimated widths of the HOMO-LUMO energy gaps in the $4n\pi$ species. This correlation applies to those systems where the HOMO's degeneracy is removed, thus resulting in a singlet configuration. The hitherto vague chain of premises related to antiaromaticity is made simple by the existence of a correlation between the ^1H NMR line shape and chemical shifts and the theoretically evaluated HOMO-LUMO gaps.

C) The study of charged π-conjugated polycycles thus affords not only extension of the scope of conjugated systems but also arrival at basic principles regarding the modes of π-electron delocalizations.

We believe that in the future, charged stacked polycycles, polymers and cyclophanes will open new avenues in applied research.

Acknowledgement: The author is deeply indebted to his coworkers (past and present), I. Willner, A. Minsky, Y. Cohen and R. Frim for their endless efforts, enthusiasm and devotion. Fruitful discussions, comments and suggestions with Professor U. Edlund, University of Umeå, Sweden and Professor K. Müllen, University of Mainz, F.R.G. are gratefully acknowledged.

10 References

1. (a) Cram DJ 1965 Fundamentals of Carbanion Chemistry, Academic Press, New York
   (b) Joner JR 1973 The Ionization of Carbon Acids, Academic Press, New York
   (c) Kaiser EM, Slocum DW 1973 McManus SP (Ed) in: Organic Reactive Intermediates, Academic Press, New York, Chapter 5
   (d) Ebel HF 1969 Die Acidität des CH-Sauren, George Thieme Verlag, Stuttgart
2. (a) Szwarc M 1974 Ions and Ion Pairs in Organic Reactions, Wiley Interscience, New York 1972, Vol. 2
   (b) Szwarc M (1968) in: Streitwieser A, Taft RW (eds) Progress in Physical Organic Chemistry, Interscience Publishers, New York, Vol. 6, pp. 323
   (c) Szwarc M (1969) Acc. Chem. Res. *2*: 87
   (d) Szwarc M (1972) ibid. *5*: 169
3. (a) Buncel E 1975 Carbanions: Mechanistic and Isotropic Aspects, Elsevier, Amsterdam
   (b) Buncel E, Durst T 1981 Comprehensive Carbanion Chemistry, Elsevier, New York
4. Thiele J (1900) Ber. *33*: 660; (1901) *34*: 68
5. (a) Schlenk W, Appenrodt J, Michael A, Thal A (1914) Chem. Ber. *47*: 473
   (b) Schlenk W, Bergmann E (1928) Liebigs Ann. Chem. *463*: 1
6. Cannon RD 1981 Electron Transfer Reactions, Butterworth, London
7. Scott ND, Walker JF, Hansley VL (1936) J. Am. Chem. Soc. *58*: 2442
8. Müllen K (1984) Chem. Rev. *84*: 603
9. Streitwieser A Jr, Suzuki S (1961) Tetrahedron *16*: 153

10. Lawler RG, Ristagno CV (1969) J. Am. Chem. Soc. *91*: 1534
11. (a) Ristagno CV, Lawler RG (1973) Tetrahedron Lett. 159
    (b) Müllen K (1976) Helv. Chim. Acta *59*: 1357
12. Gerson F, Müllen K, Vogel E (1971) Angew. Chem. Int. Ed. Engl. *10*: 920
13. Paul DE, Lipkin D, Weissman SI (1956) J. Am. Chem. Soc. *78*: 116
14. Jensen BS, Parker VD (1975) ibid. *97*: 5211
15. Jensen BS, Parker VD (1974) J.C.S. Chem. Commun. 367
16. Bard AJ, Faulkner L 1980 Electrochemical Methods, Fundamentals and Applications, Wiley, New York
17. (a) Heinze J (1984) Angew. Chem. Int. Ed. Engl. *23*: 831
    (b) Mortensen J, Heinze J (1985) Tetrahedron Lett. *26*: 415
18. (a) Hoijtink GJ, Zandstra PJ (1960) Mol. Phys. *3*: 371
    (b) Buschow KHJ, Hoijtink GJ (1964) J. Chem. Phys. *40*: 2501
    (c) Eleoranta J, Linschitz H (1963) ibid. *38*: 2214
19. Frim R, Rabinovitz M, Muszkat KA (1987) J.C.S. Chem. Commun., in the press
20. Streitwieser A Jr 1961 MO Theory for Organic Chemists, J. Wiley, New York
21. Klein J (1983) Tetrahedron Report No. 152, Tetrahedron *39*: 2733 and references cited therein
22. Cohen Y, Klein J, Rabinovitz M (1986) J.C.S. Chem. Commun. 1071
23. (a) Bates RB, Gosselink DW, Kaczynski CJ (1969) Tetrahedron Lett. 205
    (b) Klein J, Brenner S (1969) J. Am. Chem. Soc. *91*: 3094
    (c) Klein J, Brenner S (1970) Tetrahedron *26*: 5807
    (d) Klein J, Medlik A (1973) J.C.S. Chem. Commun. 275
24. (a) Heiszwolf GJ, van Drunen JAA, Kloosterziel H (1969) Recl. Trav. Chim. Pays Bas *88*: 1377
    (b) Kloosterziel H, van Drunen JAA (1970) ibid. *89*: 270
    (c) Streitwieser A, Jr, Chang CJ, Reuben DM (1972) J. Am. Chem. Soc. *94*: 5730
    (d) Streitwieser A, Jr, Murdoch JR, Hafelinger G, Chang CJ (1973) ibid. *95*: 4248
    (e) Lochmann L, Pospisil J, Lim D (1967) Tetrahedron Lett 257
    (f) Schlosser M, Hartmann J (1973) Angew. Chem. *85*: 544
25. Harvey RG, Cho H (1974) J. Am. Chem. Soc. *96*: 2434
26. (a) Young RN (1979) Progress Nucl. Mag. Reson. Spectrosc. *12*: 261
    (b) O'Brian DH (1980) in: Comprehensive Carbanion Chemistry, Buncel E, Durst T (Eds), Elsevier Amsterdam Part A, pp. 271
27. (a) O'Brian DH, Hart AJ (1975) J. Am. Chem. Soc. *97*: 4410
    (b) O'Brian DH, Russel CR, Hart AJ (1979) ibid. *101*: 633
28. Günther H (1980) NMR Spectroscopy, J. Wiley Chapter 6, pp. 67–70
29. (a) Fraenkel G, Carter RE, MacLean A, Richards JH (1960) J. Am. Chem. Soc. *82*: 5846
    (b) Farnum DG (1975) Gold V, Bethel D (Eds) Adv. Phys. Org. Chem. Vol. 11, p. 123, Academic Press, London
30. Schaefer R, Schneider WG (1963) Can. J. Chem. *41*: 966
31. Spiesecke H, Schneider WG (1961) Tetrahedron Lett. 468
32. (a) Forsyth DH, Spear RJ, Olah GA (1976) J. Am. Chem. Soc. *98*: 2512
33. Strub H, Beeler AJ, Grant D, Michl J, Cutts PW, Zilm KW (1983) ibid. *105*: 3333
34. Linder M, Hohemner A, Ernst RR (1974) J. Mag. Res. *35*: 379
35. Minsky A, Meyer AY, Rabinovitz M (1982) Tetrahedron Lett. *23*: 5351
36. Minsky A, Meyer AY, Rabinovitz M (1985) Tetrahedron *41*: 785
37. (a) Minsky A, Meyer AY, Rabinovitz M (1983) Angew. Chem. Int. Ed. Engl. *22*: 45
    (b) Minsky A, Meyer AY, Poupko R, Rabinovitz M (1983) J. Am. Chem. Soc. *105*: 2164
    (c) Cohen Y, Rabinovitz M, to be published
38. Breslow R (1973) Acc. Chem. Res. *6*: 393
39. (a) Mallion RB (1973) Molec. Physics *25*: 1415
    (b) Haigh CW, Mallion EB (1980) Progress, Nucl. Mag. Res. *13*: 303; see ref. 50
40. Karplus M, Pople JA (1963) J. Chem. Phys. *38*: 2803
41. Eliasson B, Edlund U, Müllen K (1986) J. Chem. Soc., Perkin II, 937
42. (a) Haddon RC, Haddon VR, Jackman LM (1970) Fortschr. Chem. Forsch. *16*: 103
    (b) Musher JI (1965) J. Chem. Phys. *43*: 3081

43. (a) Pople JA, Schneider WG, Bernstein HJ 1959 High Resolution Nuclear Magnetic Resonance, McGraw-Hill, New York
(b) Pople Ja, Untch KG (1966) J. Am. Chem. Soc. *88*: 4811
44. (a) Günther H (1980) NMR Spectroscopy, J. Wiley, N.Y., p. 77–79
(b) Benn R, Günther H (1983) Angew. Chem. Int. Ed. Engl. *22*: 350
45. (a) Emsley JW, Feeney J, Sutcliffe LH 1966 High Resolution NMR Spectroscopy, Pergamon Press, Oxford
(b) Jackman LM, Sondheimer F, Amiel Y, Ben Efraim DA, Gaony Y, Wolovsky R, Bothner-By AA (1962) J. Am. Chem. Soc. *84*: 4307
46. Pawliczek JB, Gunther H (1970) Tetrahedron *26*: 1755
47. Sondheimer F, Calder IC, Elix JA, Gaoni Y, Garratt PJ, Grohmann K, de Maio G, Sargent MV, Wolovsky R (1967) Chem. Soc. Spec. Publ. No. 21, 75
48. Sondheimer F (1972) Acc. Chem. Res. *5*: 81
49. (a) Boeckelheide V (1973) in: Nozoe T (Ed) Topics in Nonbenzenoid Aromatic Chemistry, Hirkawa, Tokyo
(b) Boeckelheide V (1975) Pure Appl. Chem. *44*: 751
50. Mallion RB (1980) ibid. *52*: 1541
51. (a) Vogel E (1967) Chem. Soc. Spec. Publ. No. 21, 113
(b) Vogel E (1971) Pure Appl. Chem. *28*: 355
(c) Vogel E (1980) Israel J. Chem. *20*: 215
52. (a) Schröder G (1975) Pure Appl. Chem. *44*: 925
(b) Abraham R (1961) J. Mol. Phys. *4*: 145
53. (a) Carrington A, McLachlan AD 1967 Introduction to Magnetic Resonance, Harper and Row London, p. 57
(b) Ramsey NF (1950) Phys. Rev. *78*: 699
(c) Karplus M, Pople JA (1963) J. Chem. Phys. *38*: 2802
54. (a) Gerson F 1970 High Resolution ESR Spectroscopy, Wiley, New York
(b) Wert JE, Bolton JR 1972 ESR Elementary Theory and Practical Applications, McGraw-Hill, New York
55. McLachlan AD (1961) Mol. Physics *3*: 233
56. McConnell HM (1956) J. Chem. Phys. *24*: 632
57. McConnell HM, Chesnut DB (1958) ibid. *28*: 107
58. Dewar MJS, De Llano C (1969) J. Am. Chem. Soc. *91*: 789
59. Tuttle TR, Weissman SI (1958) ibid. *80*: 5342
60. Fürderer P, Gerson F, Rabinovitz M, Willner I (1978) Helv. Chim. Acta *61*: 2981
61. Fürderer P, Gerson F, Hafner K (1978) ibid. *61*: 2974
62. Setzer WN, Schleyer PvR (1985) Advances in Organometallic Chemistry, Vol. 24, 373
63. Brooks JJ, Rhine WE, Stucky GD (1972) J. Am. Chem. Soc. *94*: 7396
64. Rhine WE, Davis J, Stucky GD (1975) ibid *97*, 2079
65. Heinze J (1984) Angew. Chem. Int. Ed. Engl. *23*: 831
66. Holy NL (1974) Chem. Rev. *74*: 243
67. Heilbronner E, Bock H 1976 The HMO Model and Its Applications, Wiley, New York
68. Hoijtink GJ (1970) Adv. Electrochem. Eng. *7*: 221
69. (a) Dietrich M, Mortensen J, Heinze J (1985) Angew. Chem. Int. Ed. Engl. *24*: 508
(b) Hebert Mazaleyrat JP, Welvart Z (1985) Nouv. J. Chim. *9*: 75
70. (a) Stevenson GR, Zigler SS, Reiter RC (1981) J. Am. Chem. Soc. *103*: 6057
(b) Stevenson GR, Schock LE, Reiter RC (1983) J. Phys. Chem. *87*: 4004
(c) Stevenson GR, Hashim RT (1986) ibid. *90*: 2896 and references cited therein
71. Cope AC, Hochstein FA (1950) J. Am. Chem. Soc. *72*: 2515
72. Winstein S, Clippinger E, Fainberg AH, Robinson GC (1954) ibid. *76*: 2597
73. Sadek H, Fouss RM (1954) ibid. *76*: 5905
74. Smid J (1972) Angew. Chem. Int. Ed. Engl. *11*: 112
75. Murdoch JR, Streitwieser A, Jr (1973) Intrascience Chem. Rep. *7*: 43
76. Hogen-Esch TE (1977) Advances in Physical Organic Chemistry, Gold V, Bethel D (Eds), Vol. 15, p. 153, Academic Press, London
77. Stevenson GR, Schock LE, Reiter RC (1984) J. Phys. Chem. *88*: 5417

78. (a) Fry AJ, Chung LL, Boekelheide V (1974) Tetrahedron Lett. 445
(b) Cox RH, Harrison LW, Austin WK (1973) J. Phys. Chem. *77*: 200
(c) Strauss HL, Katz TJ, Fraenkel GK (1963) J. Am. Chem. Soc. *85*: 2360
79. (a) Smentowski FJ, Stevenson FR (1967) ibid. *89*: 5120
(b) Smentowski FJ, Stevenson GR (1969) J. Phys. Chem. *73*: 340
(c) Allendoerfer RD, Rieger PH (1965) J. Am. Chem. Soc. *87*: 2336
(d) Hogen-Esch TE, Smid J (1965) ibid. *87*: 669
(e) Hogen-Esch TE, Smid J (1966) ibid. *38*: 307
(f) Chan LL, Smid J (1967) ibid. *89*: 4547
80. Edlund U (1977) Org. Mag. Res. *9*: 593
81. Edlund U (1979) ibid. *12*: 661
82. (a) Eliasson B, Edlund U (1983) ibid. *21*: 322
(b) Ahlbrecht H, Schneider G (1986) Tetrahedron *42*: 4741
(c) Willner I, Becker JY, Rabinovitz M (1979) J. Am. Chem. Soc. *101*: 395
83. (a) Eliasson B, Edlund U (1983) J.C.S. Perkin II, 1837
(b) Becker BCh, Huber W, Schneiders C, Müllen K (1983) Chem. Ber. *116*: 1573
(c) Eliasson B, Johnels D, Wold S, Edlund U (1982) Acta Chem. Scand. B *36*: 155
(d) Edlund U, Sygula A, unpublished results
84. (a) Streitwieser A, Jr, Swanson JT (1983) J. Am. Chem. Soc. *105*: 2502
(b) Minsky A, Meyer AY, Hafner K, Rabinovitz M (1983) ibid. *105*: 3975
85. (a) Minsky A, Rabinovitz M (1981) Tetrahedron Lett. *22*, 5341
(b) Mandler D (1986) M.Sc. Thesis, The Hebrew University of Jerusalem
86. (a) Aromaticity, Chemical Society Special Publications No. 21, London (1967)
(b) Breslow R (1965) Chem. Eng. News *43*: 90
(c) Garratt PJ 1971 Aromaticity, McGraw-Hill, London
(d) Snyder JP (Ed) Non-Benzenoid Aromatics, Vol. I, II, Academic Press, New York, 1969, 1971
(e) Lloyd DMG (1966) Carbocyclic Non-Benzenoid Aromatic Compounds, Elsevier Pub. Co. Amsterdam
(f) Rabinovitz M (Ed) (1980) Isr. J. Chem. *20*
(g) Herndon WC (1974) J. Chem. Educ. *51*: 10
87. Platt JR (1954) J. Chem. Phys. *22*: 1448
88. Randić M (1977) J. Am. Chem. Soc. *99*: 444
89. (a) Cresp TM, Sondheimer F (1975) ibid. *97*: 4412
(b) Rabinovitz M, Willner I (1980) Pure Appl. Chem. *52*: 1575
(c) Cresp TM, Sondheimer F (1977) J. Am. Chem. Soc. *99*: 194
(d) Agranat I, Hess BA, Schaad LJ (1980) Pure Appl. Chem. *52*: 1399
90. (a) Müllen K, Oth JFM, Engels HW, Vogel E (1979) Angew. Chem. Int. Ed. Engl. *18*: 229; (1979) *91*: 251
(b) Huber W (1983) Tetrahedron Lett. *24*: 3595
91. Katz TJ, Rosenberger M, O'Hara RK (1964) J. Am. Chem. Soc. *86*: 249; see ref. 61
92. Edlund U, Eliasson B (1982) J.C.S. Chem. Commun. 950
93. Cox RH, Terry HW, Harrison LW (1971) Tetrahedron Lett. *50*: 4815
94. Vogel E, Königshofen H, Wassen J, Müllen K, Oth JFM (1974) Angew. Chem. Int. Ed. Engl. *13*: 732; (1974) *86*: 777
95. Müllen K (1974) Helv. Chim. acta *57*: 2399
96. Vogel E, Engels Hw, Huber W, Lex J, Müllen K (1982) J. Am. Chem. Soc. *104*: 3729
97. Huber W, Müllen K, Schneiders C, Iyoda M, Nakagawa M (1986) Helv. Chim. Acta *69*: 949
98. Akiyama S, Iyoda M, Nakagawa M (1976) J. Am. Chem. Soc. *98*: 6410
99. Nakagawa M (1975) Pure Appl. Chem. *44*: 885
100. Lendvai T, Friedl T, Butenschön H, Clark T, de Meijere A (1986) Angew. Chem. Int. Ed. Engl. *25*: 718
101. Hafner K (1971) Pure Appl. Chem. *28*: 153
102. Dewar MJS 1969 The MO Theory of Organic Chemistry, McGraw-Hill, New York, pp. 152–190
103. (a) Elvidge JA, Jackman LM (1961) J. Chem. Soc. 856
(b) Coulson CA, Mallion RB (1976) J. Am. Chem. soc. *98*: 592
104. Longuet-Higgins HC 1967 Special Publication No. 21, The Chemical Society; pp. 109–111
105. Aihara J (1979) J. Am. Chem. Soc. *101*: 5913

106. Aihara J (1979) ibid. *101*: 558
107. Hine J (1966) J. Org. Chem. *31*: 1236
108. (a) Neumann G, Müllen K (1986) J. Am. Chem. Soc. *108*: 4105
(b) Becker BC, Neumann G, Schmickler H, Müllen K (1983) Angew. Chem. Int. Ed. Engl. *22*: 241
109. (a) Hafner K (1964) ibid. *3*: 165
(b) Hafner K (1963) Angew. Chem. *75*: 1041
(c) Edlund U, Eliasson B, Kowalewski J, Trogen L (1981) J.C.S., Perkin II, 1260
110. Katz TJ, Balogh V, Schulman J (1968) J. Am. Chem. Soc. *90*: 734
111. Stowasser B, Hafner K (1986) Angew. Chem. *98*: 477
112. Bockelheide V, Vick GK (1956) J. Am. Chem. Soc. *78*: 653
113. Huber W (1984) Helv. Chim. Acta *67*: 2582
114. Becker BCh, Huber W, Müllen K (1980) J. Am. Chem. Soc. *102*: 7803
115. Garratt PJ, Zahler R (1978) ibid. *100*: 7753
116. Boche G, Etzrodt H, Marsch M, Thiele W (1982) Angew. Chem. *94*: 141
117. Benken R, Finneiser K, v Puttkamer H, Günther H, Elliasson B, Edlund U (1986) Helv. Chim. Acta *69*: 955
118. Cremer D, Günther H (1972) Liebigs Ann. *763*: 87
119. Berris Bc, Hovakeemian GH, Lai YH, Mestdagh H, Vollhardt KPC (1985) J. Am. Chem. Soc. *107*: 5670
120. Rhine WE, Davis JH, Stucky GJ (1977) J. Organometal. Chem. *134*: 139
121. Neumann G, Müllen K (1986) Chimia *39*; 275
122. Cohen Y, Roelofs NH, Reinhardt G, Scott LT, Rabinovitz M (1987) J. Org. Chem. *52*: 4207
123. Scott LT, Reinhardt G, Roelofs NH (1985) J. Org. Chem. *50*: 5886
124. Trost BM, Kinson PL (1970) J. Am. Chem. Soc. *92*: 2591
125. Trost BM, Kinson PL (1975) ibid. *97*: 2438
126. Dagan A, Rabinovitz M (1976) ibid. *98*: 8268
127. Katz TJ, Yoshida M, Siew LC (1965) ibid. *87*: 4516
128. (a) Günther H, Günther ME, Mondeshka D, Schmikler H (1978) Liebigs Ann. Chem. 165
(b) Günther H, Shyoukh A, Cremer D, Frisch KH (1978) 150
129. Müllen K (1978) Helv. Chim. Acta *61*: 1296
130. Huber W, May A, Müllen K (1981) Chem. Ber. *114*: 1318
131. (a) Huber W, Müllen K (1986) Acc. Chem. Res. *19*: 300
(b) Heinz W, Langensee P, Müllen K (1987) J.C.S. Chem. Commun. 947
132. Paquette LA, Ewing GD, Traynor S, Gardlick JM (1977) J. Am. Chem. Soc. *99*: 6115
133. Staley SW (1981) in: Jones M, Moss RA (Eds) Reactive Intermediates, J. Wiley, New York, Vol. II, pp. 32–34
134. Willner I, Rabinovitz M (1977) J. Am. Chem. Soc. *99*: 4507
135. Rabinovitz M, Gazit A (1978) J. Chem. Res. (S) 438; (1978) J. Chem. Res. (M) 5152
136. (a) Staley SW (1981) in: Jones M, Moss RA (Eds) Reactive Intermediates, J. Wiley, New York, Vol. II, pp. 29–30
(b) Elliasson B, Edlund U, Rabinovitz M, to be published
137. (a) Hückel E (1931) Z. Phys. *70*: 204
(b) Hückel E (1931) ibid. *72*: 310
(c) Badger GM 1969 Aromatic Character and Aromaticity, CambridgePress, Cambridge
138. Clar E 1972 The Aromatic Sextet, Siely-Interscience, New York
139. Salem L 1969 The Molecular Orbital theory of Organic Chemistry, McGraw-Hill, New York
140. Ebert LB (1985) in: Ebert LB (Ed) Chemistry of Engine Combustion Deposits, Plenum Publishing Corp., New York
141. Parker VD (1976) J. Am. Chem. Soc. *98*: 98
142. (a) Breslow R (1982) Pure Appl. Chem. *54*: 927
(b) See ref. 54b, pp. 232–246
143. (a) Willigen HV, Broekhoven JAM, Boer E (1967) Mol. Phys. *12*: 533
(b) Sommerdijk JL, Boer E (1969) J. Chem. Phys. *50*: 4771
(c) Jesse RE, Bilven P, Prins R, Vroost JDW, Hoijtink G (1963) J. Mol. Phys. *6*: 633
(d) Broekhoven JAM, Sommerdijk JL, Boer E (1971) ibid. *20*: 993
(e) Glasbeek M, Vroost JDW, Hoijtink GJ (1966) J. Chem. Phys. *45*: 1852

(f) Glasbeek M, Visser AJW, Maas GA, Vroost JDW, Hoijtink GJ (1958) J. Chem. Phys. Lett. *2*: 312
144. Bauld NL, Welsher TL, Cessac J, Holloway RL (1978) J. Am. Chem. Soc. *100*; 6920
145. Mitchell RD, Bauld NL (1980) Isr. J. Chem. *20*: 319
146. (a) Hess BA, Schaad LJ (1971) J. Am. Chem. Soc. *93*: 305
(b) Hess BA, Schaad LJ (1971) ibid. *93*: 2413
(c) Randić M (1976) Chem. Phys. Lett. *38*: 68
(d) Gutman I, Milun M, Trinajstić N (1977) J. Am. Chem. Soc. *99*: 1692
(e) Illić P, Sinković B, Trinajstić N (1980) Isr. J. Chem. *20*: 258
147. Haddon RC (1979) J. Am. Chem. Soc. *101*: 1722
148. Pople JA, Untch KG (1966) ibid. *88*: 4811
149. Van Vleck JH 1932 Electric and Magnetic Susceptibilities, Oxford University Press, pp. 262–276
150. Paul DE, Lipkin D, Weissman SI (1956) J. Am. Chem. Soc. *78*: 116
151. Sternberg HW, Delle Donne CL, Pantages P, Moroni EC, Markby RE (1971) Fuel *50*: 432
152. Tuttle TR, Ward RL, Wiessman SI (1956) J. Chem. Phys. *25*: 189
153. Hendricks BMP, Canters GW, Corvaja C, de Beor JWM, de Boer E (1971) Mol. Phys. *20*: 193
154. Henrici-Olive G, Olive S (1964) Zeit, Phys. Chem. *43*: 340
155. Smid J (1965) J. Am. Chem. Soc. *87*: 655
156. Carnahan Jc, Closson WD (1972) J. Org. Chem. *34*: 4469
157. Gia HB, Jerome R, Teyssie Ph (1980) J. Organomet. Chem. *190*: 107
158. Bank S, Bockrath B (1971) J. Am. Chem. Soc. *93*: 430
159. Minsky A, Rabinovitz M (1983) Synthesis *6*: 497
160. Minsky A, Rabinovitz M (1984) J. Am. Chem. Soc. *106*: 6755
161. Fraenkel Y 1986 M.Sc. Thesis, The Hebrew University of Jerusalem
162. Glidewell C, Lloyd D (1986) Chimica Scripta *26*: 373
163. Glidewell C, Lloyd D (1984) Tetrahedron *40*: 4455
164. Glidewell C, Lloyd D (1987) J. Chem. Ed. (in press)
165. Ebert LB see reference 140; pp. 303
166. Randić M (1982) J. Phys. Chem. *86*: 3970
167. (a) Levine G, Sutphen C, Szwarc M (1972) J. Am. Chem. Soc. *94*: 2653
(b) Rabideau PW, Borkholder EG (1978) J. Org. Chem. *43*: 4283
(c) Sargent Gd, Cron JN, Bank S (1966) J. Am. Chem. Soc. *88*: 5363
(d) Sargent GD, Lux GA (1968) ibid. *90*: 7160
(e) Grast JF, Roberts Rd, Abeles BN (1975) ibid. *97*: 4925
(f) Guysten H, Horner L (1962) Angew. Chem. Int. Ed., Engl. *1*: 455
168. (a) Harvey RG (1970) Synthesis *2*: 161
(b) Harvey RG, Arzadon L (1969) Tetrahedron *25*: 4887
169. (a) Rabideau PW, Harvey RG (1970) J. Org. Chem. *35*: 25
(b) Rabideau PW, Wetzel DM, Husted CA, Lawrence JR (1984) Tetrahedron Lett. *25*: 31
170. (a) Müllen K, Huber W, Neumann G, Schneiders C, Unterberg H (1985) J. Am. Chem. Soc. *107*: 801
(b) Müllen K (1987) Angew. Chem. Int. Ed. Engl. *26*: 204
171. (a) Minsky A, Meyer AY, Rabinovitz M (1982) J. Am. Chem. Soc. *104*: 2475
(b) Minsky A, Klein J, Rabinovitz M (1981) ibid. *103*: 4586
172. Eliasson B, Lejon T, Edlund U (1984) J.C.S. Chem. Commun. 591
173. Ebert LB (1986) Tetrahedron *42*: 497
174. Lebedev YS, Sidorov AN (1981) Russ. J. Phys. Chem. (Engl. Transl.) *55*: 1220
175. Smith JG, Levi EM (1972) J. Organomet. Chem. *36*: 215
176. Kaban S, Smith JG (1983) Organometallics *2*: 1251
177. (a) Paquette LA, Hansen Jf, Kakihana R (1971) J. Am. Chem. Soc. *93*: 168
(b) Stevenson GR, Shock LE, Reiter Rc, Hanson JF (1983) ibid. *105*: 6078
(c) Buncel E 1980 in: Comprehensive Carbanion Chemistry, Elsevier, Amsterdam part A, p. 35
(d) Van Broekhoven JAM, Sommerdijk JT, De Beor E (1971) Mol. Phys. *20*: 993
178. Minsky A, Cohen Y, Rabinovitz M (1985) J. Am. Chem. Soc. *107*: 1501
179. Cohen Y, Meyer AY, Rabinovitz M (1986) ibid. *108*: 7039
180. (a) Rabinovitz M, Cohen Y (1987) Ebert L (Ed) in: ACS Advances in Chemistry Series No. 217, Chemistry of Polynuclear Aromatics

(b) Eisch JJ (1986) in: Preprints Div. of Fuel Chem., ACS *31(4)*: 798
(c) Rabinovitz M, Cohen Y (1986) in: Preprints Div. of Fuel Chem., ACS *31(4)*: 777
181. Tamarkin D, Cohen Y, Rabinovitz M (1987) Synthesis 196
182. Ebert LB, Mills DR, Scanlon JC, Symposium on Advances in Resid Upgrading, ACS Meeting Denver, Colorado, April 5–10, 1987, pp. 419
183. Sternberg HW, Delle Donne CL, Markby Re, Friedman S (1974) Ind. Eng. Chem./Process Res. Div. *13*: 433
184. Cohen Y, Klein J, Rabinovitz M (1985) J.C.S., Chem. Commun. 1033
185. Cohen Y, Klein J, Rabinovitz M (1987) J.C.S. Perkin 2 (in the press)
186. Cohen Y, Klein J, Rabinovitz M, J.C.S. Chem. Commun., in the press
187. Palmer MH, Kennedy SMF (1976) J.C.S. Perkin *2*: 81
188. Stille WC, Sreekumar C (1980) J. Am. Chem. Soc. *102*: 1201
189. Smith Jg, McCall RB (1980) J. Org. Chem. *45*: 3982
190. Müllen K, Unterberg H, Huber W, Wennerstrom O, Norinder U, Tanner D, Thulin B (1984) J. Am. Chem. Soc. *106*: 7514
191. Frim R, Raulfs FW, Hopf H, Rabinovitz M (1986) Angew. Chem., Int. Edit. Engl. *25*: 174
192. (a) Boekelhelde V (1980) Acc. Chem. Res. *13*: 65
(b) Hopf H, Klein-Schroth J (1982) Angew. Chem. *94*: 485; (1982) Angew. Chem. Int. Ed. *21*: 469
(c) Boekelheide V (1983) Top. Curr. Chem. *113*: 87
(d) Hopf H (1983) in: Keehn PM, Rosenfeld SM (Eds), Cyclophanes, Academic Press, N.Y., p. 21
(e) Heilbronner E, Yang ZZ (1984) Top. Curr. Chem. *115*: 1
(f) Gerson F (1984) ibid. *225*: 57
(g) Swann Rt, Hanson AW, Boekelheide V (1986) J. Am. Chem. Soc. *108*: 3324
193. (a) Hopf H, Mlynek C, El Tamany S, Ernst L (1985) ibid. *107*: 6620
(b) Swann RT, Hanson AW, Boekelheide V (1986) ibid. *108*: 3324
(c) Hopf H (1983) Naturwissenschaften *70*: 349
194. Haenel MW (1977) Tetrahedron Lett. 1273
195. Private communication from Professor K. Hafner — Bickert, P.: dissertation, Technische Hochschule Darmstadt 1983
196. Hopf H, Raulfs F-W, Schonburg D (1986) Tetrahedron *42*: 1655
197. Frim R, Rabinovitz M, Hopf H, Hucker J (1987) Angew. Chem. Int. Ed. Engl. *26*: 232
198. Clar E, Zander M (1957) J. Chem. Soc. 4616
199. Diedrich F, Staab HA (1978) Angew. Chem. *90*: 383
200. (a) Thulin B, Wennerström O (1976) Acta Chem. Scand. B *30*: 369
(b) Anker K, Lamm B, Thulin B, Wennerström O (1978) ibid. *32*: 155
201. (a) Huber W, Müllen K, Wennerström O (1980) Angew. Chem. *92*: 636
202. Tanner D, Wennerström O, Nordiner U, Müllen K, Trinks R (1986) Tetrahedron *42*: 4499
203. Nordiner U, Wennerström O, Wennerström H (1984) Tetrahedron Lett. 1397
204. Bieber W, Vögtle F (1977) Angew. Chem. *89*: 199
205. (a) Gerson F, Heckendorn R, Möckel R, Vögtle F (1985) H.C.A. *68*: 1923
(b) Gerson F, Huber W (1987) Acc. Chem. Res. *20*: 85

Methods of Analysis of the Relative Hydrophobicity of Biological Solutes

Boris Y. Zaslavsky[1] and Eldar A. Masimov[2]

1 Institute of Elementoorganic Compounds, Academy of Sciences of the U.S.S.R., Moscow 117813, USSR
2 Azerbaydzhan State University, Baku, USSR

Table of Contents

1 Introduction

It is generally known that the substances of natural biological origin and their synthetic analogs are the most effective biologically active agents. The choice of the best approach to the rational design of biologically active molecules is hampered by that the mechanisms of the biological activity of the substances are usually unknown. Modern drug design methods are based on the quantitative structure-activity relationship (QSAR) studies. When the QSAR analysis method is applied in drug design it is assumed that the relative biological potency of members of a set of analogs is due to the effect of the variable substituent on the chemical and physical properties of the molecule [1-3].

The major physical properties affecting biological potency of a molecule are supposed [1-3] to be: electronic, steric, and hydrophobic. Electronic properties influence the reactivities of the compounds and steric properties are clearly important in view of the steric selectivity of interactions between the compound and components of the biological system. Hydrophobic properties of a compound appear to affect its distribution throughout the body organs and tissues as well as the interaction of the compound with its target (receptor) in the biological system [1,2].

It is generally accepted at present [1,2] that the observed biological response is proportional to the number of receptor sites occupied, i.e., to the concentration of an agent in the 'receptor compartment' depending upon the selectivity of the agent distribution throughout the body regions. The interaction of the agent with its target in a given biological system can occur (and the electronic and steric properties of its molecules be realized) only if the agent gets to the target compartment. This seems to be the reason for the presence of the parameter describing the hydrophobic properties of drugs in the majority (up to 90%) of the QSAR correlation Equations [1-3].

Thus it seems obviously that the analysis of hydrophobic properties of substances of natural origin is of fundamental importance in several respects, namely with the view of development of a wider application of the QSAR approach in molecular pharmacology and biochemistry, for a more complete and better understanding of the mechanisms of the effects of drugs and natural compounds in biological systems, for development of the preparative separation techniques based on the differences in the hydrophobic properties of biological materials [5], etc.

This review deals with the possibilities and limitations of the methods available at present for studying hydrophobic properties of natural compounds as well as with some conceptual questions which have arisen recently from the results accumulated in the experimental studies of the hydrophobic character of biological compounds.

2 Main Concepts and Definitions

It is known [6] that the energy of a solute present in a solvent environment can be described as a sum of two distinct terms:

$$E = E_g + E_s \qquad (1)$$

where E_g is the energy of the inherent molecular motions of an isolated solute molecule which can be calculated by some quantum chemistry method; E_s is the energy of solvation which in turn (in the absence of specific interactions) can be presented as a sum of a number of contributions:

$$E_s = E_e + E_r + E_p + E_{vdW} + E_{cav} \quad (2)$$

where E_e is the energy of electrostatic interactions between the solvent and solute; E_r the energy of repulsion; E_p the polarisation energy; E_{vdW} accounts for the energy of van der Waals interactions between the solute and solvent; E_{cav} expresses the energy of formation of a cavity in the solvent to accommodate the solute molecule.

It should be emphasized that the above resolution of the E_s parameter according to Eq. [2] is rather arbitrary; the other formulations can also be used [6]. For example, E_r and E_p in some cases are included into the E_{vdW} term, in other cases E_p is combined with E_e, and so forth. The most generally used form of Eq. [2] seems to be the following simplified one:

$$E_s = E_e + E_{vdW} + E_{cav} \,. \quad (3)$$

According to the modern conceptions the E_s value is a measure of the lyophilic or lyophobic character of the solute, i.e., an index of the intensity of the solute-solvent interactions. Hydrophobicity and hydrophilicity of a solute are the particular case of the lyophobic and lyophilic character of the solute. The hydrophilicity and hydrophobicity are the measures of the intensity of molecular interactions of a solute or the surface of a solid phase with water in dispersed systems, the dispersing medium in which is water [7].

Hydrophilicity (as the lyophilicity in general) is specified by the value of the free energy of hydration (solvation) of a given compound or of a solid phase surface [7]. Hydrophobicity should be regarded as a small extent of hydrophilicity, since all substances possess the latter property to a certain degree [7]. Actually even the most hydrophobic pure hydrocarbon surface of paraffin absorbs water, i.e., it is hydrophobic only in the sense of being of the very slightly hydrophilic character [7]. The concept of hydrophilicity and hydrophobicity is applicable not only to the solid phases, for which it is the property of a surface, but also to single molecules, their fragments, atoms, and ions. Electrostatically charged and polar groups having a dipole moment are usually hydrophilic [7]. These groups increase the aqueous solubility of the molecules possessing such groups, whereas the hydrophobic fragments incorporated in the molecules decrease their solubility in water. Thus, solubility of a compound in water and nonpolar organic solvents is an overall result of the interactions of hydrophobic and hydrophilic groups of the compound molecule with a given solvent environment.

It should be emphasized that the value of the E_s term in Eq (1) can be calculated only within the framework of models based on the approximations of the classic or quantum mechanics; the E_s value cannot be determined experimentally. Therefore, in order to estimate the hydration energy of a given solute experimentally the free energy change for transferring the solute molecule from the pure solute phase to water [8], from the gas phase into water [9], or from one solvent to another one [10, 11] is examined.

As a result of an analysis of thermodynamic characteristics of the above types of transfer, some oversimplified definitions of hydrophobicity and hydrophilicity have appeared in the literature. For example, the term hydrophobic is often used for the compounds which are readily soluble in many nonpolar organic solvents and only sparingly soluble in water [12]. According to Tanford [12] the hydrophobicity of a solute is represented by the free energy of transfer of the solute between water and a nonpolar organic solvent. The sign of the corresponding value of the free energy change of transfer is indicative of the hydrophobic or hydrophilic character of the solute under study.

3 Water Structure and its Effect on Solutes Hydration

It follows from Eq. (2) or from its simplified form of Eq. (3) that the energy of solvation (hydration) E_s should be dependent, on the one hand, upon the properties of the solute and, on the other hand, upon the properties of the solvent (water). This obvious fact is for no apparent reason often ignored in the literature on the hydrophobic-hydrophilic properties of chemical compounds. In some cases it leads to an inadequate interpretation of the results obtained. This question will be dealt with below.

The literature on the properties of water is so abundant that a brief discussion of these properties in the present review would be quite superfluous but for one consideration, that is, the water properties in biological systems appear to differ depending on the type of the particular system and seem to differ from those of pure water [13]. This fact is fundamentally important when the methods for studying the hydrophobic character of biological solutes are considered and therefore it calls for a more thorough discussion.

It is known that the main specific property of pure liquid water as compared to the other solvents consists in its being highly structured. Innumerable models of the water structure, enumeration, and classification of these models are proposed in the literature (see, e.g., [14]). Some of these models such as the one suggested by Samoilov [15] or the one advanced by Nemethy and Scheraga [16] have greatly influenced the concepts of the water structure; some of the other models, as Naberukhin [17] wittily puts it, have demonstrated rather inexhaustible imagination of their authors. The basic limitation of all of these models is due to their qualitative character. It means [17] that the main statements and concepts of these models are in requirement of quantitative specification and that their quantification remains basically unsupported by the primary principles of statistical physics.

The main types of the water structure models according to the classification given by Angell and Rodgers [18] include:

a) simple, unrepentant, two-state models which imply two distinct species, ideally mixing;
b) generalized two-state models, according to which there exist two classes of oscillator, strongly hydrogen-bonded and weakly hydrogen-bonded, respectively;

c) quasilattice with broken bonds, i.e., an effective two-state model with bond states replacing molecular states;
d) continuum models with preferential exchange of strong bonds for weak bonds on increasing temperature (leaving intermediate strength bonds constant in population);
e) continuum models with continuous bond weakening on increases of temperature.

Various authors are giving more or less convincing arguments in favor or different models [13-18]. To us the most adequate seems to be the concept proposed by Naberukhin according to which "the main pattern of water structure is a uniform continuous four-coordinated irregular network of hydrogen bonds with considerable fluctuations of such features as the length of the O ... O bonds, bond angles, and the bonds' energy parameters" [17].

This concept agrees in particular with the main conclusions drawn by Beall [13] from numerous data on the state of water in biological systems. These conclusions [13] are as follows:

a) water in cells does not behave as if it was all a dilute solution; there is no measurable "ice-like" crystalline solid water in normal functioning cells; and the simplified model of a small fraction of bound water and a large fraction of bulk water in cells is not supported by current biophysical evidence; and
b) the most appropriate view of cell water, consistent with experimental data, is a distribution of states the shape of which (distribution) will be a function of the type of cell, its macromolecular composition, and its physiological state.

In a simplified form the above conclusions [13] can be summed up as follows: the structure and/or the state of water in a complex multicomponent aqueous system (which is an oversimplified physico-chemical concept of a biological system) differ from those of pure liquid water assumingly due to an influence of the system components on these water features. This conclusion seems to agree with the above concept of the water structure [17] since an introduction of a certain amount of an additive into water may distort not only local parameters of the hydrogen bonds network but the averaged ones as well due to the space continuity of the network.

In the literature on the effects of different solutes on water structure the nearest order of perturbation of the structure induced by ions [15,19,20], nonpolar [21,24], and amphiphilic [25] compounds is emphasized. To describe the process of structure promoting in aqueous solutions of nonpolar molecules and some ions, the terms of hydrophobic hydration [26] and positive hydration [15] have been proposed. To describe the reverse process, i.e., the destabilizing effect of ions and hydrophilic molecules on water structure, the terms of negative [15] and hydrophilic hydration [26] are used. Solvation interactions bringing about hydrophobic or hydrophilic (positive or negative) hydration can be described to a certain degree of approximation by means of the computer simulation techniques [19-24]. The structure-changing influence of solutes on the bulk water properties cannot be thoroughly examined at present by computer simulation studies. The aforementioned influence is experimentally studied by means of infrared spectroscopy [18], nuclear magnetic relaxation measurements [13], analysis of solubility of nonpolar solutes in aqueous solutions of various composition [26], partitioning in water-organic solvent biphasic systems [27-29], analysis of salt effects on solubility of macromolecules in aqueous solutions [30], analysis of the effect of various solutes on the structure temperature of water [31], etc. The main result of these studies seems to be the conclusion that the solute-water interactions appear to alter not only the structure of water vicinal to the solute molecule but that of bulk water as well. The correlation relationship established between the relative intensity of the macromolecule-water interactions and the relative affinity of an aqueous solution of a given macromolecule for a CH_2-group [32] seems to serve to illustrate the above conclusion.

It seems reasonable to assume that the water structure-changing influence of solutes should affect the energy of formation of a cavity in the solvent to accommodate the solute molecule, i.e., the value of the term E_{cav} in Eq. (2). Melander and Horvath [30] examined particularly the relationship between the effect of inorganic salts on the surface tension of the aqueous medium and the free energy change for formation of a cavity of a given size in the solvent. Experimental results obtained by Masimov et al. [27–29,32] on the effect of macromolecules on the relative affinity of their aqueous solutions for a CH_2-group (i.e. on the relative hydrophobic character of the aqueous medium) appear to imply that the energy of formation of a cavity and the ability of water to participate in van der Waals interactions with a solute, i.e., the values of the terms E_{cav} and E_{vdW} in Eq. (2), depend upon the chemical composition of an aqueous solution. The contribution of the energy of electrostatic interactions between the solvent and solute E_e to the overall hydration energy E_s in Eq. (2) (or in Eq. (3)) seems to depend upon the presence of electrolytic additives [30] and upon the thermodynamic state of water dipoles [20] which is presumably dependent upon the components of the solution modifying the structure and/or the state of water in the solution.

It follows, therefore, that the hydration energy E_s (or the hydrophobicity or hydrophilicity, which is one and the same) of a solute depends not only on the chemical nature and structure of the solute molecule, but also on the structure and state of water in a given aqueous medium, the latters being governed at a given temperature by the chemical composition of the medium.

Since the present review deals with the methods of analysis of the relative hydrophobicity of biological molecules, the role of the above dependence of the state of water in biological systems on the composition of these systems in biological processes should be briefly considered.

According to Tanford [33] the dynamic processes of life occur within an organized structural framework that turns over slowly or not at all. It is suggested [33] to consider this framework as being essentially at equilibrium. The hypothesis advanced by Tanford [33] consists of that distribution of a biological molecule between various compartments (phases) or places in the biological system is governed by the difference in the chemical potential values of the molecule in these compartments. In other words, the concept suggested [33] amounts to that the concentration of a solute in a given compartment of the biological system in the absence of specific interactions depends on the affinity of the solute for the compartment medium, i.e., for the aqueous medium of the specific chemical composition. The concept under discussion implies that the differences in the state and/or structure of water in various compartments (phases) of the biological system caused by their different chemical composition [13] would affect the affinity of biological solutes for the compartments and hence, distribution of the solutes between organs, tissues, and subcellular structures of the living body.

This principle as considered in detail by Zaslavsky et al. [34] is likely to be at the root of an arrangement and functioning of blood-tissue barriers of a living organism, to be central to the mechanism of the counterpoison effects of such synthetic polymers as polyvinylpyrrolidone, polyvinyl alcohol, etc. Similar considerations, although not explicitly formulated seem to be underlying the thermodynamic equilibrium concept of the relationship between the biolocical potency and the relative hydrophobicity of drugs proposed by Higuchi and Davis [35]. The above principle was used by Zaslavsky et al. [36] in order to formulate the physical meaning of the term "the hydrophobic character of a biological liquid or tissue" and to evaluate the difference in the relative hydrophobic character of several tissues and fluids of rat.

The main conclusion from the above experimental facts and theoretical concepts to be emphasized is that the hydration energy or the hydrophobicity (hydrophilicity) of a solute depends upon the chemical composition of a given aqueous medium. It seems reasonable to assume that this dependence plays a regulating role in living systems and therefore it should be taken into account while studying the hydropobicity of biological molecules.

4 Study of Transfer of a Solute from Water to an Organic Solvent

It has been noted above that according to simplified definition of hydrophobicity (or hydrophilicity) of chemical compounds [12] the hydrophobic character of a solute can be measured by the free energy of transfer of the solute from water to a nonpolar environment of an organic solvent. To quantify the free energy value, one of the three following methods are usually employed: analysis of comparative solubility of compounds in water and in organic solvent; partitioning of compounds in water- organic solvent biphasic systems; and partition chromatography.

4.1 Analysis of Comparative Solubility of a Solute in Water and in Organic Solvents

An employment of measurements of comparative solubility of a substance in water and in an organic solvent to estimate the relative hydrophobicity of the substance is based on the general Gibbs equilibrium condition which is that the chemical potential μ_i of a given solute must have the same valuc in the saturated solution of the solute and in the phase of the pure solute. It follows from this condition that the equation describing the free energy change for transferring the solute from water into an organic solvent is:

$$\Delta G_{w \to s} = \mu_s^0 - \mu_w^0 = RT \ln (f_w/f_s) - RT \ln (C_s/C_w) \qquad (4)$$

where μ^0 is the standard chemical potential of a solute; C the solubility of the solute, i.e., the molar concentration of the solute in the saturated solution; f the activity coefficient of the solute; indices "w" and "s" denote water and organic solvent, respectively.

When the solubility of a compound in water and in the organic solvent is sufficiently low, the activity coefficients f_w and f_s are close to unity, and Eq. (4) becomes:

$$\Delta G_{w \to s} = -RT \ln (C_s/C_w) \tag{5}$$

The simplest case of the solution process is the dissolution of a nonpolar solute in water. Theoretically this process can be divided into three hypothetical steps [37-40]: the removal of the solute molecule from its initial environment; the formation of a cavity in water to accommodate the solute molecule; and the introduction of the solute molecule into the cavity.

When the factors affecting each of the above steps of the solution process are considered, it seems clear that the two latter steps should depend on the size and the effective surface area (or volume) of the solute molecule, and on the magnitude of the molecular solute-water and water-water interaction energies. The water structure-perturbing effects of various additives have been discussed above. It is therefore evident that the solubility of a compound in water and in an aqueous solution of a salt or some other solute may differ. This should be particulary taken into account while studying the water solubility of readily solube compounds [41] as the saturated aqueous solution of such a compound should be regarded as the aqueous medium, the structure of water in which has been modified by the dissolved compound (even assuming the absence of the solute-solute interactions).

The first step of the solution process, i.e., the removal of a solute molecule from its original environment appears to depend upon the intensity of intermolecular interactions in the pure phase of the solute [42-44]. Amidon et al. [42] have examined the aqueous solubilities of various aliphatic hydrocarbons, olefins, alcohols, ethers, ketones, aldehydes, esters, and fatty acids together with their molecular surface areas. The results obtained [42] indicate in particular that the functional group contributions to the free energy of solution in water are nearly equivalent from the pure solute standard state while being significantly different when the gas phase (1 mm Hg) standard state is chosen. For this and other aforementioned reasons the parameters characterizing the difference in the solubility of solutes in water and in an organic solvent and not just the aqueous solubilities are used to estimate the hydrophobic character of solutes.

The estimates of the relative hydrophobicity of solutes obtained by measurements of their comparative solubility in water and in organic solvents are usually in agreement with those obtained by the partition technique [45,46] (see below).

Numerous efforts [38,40,42,46-51] were undertaken to find out in what manner the aqueous solubility and the partition coefficients of different solutes in water-organic solvent biphasic systems are related to the size of the solute molecules. Since the solute packing into the solvent clearly depends on the solute surface, a relationship between surface area and solution thermodynamics seems to be reasonable.

The relationship between the aqueous solubility of a homological set of solutes and their molecular surface area is described as:

$$-\ln C_w = b_0 \cdot F_m + w_0 \tag{6}$$

where F_m is the solute molecular surface area accessible for the solvent; C_w the aqueous solubility of the solute; b_0 and w_0 are constants.

Hermann[38)] has estimated the coefficient b_0 values for a series of alkanes and cycloalkanes and for a number of alkylbenzenes. The b_0 values are 33 $Å^{-2}$ for alkanes and cycloalkanes and 30 $Å^{-2}$ for aromatic systems[38)]. From the data obtained by Amidon et al.[42)] it appears that the b_0 value for monofunctional aliphatic alcohols, ethers, aldehydes, ketones, and fatty acids is constant and amounts to 22.6 $Å^{-2}$. A similar b_0 value of 22.0 $Å^{-2}$ has been found by Chotia[47)] for the side chains of nonpolar amino acids — those of alanine, valine, leucine and phenylalanine. For the side chains of serine, threonine, histidine, methionine, and (for no apparent reason) tryptophan, the coefficient b_0 value is ca. 13–15 $Å^{-2}$[47)].

Two essential issues from the above results should be emphasized here. Firstly, the aqueous solubility of solutes of similar chemical nature, i.e., the free energy of transfer of a solute from the pure solute phase into water, is linearly related to the solute molecular surface area. Secondly, it appears that the value of the b_0 coefficient decreases with an increase in the hydrophilicity of the solutes under study. It seems to indicate that hydration interactions accompanying transfer of an amphiphilic molecule into water[52)] are likely to oppose to some degree the stabilizing effect of the hydrophobic fragment of the molecule on the vicinal water structure. This assumption agrees well, among other things, with the behavior of amphiphilic solutes in water-organic solvent biphasic systems[53)] (see below). It remains unclear, however, whether the differences in the b_0 values are due to the hydration effect or to the different intensity of the intermolecular interactions in the pure solute phase.

The additional complication involved in the employment of the method under discussion in the study of the hydrophobicity of solutes is due to the problem of the appropriate choice of an organic solvent to serve as a nonaqueous medium. The data reported by Nozaki and Tanford[41)] and those by Fendler et al.[54)] on the solubility of various amino acids in water, aqueous dioxane and ethanol solutions, and in *n*-hexane provide a typical example of the difficulties accompanying the aforementioned choice. Nozaki and Tanford[41)] measured the solubilities of different amino acids in water and in progressively increasing concentrations of ethanol and dioxane in water. The solubilities of the amino acids were extrapolated to pure organic solvents and the free energy of transfer for the amino acid from pure solvent to water was calculated. Using glycine as a reference, and subtracting its free energy of transfer from that of all the other amino acids, the relative hydrophobicities of the side chains of the amino acids were estimated[41)]. A similar approach was employed by Fendler et al.[54)] using *n*-hexane as the organic solvent. The values of the free energy of transfer from ethanol and from dioxane to water were reported for five amino acids[41)]. For three of these amino acids (tryptophan, tyrosine, and histidine) the free-energy-of-transfer values for the side chains appear to be independent of the organic solvent used[41)]. For the phenylalanine side chain the values found[41)] differ for ethanol and dioxane within the experimental error range (ca. 100 cal/mole), whereas for the leucine side chain the values reported differ more than by 800 cal/mole. When the estimates of the relative hydrophobicity for the amino acid side chains reported by Nozaki and Tanford[41)] are compared with those for the same side chains reported by Fendler et al.[54)], it appears that the values in question agree within the experimental error range only for the side chains of three amino acids (valine, histidine, and phenylalanine) and differ considerably for those of five other amino acids (alanine, leucine, isoleucine, serine, and threonine). It seems likely that the above agreement of the estimates obtained using different organic solvents is accidental. The data on the solubility of adenine and thymine in water and in ethanol and *n*-propanol reported by Herskovits et al. [55] seem to support this conclusion. The difference between the hydrophobicity values of adenine and thymine measured by the free energies of transfer of the solutes from water into an organic solvent is 105 cal/mole when ethanol is used, but the same difference amounts to 250 cal/mole if *n*-propanol is used as a nonaqueous medium [55]. Shruggs et al. [56] observed particulary that the solubility of adenine in chloroform is markedly affected by the presence of water in the solvent.

Thus, the main limitations of the method of study of the hydrophobicity of solutes based on measurements of the solutes comparative solubility in water and in an organic solvent are:

1. The method can be employed only for the study of the compounds possessing moderate solubility both in water and in organic solvents.
2. Validity of a choice of a given organic solvent to be used as nonpolar medium is usually open to objection.
3. The method cannot be used to study labile biological solutes (proteins, nucleic acids, etc.), the intact features of which are affected by organic solvents.
4. The method can be used to study only the relative hydrophobicity of solutes of similar chemical nature.

Since the first of the above limitations can be bypassed with an approach based on the study of partitioning of solutes in water-organic solvent biphasic systems, most of the hydrophobicity estimates for chemical compounds have been obtained by this method.

4.2 Partitioning of Solutes in Water-Organic Solvent Biphasic Systems

When a solute distributes at constant temperature between two solvents, which are immiscible or partially miscible, there exists the equality of the chemical potentials of the solute in the two phases. This situation is described by Eq. (5) in the form:

$$\Delta G_{w \to s} = -RT \ln (c_s/c_w) = -RT \ln P \qquad (7)$$

where P is the partition coefficient of the solute in a given water-organic solvent biphasic system, c_s and c_w represent the equilibrium concentrations of the solute in the organic solvent phase and in the aqueous phase of the system, respectively.

When the partitioning of a solute in a biphasic system is to be considered in terms of thermodynamics, two conditions are postulated [57]: a) both phases of the system are regarded as immiscible liquids; and b) the behavior of the solute partitioned in the system is regarded as the ideal one. Naturally, neither of the two conditions is fulfilled in reality.

Since the estimation of the hydrophobicity of a solute requires the determination of the difference between the chemical potentials of the solute in both phases, it is clear that the partition coefficient value should be measured for the solute molecules being in the same form in both phases, i.e., for the nonprotonated or the ionized monomeric species. The basic methods of corrections for the degree of association of solute molecules in the phases of the partitioning systems are discussed in detail in review [53]. Various methods of corrections for the effect of the ionization degree of a solute on its partition coefficient have been suggested by Alhaider et al. [58] and by Martin [59].

The effect of the pH and ionic composition of the aqueous phase of the solvent system on the estimates of the hydrophobic character of a number of solutes have been shown by Wang and Lien [60]. It has been particularly shown that even the partition coefficients of nonionic solutes depend upon the type of buffer used as an aqueous phase in the *n*-octanol-buffer system. It seems possible to explain the effects of the ionic composition of the aqueous phase on the partitioning of solutes in the above biphasic system [60] by the influence of the ions on the state and/or structure of water in the phase of the system.

When the relative hydrophobicity of solutes is estimated by the partitioning in the solvent system, one is faced with the aforementioned problem arising when measuring the comparative solubility of substances in water and in an organic solvent, that is which solvent should be used to simulate a nonpolar medium.

In terms of the discussion in the preceding section, ideally a hydrocarbon solvent such as *n*-hexane should be used to measure the relative hydrophobicity of a solute [30]. Such solvents, however, suffer from the disadvantage that most polar organic compounds are essentially insoluble in them with the result that partition coefficients cannot be measured with sufficient accuracy to be useful. Additionally, when a polar molecule does dissolve in such solvents, it brings with it water molecules. Dissolved molecules also tend to associate with each other rather than with the solvent in hydrocarbon solvents. The net result is that if hydrocarbon solvents are used for partition coefficient measurements, the organic phase contains several different species of solute. Hence, the measured partition coefficient cannot be easily interpreted in terms of fundamental molecular interactions [2,61]. One should also bear in mind that the features of nonaqueous phases in biological systems differ from those of hydrocarbon solvents. These phases as a rule contain considerable amounts of water linked with the polar and ionized groups of biological molecules present in the phases [61].

A number of more polar organic solvents have been used as a model nonaqueous phase: diethyl ether, chloroform, olive oil, oleyl alcohol, *n*-octanol, *n*-butanol, etc. [10,43,53,61]. When choosing an organic solvent to simulate a nonpolar medium in a partitioning system, one should take into account the following [10,53,61]: a) the mutual solubility of water and the solvent; b) the solvation capacity of a solvent in relation to the solute being partitioned; c) the hydrogen bond-donating and acceptor properties of a solvent.

The most generally used solvent at present seems to be 1-octanol [1-3,10,53,61]. Both because of its hydroxyl group and the relatively high concentration of water (2.3 M at saturation), octanol appears to be an appropriate solvent for most organic compounds. Water-saturated octanol is sufficiently polar so that dissolved molecules tend to associate with the solvent rather than with each other. It has a regular structure which is not changed by the addition of solute [61]. Additionally, octanol is chemically stable, commercially available, non-volatile, and it does not absorb ultraviolet light. All these characteristics are of practical importance. The use of *n*-octanol is preferable as compared to that of other alcohols primarily owing to the fact that historically it has been the *n*-octanol-water biphasic system in which the partition coefficients of a vast number of chemical compounds has been measured in order to study their relative hydrophobicity [10,53].

It should be noted that the very range of parameters considered when choosing an organic solvent for a partitioning system [53,61] indicates that the aforementioned ideal conditions [57] are completely ignored as unrealistic. The parameters considered are essential for postulating some physical model of the partitioning process. Such a model [10] agrees with the above theoretical concept of the solution process, but differs in that instead of the pure solute phase the solution of the solute in an organic solvent is considered. According to this model [10], transfer of a solute molecule from a nonaqueous phase into an aqueous one is simulated by a cavity-to-cavity (or "hole-to-hole") transfer, the process being dependent upon the difference in the free energies required to form an appropriate cavity in both phases of the system. Rekker [10] suggests that an adequate description of the solute partitioning in a solvent biphasic system should take into account not only the size of the solute molecule but primarily the structural features of the media in both phases of a given system.

It should be acknowledged that the "hole-to-hole" concept advanced by Rekker [10] seems to be the most adequate one to account for the relationship known to exist between the aqueous solubility of solutes and their partition behavior in the solvent systems [8] as well as for the differences observed in the partitioning of solutes in various solvent biphasic systems [10,43,53,60,61]. The validity of this concept appears to be supported by the data reported by Leo et al. [51] on the effect of the chemical nature of nonpolar solutes on the relationship between their partition coefficients in the octanol-water system and the molecular surface area of the solutes. The above concept [10] also seems to be in line with the results obtained by Harris et al. [40] indicating that the value of the free energy of transfer of polar solutes from water into an organic solvent depends upon the specific solute-solvent interactions to a much greater degree than it does upon the surface area of the solute molecule.

It is known [10,43,53,60–62] that the partition coefficients of the same solutes in various solvent biphasic systems are interrelated according to the so-called solvent regression equation:

$$\log P_i = a_i \cdot \log P_0 + b_i \tag{8}$$

where P_i and P_0 are the partition coefficients of a given solute in the i-th biphasic system and in the system chosen for reference, respectively; a_i and b_i are constants. The *n*-octanol-water system is usually chosen as the reference system.

Leo et al. [53,62] compared twenty different partitioning systems with the octanol-water system and derived correlations of Eq. (8) type from the results obtained. It was found [53,62] that Eq. (8) is of the same form in all the cases, but the a_i and b_i coefficient values differ for a given pair of the systems under comparison depending upon the nature of the solutes being partitioned. It has been suggested [53,62] to divide all the solutes into so-called donor and acceptor compounds. The differentiation of the solutes into donor and acceptor compounds improves the statistics and two equations of Eq. (8) type are derived [10,53,62].

It should be noted that an examination of the data on the a_i and b_i coefficient values reported by Rekker [10] and by Leo et al. [53,62] indicates that the discrepancy between the a_i values for donor and acceptor solutes in a given pair of solvent system does not exceed experimental errors in the corresponding a_i values for the two 'classes' of solutes. The discrepancy between the b_i values in the same equations seems to be much greater. Regarding the physical meaning of the a_i and b_i coefficients, Leo et al. [53,62] suppose that both coefficients represent the difference in the relative hydrophobic character of the organic solvents, but "to a different degree". The vagueness of such a conclusion seems to be self-evident.

Leo et al. [53,62] stress that the intercept b_i values of the solvent regression equations are clearly related to the extent to which water is dissolved in the organic phase of the partitioning system. They found [53,62] that the b_i coefficient value is related to the above parameter as follows:

$$\log [H_2O]_i = 1.077 \cdot b_i + 0.249$$

$$N = 17, \quad r = 0.979, \quad s = 0.217, \tag{9}$$

where $[H_2O]_i$ is the water concentration in the i-th organic solvent at saturation; N the number of data point used in deriving the equation; r the correlation coefficient; s the standard deviation from the regression.

When deriving Eq. (9), the authors [53,62] have used the b_i values from Eq. (8) type equations for solutes of basic nature. It should be emphasized that the solubility of water in an organic solvent represents only the features of the nonaqueous phase of the partitioning system. For a more complete description of the partitioning systems Recker [10] suggested to use the so-called "discriminating power" of the system which denotes the spread any given solvent system imparts to the partition coefficient values of a set of molecular structures presented to that system for partitioning. Davis et al. [43] suggested to measure the relative hydrophobic character of a solvent system by the free energy of the hypothetical transfer of a CH_2 group between the phases of the system, the value of which appears to vary from ca. 1000 cal/mole to 450 cal/mole CH_2 depending upon the type of the solvent system.

The differences in the discriminating power and in the free energy of interphase transfer of CH_2-group values observed in the different partitioning systems [10,43] seem to be due to the nonideal character of any such system, i.e., to mutual solubility of water and the solvent used in the system. The water-structure changes in the aqueous phase caused by an organic solvent are accompanied by some alterations of the relative hydrophobicity of this phase, and also the relative hydrophobic character of the nonaqueous phase is altered due to the presence of water in the latter. The data obtained by Wang and Lien [60] cited above seem to illustrate the changes in the properties of the octanol-water system induced by the presence of buffer salts in the aqueous phase of the system. This should be borne in mind when analyzing extensive reports [3,10,53] on the partition coefficients for various solutes in different solvent systems, since these reports do not usually distinguish between water-organic solvent and buffer-organic solvent systems.

An analysis of the published partition coefficients measured in various solvent systems for homologous series of fatty acids, aliphatic alcohols, and amines was performed by Zaslavsky et al. [63] in order to clarify the physical meaning of the a_i and b_i coefficients in the solvent regression Equation (8). The partition coefficients for each homologous series of solutes in a given solvent system were treated according to equation:

$$\ln P_i = A_i + E_i \cdot n \qquad (10)$$

where P_i is the partition coefficient of a given solute; n the number of CH_2 and CH_3 groups in the solute molecule alkyl chain; the subscript i denotes the solvent system; A_i and E_i are constants which represent the contributions of a hydrophilic polar group and of a CH_2 group to the $\ln P_i$ value, respectively. These constants are related to the free energies of the interphase transfer of the above groups of the solute according to equation identical to Eq. (7) (see above).

An analysis of the values of the constant E_i and of the P_i values in accordance to Eq. (8) in the modified form (using ln P instead of log P values) has shown [63] that:

$$a_i = E_i/E_0 = \Delta G_i^{CH_2}/\Delta G_0^{CH_2} \qquad (11)$$

where a_i and E are defined above; ΔG^{CH_2} is the free energy of the interphase transfer of a CH_2 group; indices "i" and 0" denote the i-th system and the reference system, respectively.

From Eqs. (8), (10), and (11) it follows that the b_i coefficient in the Eq. (8) type Equation is described as:

$$b_i = E_i(A_i/E_i - A_0/E_0) = A_i - a_i \cdot A_0 \qquad (12).$$

The relationship between a_i and b_i described by Eq. (12) seems to clarify why both coefficients are considered [10,53,62] to be slightly different measures of the relative hydrophobic character of the organic solvent in a given partitioning system.

The above considerations show that it is the values of the ratio A_i/E_i and not just the A_i values that should be compared to examine the estimates of the relative hydrophobicity of polar groups of solutes obtained when using different partitioning systems [63]. An analysis of the ratio A_i/E_i values for aliphatic alcohols, amines, and fatty acids, calculated from the known partition coefficients for these solutes in various solvent systems, indicated [63] that there is a relationship between the ratio value and the solubility of water in a given organic solvent. This relationship is described as:

$$(A_i/E_i)_j = q_{ij} + ß_{ij} \log [H_2O]_i \qquad (13)$$

where $[H_2O]_i$, A_i, and E_i are as defined above; q and ß are constants; index "i" denotes the organic solvent; index "j" denotes the type of the solutes being partitioned.

It has been found [63] that the relationship described by Eq. (13) for fatty acids, aliphatic alcohols, and amines when plotted as the A_i/E_i value against $\log [H_2O]_i$ are represented by the mutually intersected straight lines. The fact that the obtained lines intersect [63] means that the apparent hydrophobic character of one polar group with respect to the other may be reversed, depending on the particular partitioning system used. The results obtained by Zaslavsky et al. [63] imply that the relative hydrophobicity only of the solutes of the same chemical nature can be examined by the partitioning in water-organic solvent biphasic systems.

Thus, an employment of the partitioning of solutes in the water-organic solvent-system technique in studies of the hydrophobic character of solutes is .limited by that:

a) the estimates of the hydrophobic character of solutes appear to depend upon the choice of a particular partitioning system;
b) the estimates obtained can be used only for a comparative characterization of solutes of the same chemical nature; and
c) the method cannot be used to study labile biological solutes which are liable to denaturation or conformation changes induced by organic solvents.

4.3 Partition Chromatography

The use of paper and thin-layer chromotography (TLC) and high-pressure liquid chromatography (HPLC) for the estimation of the hydrophobis character of solutes has been extensively reviewed by Tomlinson [64] and other authors [65-68]. The use of the different versions of partition chromatography in the hydrophobicity studies is based on the fact that the chromatographic behavior of many solutes (characterized by retention time, R_M value, etc.) correlate well with the logarithms of the partition coefficients of the solutes in the octanol-water biphasic system [67,68]. The main advantages and limitations of the use of the chromatography in the study of the hydrophobic character of chemical compounds have been discussed at length in a number of reviews [64-68]; therefore, only the most important of these are briefly mentioned here.

Experimental advantages of the partition chromatography method over the methods discussed above are as follows: it is relatively fast and labor saving, and it allows one to work with rather impure compounds when only very small amounts of sample are available. The main drawback of the method is, in our opinion, the need to use an organic solvent or a mixture of solvents which restricts the applicability of the method for the study of many biological solutes and, as indicated above, leads to an ambiguity of the estimates of the relative hydrophobic character of polar organic compounds.

In summing up the present section of the review, it should be noted that transfer of a solute from water into a nonpolar solvent medium simulates roughly such biochemical processes involving transfer of a solute from an aqueous environment to a nonaqueous one as transfer of protcins from blood plasma to cellular membranes, processes of penetration of drugs through skin, binding of ligands with nonpolar sites in protein macromolecules, etc. From this point of view the studies of the thermodynamic quantities of transfer of solutes, particularly of biological origin, from water into an organic solvent are of both theoretical and practical importance.

It should be taken into account, however, that the approximation used is extremely rough. Firstly, the medium of a nonpolar organic solvent appears to be very inadequate model of the hydrophobic nonaqueous compartments (phases) in biological systems. Secondly, organic solvents employed for the commonly used partitioning systems are far from being inert to the solutes being partitioned. The effect of the solute-solvent interactions on the partition coefficient of the solute cannot be usually quantified.

Hence, the estimates of the hydrophobic character of solutes can be used only for a relative rating of the solutes of the same or very similar chemical nature. Additionally, the experimental methods considered in this section of the review cannot, as has been repeatedly noted above, be used in the studies of biological solutes, the intact properties of which may be altered by an organic solvent.

The above considerations prompted development of a number of special methods for studying the hydrophobic properties of biological compounds.

5 Study of Transfer of a Solute from Water to a Gaseous Phase

Since most biochemical processes in living systems occur in aqueous media, attempts have been made to develop a method to characterize "the absolute tendencies of solutes to leave water and enter a 'naive' or featureless cavity of unit dielectric constant that neither attracts nor repels the solutes" [9]. It is clear that these attempts are aimed at creating a method of direct estimation of the hydration energy E_s in Eq. (1) (see above).

The method considered extensively by Wolfenden [9] is based on measuring the dimensionless equilibrium constant for transfer of a substance from the dilute vapor phase, in which each molecule exists in virtual isolation, to an aqueous solution so dilute that each solute molecule is completely surrounded by water, and solute-solute interactions can be neglected. This can be accomplished by measuring solubilities of a gas under known pressure or, for less volatile compounds, by determining concentrations of solute in the gas space over solutions of known concentrations. In the cases of highly hydrophilic solutes, measured volumes of an inert carrier gas can be bubbled through an aqueous solution of known concentration, and then through an efficient trap that recovers the solute quantitatively from the vapor phase [9]. Specific methodical details of this technique are presented in the papers by Wolfenden et al. [69,70]. Some advantages and limitations of this method should be noted here.

From the viewpoint of studying the hydrophobic character of biological compounds, the important advantage of the approach under discussion seems to be that it excludes the need for an organic solvent. The possibility to use the method for the study of the relative hydrophobicity of highly hydrophilic biological macromolecules is, however, open to question. The correlation observed between the relative hydrophobicity estimates of solutes obtained by partitioning of the solutes in water-organic solvent systems and the "vapor-to-water" partition coefficients for the same solutes is concluded by Wolfenden to be "not bad" [9] it does not seem to be sufficiently convincing at present. The important drawback of the approach seems to us to be that the processes characterized by the method under discussion are not realized in biological systems. It should be noted, however, that the authors [9,69,70] do not claim that the method is applicable for studying the relative hydrophobicity of biological molecules. They advanced the term "hydration potential" of a solute [71] which seems to be related to the hydrophobic character of the solute. The outlook for the applicability of the approach to study the relative intensity of the hydration interactions of biological molecules seems to us to be rather uncertain at present.

6 Special Methods for Studying Hydrophobic Properties of Biological Macromolecules

The importance of the hydrophobic properties of biological macromolecules with regard to their function and structural organization has long been recognized. It is generally believed that the genetic code in its most primitive form could only differentiate between two classes of amino acids, i.e., hydrophilic and hydrophobic, and the grouping of codons and amino acids by similar hydrophobicity criteria has been advocated [72,73]. All of the known "signal" amino acid sequences of secreted or membrane-bound proteins are highly hydrophobic which seems to be essential for the membrane transfer of the proteins [74,75]. The protein-aqueous medium interactions are suggested [76] to affect the particular localization of a given protein macromolecule in the structure of a biological membrane, etc.

The most generally used methods employed in the study of the hydrophobic character of biological macromolecules have been extensively reviewed by Ochoa [5]. Therefore, only the main merits and limitations of these methods are considered below.

6.1 Calculating Methods

All calculating methods for estimating the hydrophobic character of biological macromolecules, primarily of proteins, are based on the hydrophobicity-hydrophilicity classification of amino acids. The earliest attempts to classify amino acids according to this property were based upon considerations of their chemical nature and steric structure of their side chains. According to the first classification, all the amino acids have been grouped into two classes — hydrophobic and hydrophilic ones. Capaldi et al. [77] suggested to divide amino acids not into two but into three groups — hydrophilic, hydrophobic, and intermediate ones. For many amino acids, attribution to one or the other group varies depending on the authors' opinion [77,78]. Various qualitative classifications have been used to estimate the differences in the relative hydrophobicity of membrane-bound proteins and lipoproteins as compared to globular proteins [77,79]. As a quantitative measure of the relative hydrophobicity of a given protein, the sum of the residue mole percentages of hydrophilic amino acids in the protein macromolecule (the so-called polarity index) has been proposed [77]. The attempts to employ this criterion to estimate the relative hydrophilic (or hydrophobic) character of proteins [77,79] have failed mainly due to two major drawbacks of the approach under discussion, namely to a qualitative and somewhat incorrect division of amino acids into hydrophilic and hydrophobic and to neglect of that the interaction of a protein macromolecule with the aqueous medium (i.e., the relative hydrophilicity or hydrophobicity of the protein) depends upon the tertiary structure of the macromolecule.

Rather more sound seem to be numerous attempts to analyze the hydrophobic character of proteins and polypeptide chain fragments on the basis of the aforementioned quantitative estimates of the relative hydrophobicity of the amino acids side chains obtained by the experimental methods considered above [41,54,71,80,81]. It should be noted that many of these estimates seem to be questionable due to the above drawbacks of the methods based on the study of transfer of a solute from water into an organic solvent. As an example it should be noted that the values of the free energy of transfer of an amino acid side chain from a nonpolar medium into water reported in the literature vary: for alanine from 0.2 to 0.83 kcal/mole (41,80) and to —0.42 kcal/mole [54]; for threonine from 0.40 to 0.83 kcal/mole [41,80] and to —0.04 kcal/mole [54]. The estimates reported for amino acids with ionizable side chains such as Glu, Asp, Lys, and Arg [41] seem to be particularly inadequate.

Segrest et al. [82] suggested to use as a measure of the hydrophobic character of an amino acid sequence the so-called hydrophobicity index (HI) calculated by dividing the sum total of the hydrophobicity values for the amino acid sequence by the number of the residues in the sequence. The scale, which assigns a hydrophobicity value for all nonpolar and neutral amino acids residues, advanced by Segrest [82] is a modification of the scale established by Nozaki and Tanford [41]. To estimate the mean hydrophobicity index (HI) of a given protein, Segrest et al. [82] suggested to calculate the HI values for the polypeptide backbone fragments including not less than ten residues disposed between the residues with ionizable side chains. The HI values for such fragments are summed up and the mean protein hydrophobicity index is calculated as a ratio of the total HI value to the number of the fragments. Segrest et al. [82] have shown that the mean hydrophobicity index values for the membrane-bound proteins and lipoproteins exceed significantly those specific for globular proteins. The results obtained by Segrest et al. [82] are of interest, although the approach under consideration is based on an analysis of the protein amino acid sequence and does not allow one to estimate the actual hydrophobic character of a protein macromolecule which is governed by the residues exposed to an aqueous environment.

An altogether different, though also strictly formal approach discussed in papers [47–49,83–86] is used to estimate the hydrophobic surface area of protein macromolecules. By means of this approach using a computer-based analysis of known proteins structures, it has been shown that as much as 40 to 60 percent of the surface area of many globular proteins are taken up by nonpolar amino acids residues [83,84]. These data dispel older concepts according to which most of the nonpolar residues are buried in the interior of the macromolecule. The nonpolar amino acids found in the protein surface are assumed to account for the biospecific conformation of the protein as well as for its ability to complex or aggregate with other types of biological molecules. This line of investigation of hydrophobic properties of proteins seems to be of a great interest. Its limitations, in our opinion, lie in its applicability only to proteins with a known tertiary structure as well as in a strictly formal manner of analysis. For instance, all N, O, and S atoms as well as amino acid residues with these atoms in a side chain are regarded as hydrophilic.

An exception is made for polar groups which form intramolecular hydrogen bonds in the protein which together with all C atoms and amino acid residues with side chains of hydrocarbon nature are regarded as hydrophobic. These features seem to prevent the method from being applied to studying the relative hydrophobicity of proteins. It should be emphasized, however, that the results obtained by this method unambigously invalidate the using of such measures as a polarity index and the mean hydrophobicity index to quantify the hydrophobic character of biological macromolecules.

6.2 Analysis of Interactions of Biological Macromolecules with Nonpolar and Amphiphilic Probes

Experimental methods based on an analysis of interactions of biological macromolecules with free or immobilized nonpolar or amphiphilic ligands are most commonly used at present to study the relative hydrophobicity of biopolymers.

The effect of hydrophobic binding of biopolymers to nonpolar ligands coupled to an inert insoluble matrix underlies the method for the isolation and purification of macromolecules called the hydrophobic chromatography method [5]. This method seems to appear similar to the reversed-phase liquid chromatography used successfully in studies of the relative hydrophobicity of chemical compounds (see above). It is supposed that the volume of an eluent needed to remove the protein under study from a given matrix can be used as a measure of the relative hydrophobicity of the protein [5]. It should be pointed out, however, that while investigating biopolymers by means of hydrophobic chromatography, only some nonpolar sites or "pockets" of the surface of a protein but not the whole macromolecule surface are involved in the interactions of the protein with the immobilized ligand.

Several approaches [87–89] developed for the study of the hydrophobicity of proteins are based on an analysis of interactions of proteins with free hydrophobic or amphiphilic probes. One of the approaches [87] consists of the study of binding of a nonionic surfactant, Triton X-100, to proteins in order to distinguish between integral and peripheral membrane proteins. The differences between the two types of proteins are revealed by means of electrophoresis in agarose gel [87]. The binding of the surfactant to the more hydrophobic integral membrane proteins exceeds that specific for the peripheral proteins due to which a marked change in the electrophoretic mobility of integral proteins as compared to that of the peripheral ones is observed [87].

The other approach to the study of the proteins hydrophobicity by means of free probes is based on the assumption that the process of binding of a probe molecule to a nonpolar site in a given macromolecule can be regarded as transfer of the probe molecule from an aqueous medium to a hydrophobic one [88]. Actually the probe-protein binding is usually a cooperative process limited by the type and dimensions of the probe-binding sites [88, 89].

It should be emphasized, however, that whatever experimental technique is used, the data on the binding capacity of a macromolecule in relation to a given nonpolar or amphiphilic probe does not provide a means for estimating the hydrophobic character of the macromolecule which is a measure of the intensity of its interactions with an aqueous medium.

6.3 Analysis of Effects of Salts on the Aqueous Solubility of Proteins

The conception that the hydrophobicity of a protein is the property of its surface is used as the basis of the technique suggested by Melander and Horvath [30]. This technique consists of an analysis of the effects of inorganic salts on the aqueous solubility of proteins [30]. According to the model considered by Melander et al. [30], the free energy of solvation of a protein macromolecule in aqueous solution is described by Eq (3) (see above). The presence of a salt alters the protein solubility due to the concentration-dependent effect of the salt on the free energy of formation of a cavity in water, E_{cav}, and the free energy of electrostatic interactions of protein with the solvent, E_e [30]. Melander et al. [30] have quantified the structure-changing effect of salts on water by a molal surface tension increment of a given salt.

It is shown [30] that the analysis of a relationship between the salting-out constants of a given protein and the molal surface tension increments makes it possible to estimate "the relative surface hydrophobicity" which is calculated as the ratio of the nonpolar surface area to the molecular weight of the protein. According to the concept developed by Melander et al [30], the hydrophobic character of a protein is believed to be constant at both high and "physiological" concentrations of different inorganic salts, which seems to be untrue in most cases (see below).

The approach under discussion, although not commonly used, deserves special attention as it seems to be the only one taking into account the fact that hydration interactions of a biological macromolecule depend upon the concentration of the components of an aqueous solution affecting the structure and/or thermodynamic state of water in the solution [30]. At the same time the authors of this model [30] assume the constancy of the hydrophobicity of a protein which is a measure of the above interactions varying with the composition of an aqueous medium. This example seems to be typical of that even in the case of an obvious discrepancy between experimental results and interpretation of the results from the conventional point of view, the conventional ideas take the upper hand which in this particular case [30] means that the hydrophobicity of a solute is considered as an intrinsic property of the solute and not as a measure of the intensity of the interaction between the solute and the solvent which clearly depends on both properties of the solute and of the solvent.

7 Study of Transfer of a Solute from Water-I into Water-II (Partition in Aqueous Polymer Biphasic Systems as a General Method for Estimation of the Relative Hydrophobicity of Biological Molecules)

7.1 Theory of the Method

When aqueous solutions of two incompatible polymers are mixed above certain concentrations they form two immiscible aqueous phases, one of which is rich in one of the polymers, the second one — in the other. These solutions can be buffered, made isotonic, etc., the salt composition of the phases can be varied within a wide range [90]. Biological objects added to such systems are partitioned between the two phases. Partition behavior of a solute in an aqueous polymer biphasic system is governed by the properties of the solute as well as by pH, and by the polymeric and ionic composition of the phases [90–92].

The method of partitioning in aqueous polymer biphasic systems was developed for fractionation of different biological materials: proteins, nucleic acids, viruses, cells, etc. [90], but lately it has been applied to studying the properties of biological particles and solutes [11, 91, 92]. Properties of aqueous polymer biphasic systems have been considered extensively in a monography by Albertsson [90] and in several reviews (see, e.g. Ref. [91]). The salient features of the systems important from the standpoint of using these systems to estimate the relative hydrophobicity of biological solutes [11] are as follows.

1. Both phases of an aqueous polymer biphasic system contain not less than 70 to 80% (w/w) of water. Separation into two phases occurs only above certain concentrations of the phase polymers and the curve relating these concentrations in a phase diagram is called the binodial [90]. The position of the binodial depends on the chemical nature of the phase polymers and on their molecular weights as well as on the nature and concentration of inorganic salts present in the system [93, 94].

2. The most commonly used nonionic phase polymers, dextran-500 and poly(ethylene glycol)-6000 (PEG-6000) [90, 91], are not inert to biological structures. Dextran-500 with a molecular weight of $5 \cdot 10^5$ decreases the aqueous solubility of proteins [95], enhances *in vitro* mixed lymphocyte-type reactions between lymphocytes from unimmunized animals and tumor cells [96], increases pulmonary metastases of the carcinosarcoma of Walker 256 in rats [97], etc. Poly(ethylene glycol) (PEG) with a molecular weight of $6 \cdot 10^3$ induces precipitation of proteins in aqueous solutions [98] and formation of compact forms of nucleic acids and polynucleotides [99], induces fusion of biological membranes [100], etc. An employment of the aqueous dextran-500-PEG-6000 biphasic systems for preparative purposes is based on that the possible changes of the objects being partitioned seem to be reversible [90]. Even reversible changes of the objects in the course of the partitioning, however, makes it impossible to use the method for analytical purposes. For this reason we believe that aqueous biphasic systems formed by polymers which are more inert to biological structures, as e.g., ficoll-400 and dextran with molecular weights in the range of $4 \cdot 10^4$ to $7 \cdot 10^4$, are better suited for analytical purposes [93].

3. Partition coefficient values of a given solute in the aqueous ficoll-dextran biphasic systems prepared with various polymers samples particularly differing in the molecular weight distribution, are interrelated as described by the above solvent regression Eq. (8) [101]. The coefficient a_i in Eq. (8) represents the difference in the hydrophobic properties of the phases similarly to the case of water-organic solvent systems [102]. The coefficient b_i in Eq. (8) depends on the particular systems compared, but in contrast to the water-organic solvent systems, it is independent of the chemical nature of the solutes being partitioned [103]. This fact indicates that the solutes being partitioned in the aqueous ficoll-dextran biphasic systems do not interact directly with the phase polymers. Hence, the solute partitioning in such systems should be considered as transfer of the solute molecule from one aqueous medium (phase) into another.

4. The properties of an aqueous medium in the two phases of an aqueous-polymer biphasic system differ and depend on the nature and concentration of the phase polymers as well as on the nature and concentration of inorganic salts and nonelectrolytes present in the system [11, 90, 102, 104–106]. The salient properties of the medium of the phases of a given system, which seem to govern the partitioning of a solute in the system, insofar as it is known at present [11, 105, 106], are: a) the affinity of the medium of a given phase for nonpolar groups, e.g., for a CH_2 group, relative to that of an organic solvent such as *n*-octanol chosen as a reference solvent [102]; b) the difference in the relative affinity of the two phases of a given biphasic system for nonpolar

groups, e.g., for a CH_2 group [104, 105]; c) the difference in the capacity of the medium in both phases of the system to participate in the electrostatic interactions with a solute being partitioned [11, 106].

According to the above conceptions, the free energy of transfer of a solute from the aqueous phase I into the aqueous phase II can be described by an equation identical to Eq (3):

$$\Delta G_{I \to II} = \Delta G_e + \Delta G_{vdW} + \Delta G_{cav} \tag{14}$$

where ΔG_e is the free energy change due to the difference in electrostatic interactions between the solute and solvent in the two phases of the system; ΔG_{vdW} the free energy change due to the difference in van der Waals interactions between the solute and solvent in the two phases; and ΔG_{cav} the free energy change due to the difference in the energies for formation of a cavity in both phases.

It is clear that all the terms on the right-hand side of Eq (4) depend, on the one hand, upon the chemical nature and structure of the solute being partitioned and, on the other hand, upon the chemical composition of both phases of the system which governs the structure and/or the state of water in the phases.

In order to estimate the state of the aqueous medium in a given phase, it has been suggested [102, 105] to examine the affinity of the medium for a nonpolar group arbitrarily chosen as a standard group, e.g., for a methylene group. For this purpose, partition coefficients of a homologous series of solutes with varied length of the aliphatic chain in the *n*-octanol-aqueous phase of the polymeric biphasic system are examined. The partition coefficients obtained are described by Eq (10) (see above) which makes it possible to estimate the E parameter value [102, 105] which is related to the free energy of transfer of a CH_2 group from *n*-octanol into the aqueous phase in question according to Eq (7). On the basis of the experimentally obtained free energy values for transferring a CH_2 group from *n*-octanol into the phases of several aqueous polymer biphasic system [102, 105], a general hydrophobicity scale for both organic solvents and aqueous polymeric biphasic systems has been formulated [63].

Positions of the phases of aqueous polymeric biphasic systems in relation to the above general solvent hydrophobicity scale [63, 102] show that the relative hydrophobic character of these phases is close to that specific for a medium of biological systems. It should be noted that the free energy of transfer of a solute between the two phases of a given aqueous polymeric system depends on the difference in the state of water in these phases. To quantify this difference, it has been suggested to use the value of the free energy of transfer of a CH_2 group between the two phases [11, 104, 105]. In order to estimate this value, partition coefficients for a homologous series of solutes in a given aqueous polymeric biphasic system are examined as described above. Depending on the particular aqueous biphasic system studied the free energy value for a CH_2-group interphase transfer is varied from ca. 10 to 80 cal/mole of CH_2 [104, 105]. It should be noted here that the same parameter varies from ca. 400 to 1000 cal/mole of CH_2 in the case of water-organic solvent systems [63].

It follows from Eq. (14) that the free energy of interphase transfer of a CH_2 group provides a measure of the difference in the energies for formation of a cavity together with that in the capacities of water in both phases to participate in van der Waals interactions with a solute being partitioned.

In order to describe partitioning of a solute capable to participate in electrostatic interactions with the solvent in a given aqueous biphasic system, however, proper allowance must be made for the difference in the capacities of the solvent in the two phases to participate in this type of interactions with the solute.

It has been suggested [90,91] that partitioning of a charged solute in an aqueous biphasic system is governed by an electrostatic potential difference (an interfacial potential) between the phases caused by unequal distribution of some inorganic salts present in the system. De Ligny and Gelsema [107] have reasoned this suggestion to be misleading. An interfacial potential resulting from an unequal partitioning of a salt in a given biphasic system which is called distribution potential has been analyzed extensively in the literature dealing with electro-chemical phenomena in water-organic solvent systems (see, e.g., Ref. [108]). It is generally accepted [108] that the distribution potential represents the difference in the solvation energies of the potential-determining ions in the phases of the system. This parameter cannot be used as a measure of the difference in the properties of the phases of an aqueous polymeric biphasic system, since the salts present in the system not only distribute between the phases but seem to govern formation of the system (see above) and therefore are likely to affect the properties of both phases.

It has been suggested by Zaslavsky et al [106] to use the logarithm of the partition coefficient of a solute (or its fragment) with a single ionized group as a measure of the difference in the capabilities of the aqueous medium of the two phases to participate in electrostatic interactions with a charged solute being partitioned. The choice of such a reference solute is rather arbitrary and it has been established by us that sodium salt of dinitrophenylated (DNP-) glycine is an adequate choice.

Thus, the concept of partitioning of a solute in an aqueous polymeric biphasic system suggested by Zaslavsky et al. [11,105,106] is as follows:

a) both phases of a given aqueous biphasic system comprise two aqueous polymeric media of different chemical composition;
b) the different chemical composition of the two phases causes the difference in the properties of the phases;
c) partitioning of a solute between the two phases should be regarded as transfer of the solute from the aqueous medium with one set of properties into the aqueous medium with a different set of the properties;
d) the difference in the properties of the medium of the two phases consists in the different structure and/or thermodynamic state of water in the phases causing the difference in the free energies of formation of a cavity and the difference in the capabilities of water to participate in van der Waals and electrostatic interactions with the solute being partitioned.

To quantify the difference in the free energies of formation of a cavity and in capabilities of water to participate in van der Waals interactions with a solute it has been suggested [104,105] to use the free energy of interphase transfer of a CH_2 group. To quantify the difference in the capability of water in the phases to participate in electrostatic interactions with a solute, it has been proposed [11,106] to use the logarithm of the partition coefficient of Na-salt of DNP-glycine.

Thus, according to the above concept the partition coefficient of a solute in an aqueous polymeric biphasic system can be described as:

$$\ln K = n_0 \cdot E + m \cdot C \quad (15)$$

where E accounts for the difference in the relative hydrophobic character of the phases (see Eq. (10) above); C represents the difference in the capabilities of water in the phases to participate in electrostatic interactions with the solute; n_0 expresses the difference in the intensity of van der Waals interactions between the solute and solvent in the two phases relatively to that for a methylene group; m represents the difference in the intensity of the solute-solvent electrostatic interactions in the two phases relatively to that for Na-salt of DNP-glycine.

In order to estimate the relative hydrophobicity of a solute it has been suggested [11, 109] to use the ratio of the free energy of the interphase transfer of the solute to that of a CH_2 group:

$$n^{CH_2} = \Delta G_{I \to II} / \Delta G_{I \to II}^{CH_2} = \ln K/E \quad (16)$$

where $\Delta G_{I \to II}$ is as defined in Eq. (14); $\Delta G_{I \to II}^{CH_2}$ the free energy of transfer of a CH_2 group from phase I into phase II; K and E are as defined above.

The parameter n^{CH_2}, i.e. the equivalent number of CH_2 groups [11, 109], characterizes the relative hydrophobicity of the solute under study. The positive value of the n^{CH_2} term means that the solute is hydrophobic and its relative hydrophobicity is equivalent to that of n methylene groups. The negative value of n^{CH_2} means that the solute is hydrophilic and its hydrophobicity is equal by value and opposite by sign to the total hydrophobicity of n CH_2 groups. The distinctive feature of the above expression of the relative hydrophobicity of a solute consists in that the hydrophobic character of the solute is considered not as the affinity of the solute for water as compared to that for nonpolar environment [12] but as the difference in the solute affinity for the aqueous media with various properties relative to that for a methylene group.

It has been shown that the above approach when applied to solutes of similar chemical nature such as a series of amino acids [109] or a series of mononucleosides [110] provides the estimates of the relative hydrophobicity of the solutes which are in a good agreement with those obtained by studying of transfer of the solutes from water into an organic solvent. At the same time the approach under consideration has been shown [103, 109–111] to have many points in its favor which are to be discussed below.

7.2 Some Examples of Application of the Method

It should be emphasized that the conditions found in aqueous polymeric biphasic systems seem to simulate those in biological systems to a good approximation [33]. The properties of both phases of such systems are only slightly different as compared to those of the phases of water-organic solvent systems. Hence, the partition coefficient of a solute in an aqueous polymeric biphasic system is much more responsive not only to the modifications of the molecule structure but to the alterations in the conformation of the molecule as well. The application of the partition technique to the studying of the relative hydrophobicity of biological solutes and their synthetic analogs has shown that: i) the generally accepted principle of additivity of hydrophobic increments of substituents is invalid for the absolute majority of the solutes studied so far; ii) the relative hydrophobicity of biological molecules depends to a considerable degree upon the steric structure and specific conformation of a molecule in the solvent; iii) the relative hydrophobicity of the majority of biological solutes depends upon the chemical (ionic, in particular) composition of an aqueous medium.

The experimental data substantiating the above conclusions and the biological significance of the latter are considered below.

It should be noted here that all the studies of the relative hydrophobicity of biological solutes discussed below have been performed using the aqueous ficoll-dextran biphasic system.

In the study of the relative hydrophobicity of amino acids [109] it was particularly found that the relative hydrophobicity of ionized side chains of such amino acids as lysine, arginine, glutamic and asparagic acids depends upon the ionic composition of the aqueous medium at pH 7.4. In particular, the relative hydrophobicity of the lysine side chain appears to be equivalent to that of $-4.2\,CH_2$ groups in the presence of 0.11 M phosphate buffer at pH 7.4 and to that of $-0.4\,CH_2$ groups in the presence of 0.15 M NaCl in 0.01 M phosphate buffer at pH 7.4 [109]. An even more pronounced effect of the ionic composition of the aqueous medium at pH 7.4 on the relative hydrophobicity of mononucleotides is observed [110]. The relative hydrophobicity of AMP is equivalent to that of $-2.4\,CH_2$ groups in the presence of 0.11 M phosphate buffer, and to that of $-10.8\,CH_2$ groups in the presence of 0.15 M NaCl in 0.01 M phosphate buffer (pH 7.4 in both cases), to give only one typical example.

The probable biological significance of the influence of the ionic composition of an aqueous medium on the relative hydrophobicity of biological solutes will be considered below. Here only one important conclusion drawn from the above data should be stressed. It seems that there is no absolute hydrophobicity scale for solutes and for hydrophobic increments of polar and ionizable substituents as the relative hydrophobicity of solutes and the substituents depends upon the chemical composition of an aqueous medium.

No less important for the understanding of the hydrophobic properties of chemical compounds is the matter of additivity of hydrophobic increments of substituents. On the one hand, many examples of successful employment of the additivity approach in calculating the relative hydrophobicity of various chemical compounds are reported [1,3,10]. On the other hand, quite a few examples of significant deviations of the experimental data from the calculated values are known (see, e.g., Ref. [112] and [113]). The results obtained in the study of the relative hydrophobicity of a number of peptides [11,111] have shown that the above additivity principle is often invalid for amino acid residues under the conditions employed. Comparison of the estimates of the relative hydrophobicity of such mononucleotides as AMP, ATP, and c-AMP [110] indicates that formation of the intramolecular covalent bond affects the relative hydrophobicity of AMP much more than introduction of two additional phosphate groups. Such a result would be quite unpredictable on the basis of the principle of additivity of hydrophobic increments of substituents. The infringement of the additivity principle was demonstrated most distinctly in the study of the relative hydrophobicity of dinucleotides and their synthetic analogs [114]. For example, the relative hydrophobicity of (3′–5′)- and (2′–5′)-ApA isomers in the presence of 0.11 M phosphate buffer at pH 7.4 amounts to 12.9 equivalent CH_2 and to —2.3 equivalent CH_2 groups, respectively [114]. The data obtained [114] show in particular that the substitution of the ribose residue in a dinucleoside phosphate molecule for the 2-oxyethyl fragment in the cases of ApA and UpU is accompanied not by an expected increase but by a decrease of the relative hydrophobicity of the molecule. The above examples clearly show that at least in the case of biological molecules one should be very careful when employing the additivity principle. It seems highly desirable to check experimentally the validity of the principle in each particular case. The causes of the possible invalidity of the principle are discussed by Draffehn et al. [113].

The results from the studies of the relative hydrophobicity of peptides [11,111], nucleic compounds [110,114], and proteins [103] indicate that the intensity of the hydration interactions of the solutes depends upon the conformation of the solute molecule. For example, the relative hydrophobicity of γ-endorphine (16 amino acid residues) is equivalent to that of 2.2 CH_2 groups in the presence of 0.11 M phosphate buffer at pH 7.4 which is equal within the experimental error range to the relative hydrophobicity of 1.5 equivalent CH_2 groups of Phe-Leu-Arg peptide under the same conditions [11,111]. The results on the relative hydrophobicity of proteins [103] seem also to indicate that the macromolecules conformation affects the hydrophobic character of the protein. For instance, different hydrophobic properties are displayed by two samples of human serum albumin which differ with regard to the nature of their lipid contaminants [115]. It should be emphasized that various methods of protein analysis, e.g., electrophoresis, electrical focussing, sedimentation analysis, etc. could not detect any difference in the above albumin samples [115]. The results obtained in the study [103] of the pH dependence of the relative hydrophobicity of human serum albumin and human oxyhemoglobin seem to support the hypothesis that the hydrophobic character of a solute is affected by the conformation of the solute molecule.

Referring once more to the effect of the aqueous medium composition upon the relative hydrophobicity of biological solutes, the correlation relationship established [103,116] between the effects of the medium ionic composition on the relative hydrophobicity of serum albumins of various origin and on that of erythrocytes from the same species should be noted. The ionic strength value of the medium in the aqueous ficoll-dextran biphasic system has been used as a quantitative index of the ionic composition of the system, and the ionic composition was varied from 0.11 M phosphate buffer to 0.15 M NaCl in 0.01 M buffer at pH 7.4 [116]. Partition coefficients of cells in an aqueous polymeric biphasic system are determined as the ratio of a number of the cells in the phase to that of the cells present at the interphase [90]. Specific features of the partition behavior of cells in aqueous biphasic systems are discussed in detail elsewhere (see, e.g., Ref. [90,91]). It has been established [116] that erythrocytes of different species are distributed in the aqueous ficoll-dextran biphasic system according to the following equation:

$$\ln K_{cell} = A_{cell} + B_{cell} \cdot I \tag{17}$$

where K_{cell} is the partition coefficient of a given population of cells; I the ionic strength value used as a quantitative index of the ionic composition of the system under the conditions employed; A_{cell} and B_{cell} are constants. It is clear that the value of the B_{cell} coefficient reflects the degree to which the ionic composition of the medium affects the partitioning (and the relative hydrophobicity) of cells.

It has been shown [103] that the effect of the ionic composition of the medium on the relative hydrophobicity of serum albumins follows the relation similar to that described by Eq. (17), and there is a correlation between the coefficients B_{cell} and B_{alb} both of which represent the degree of the influence of the ionic composition of the aqueous medium on the hydrophobic character of erythrocytes and albumins, respectively:

$$B_{cell} = 2.28 \cdot B_{alb} + 26.63 \quad (18)$$

$$N = 4, \qquad r^2 = 0.943,$$

where N is the number of species studied (human, rabbit, rat, sheep); r^2 being the correlation coefficient.

The relationship described by Eq. (18) cannot be regarded as statistically valid due to a limited number of species studied. The existence of such a correlation, however, seems to support the assumption [115] that the effect of the composition of an aqueous medium upon the hydrophobic character of biological solutes and particles is important for regulation of their biological functions. According to this assumption, the changes in the chemical composition of an aqueous environment of components of biological fluids (blood, lymth, tissue fluid, etc.) may affect the relative hydrophobicity of these components inducing, e.g., adsorption or desorption of proteins on membranes of blood cells, blood vessel surfaces, and so forth, affecting aggregation processes and complex formation, controlling enzyme reactions, etc. The results of the studies of the relative hydrophobicity of liposomes [117] and of neuraminidase-treated erythrocytes [118] seem to indirectly support this hypothesis. The data obtained in the study [111] of the correlations of the hydrophobic character of opioid peptides with their biological activity measured in various bioassay systems also appear to indirectly confirm the above hypothesis. These data sould be considered in greater detail.

The data obtained on the relative hydrophobicity of opioid peptides [111] indicate that the relative hydrophobicity of a given peptide may be affected by the ionic composition of the aqueous medium or not affected (at pH 7.4 in the presence of NaCl and phosphate buffer) depending on the structure of the peptide molecule. The biological activity of the peptides determined in several tests was correlated with their relative hydrophobicity at different ionic composition of the aqueous medium [111].

It has been established [111] that the affinity of the opioid peptides for the naloxone binding sites in rat brain homogenate is related to the hydrophobic character of the peptides if the composition of an aqueous medium provides the capacity of the medium to participate in electrostatic interactions which amounts to 0.086 in terms of the parameter C value (see Eq. (15)). This correlation seems to be of either parabolic of bilinear type [1-3] but for the most peptides studied it fits mainly the left-hand side of the curve [111] which is described as:

$$\log (1/IC_{50})_{RBR} = 7.065 + 0.345 \cdot n_{0.086}^{CH_2} \quad (19)$$

$$N = 12; \qquad r^2 = 0.996; \qquad s = 0.044,$$

where $(IC_{50})_{RBR}$ is the peptide concentration (in M) which under standard conditions [111] displaces 50% of naloxone bound to rat brain homogenate; $n_{0.086}^{CH_2}$ is the relative hydrophobicity of peptide (expressed in terms of equivalent CH_2 groups) at the composition of an aqueous medium providing the capacity of the medium to participate in electrostatic interactions which amounts to 0.086 in terms of the parameter C value; N r^2, and s have the usual meaning.

The number of the peptides fitting the hypothetical right-hand side of the above curve [111] was insufficient in order to determine the so-called "optimal value" of the relative hydrophobicity corresponding to that of the peptides with the highest affinity for rat brain homogenate. An approximate estimate of this value amounts to 6–8 equivalent CH_2 groups. It should be stressed that the results obtained [111] indicate that the relationship between the relative hydrophobicity of morphine and a number of similar drugs and their affinity for rat brain homogenate features the "optimal value" of the hydrophobic character of drugs which amounts to 3–4 equivalent CH_2 groups. The difference in the observed "optimal values" of the relative hydrophobicity of opiates and opioid peptides [111] seems to be in line with the generally accepted opinion that opioid peptides and opiates interact with different rat brain receptors. The assumption that an existence of two different "optimal values" of the relative hydrophobicity of drugs plotted against their biological potency implies two different types of receptors with which these drugs interact in a given biological system has been put forward by Dearden et al [119]. The results obtained by Zaslavsky et al [111] seem to support this assumption.

A quite good correlation was also found [111] between the potency of the peptides in the mouse vas deferens assay and the relative hydrophobicity of the peptides, $n_{0.322}^{CH_2}$, at the composition of an aqueous medium providing the capacity of the medium to participate in electrostatic interactions which amounts to 0.322 in terms of the parameter C [111]. The correlation is described as [111]:

$$\log (1/IC_{50})_{MVD} = 5.755 + 0.336 \cdot n_{0.322}^{CH_2} \qquad (20)$$

$$N = 14; \quad r^2 = 0.866; \quad s = 0.540,$$

where $(IC_{50})_{MVD}$ is the peptide concentration (in M) which under standard conditions (see Ref. [111]) ensures 50% depression of electrically evoked contraction of the mouse vas deferens; $n_{0.322}^{CH_2}$, N, r^2, and s are as defined above.

The observed similarity of the relationships described by Eqs. (19) and (20) seems to be not accidental if one takes into account that the relative hydrophobicity of some of the peptides under study is diametrically opposite in the aqueous media with the different values of parameter C. The difference in the C values reflecting that in the capacity of the aqueous medium to participate in electrostatic interactions with a solute observed when Eqs. (19) and (20) are compared seems to indicate an important role of the chemical composition of an aqueous medium probably regulating the interactions of peptides with receptors [111].

Thus, using the method of partitioning of solutes in the aqueous ficoll-dextran biphasic systems for estimating the relative hydrophobicity of biological molecules it was established that:

i) the conformation and steric structure of a solute molecule affects its hydrophobic character;
ii) the principle of additivity of hydrophobic increments of substituents seems to be valid for nonpolar solutes and only for a very limited number of polar and amphiphilic solutes;
iii) the relative hydrophobicity of many biological (and synthetical) solutes and of their ionizable fragments depends upon the chemical composition of an aqueous medium, and this dependence is likely to regulate transport and functions of these solutes in biological systems.

8 Conclusions

The above examples imply that the modern concepts of the hydrophobic properties of both biological and synthetical solutes are far from adequate. As is usually the case, an advancement of the concepts seems to depend upon the development of new experimental techniques. We believe that the most important question at present is the one of the influence of the chemical composition of an aqueous medium on the hydrophobic character of a solute. This question is, on the one hand, essential for the progress of the concepts of hydrophobic properties of solutes and of fundamentals of water structure and solution theory and, on the other hand, it seems important for better understanding the factors regulating biological processes and biological potency of chemical agents.

The aforementioned examples of deviations from the principle of additivity of hydrophobic increments of constituents compel us to take a "retrograding" position. It is generally accepted [1,2] that one of the main achievements provided by the Hansch approach [3] is the possibility to calculate the relative hydrophobicity of solutes which takes a lot of burden from the experimenter. On the strength of the aforementioned experimental data and theoretical concepts, however, the authors of the present review are forced to stress the necessity to return to experimental measurements of the relative hydrophobicity of solutes. It seems to us that this is essential in order to turn the QSAR studies of drugs and biological molecules from the correlation analysis field into that of the study of the mechanism of biological action of chemical compounds. We believe that the possibility to quantify the relative hydrophobicity of solutes of different chemical nature provided by the method of partitioning in the aqueous ficoll-dextran biphasic system serves as the basis for this perspective.

9 References

1. Franke, R.: Theoretical Drug Design Methods, Amsterdam, Elsevier 1984
2. Martin, Y. C.: Quantitative Drug Design: A Critical Introduction, New York, Marcel Dekker 1978
3. Hansch, C., Leo, A.: Substituents Constants for Correlation Analysis in Chemistry and Biology, New York, Wiley 1979
4. Kaiser, E. T., Kezdy, F. J.: Science 223, 249 (1984)
5. Ochoa, J.-L.: Biochimie 60, 1 (1978)

6. Gorb, L. G., Abronin, I. A. Harchevnikova, N. V., Zhidomirov, G. M.: Rus. J. Phys. Chem. 58, 9 (1984)
7. Rebinder, P. A.: Hydrophilicity and Hydrophobicity, in: Kratkaya Chim. Enciclopedia, Vol. 1, p. 937, Moscow, Sov. Enciclopedia 1961
8. Yalkowsky, S. H., Valvani, S. C., Roseman, T. J.: J. Pharm. Sci. 72, 866 (1983)
9. Wolfenden, R.: Science 222, 1087 (1983)
10. Rekker, R. F.: The Hydrophobic Fragmental Constant: Its Derivation and Application. A Means of Characterizing Membrane Systems, Amsterdam, Elsevier 1977
11. Zaslavsky, B. Yu., Mestechkina, N. M., Miheeva, L. M., Rogozhin, S. V.: J. Chromatogr. 256, 49 (1983)
12. Tanford, C.: The Hydrophobic Effect: Formation of Micelles and Biological Membranes, New York, Wiley 1973
13. Beall, P. T.: Cryobiology 20, 324 (1983)
14. Structure of Water and Aqueous Solutions, (ed.) Luck, W. A. P., Weinheim, Verlag Chemie 1974
15. Samoilov, O. Ya.: Structure of Aqueous Electrolyte Solutions and Hydration of Ions, Moscow, Acad. Sci. USSR 1957
16. Nemethy, G., Scheraga, H. A.: J. Chem. Phys. 36, 3382 (1962)
17. Naberukhin, Yu. I.: J. Struct. Chim. (Rus.) 25, 60 (1984)
18. Angell, C. A., Rodgers, V.: J. Chem. Phys. 80, 6245 (1984)
19. Blum, L., Hooye, J. S.: J. Phys. Chem. 81, 1311 (1977)
20. Planche, H., Renon, H.: J. Phys. Chem. 85, 3924 (1981)
21. Geiger, A., Rahman, A., Stillinger, F. H.: J. Chem. Phys. 70, 263 (1979)
22. Pratt, L. R., Chandler, D.: J. Chem. Phys. 67, 3683 (1977)
23. Okazaki, S., Nakanishi, K., Touhara, H., Adachi, Y.: J. Chem. Phys. 71, 2421 (1979)
24. Lee, C. Y., McCammon, J. A., Rossky, P. J.: J. Chem. Phys. 80, 4448 (1984)
25. Mirgorod, Yu. A.: J. Struct. Chim. (Rus.) 24, 93 (1983)
26. Ben-Naim, A.: Hydrophobic Interactions, New York, Plenum Press 1980
27. Zaslavsky, B. Yu., Masimov. E. A., Miheeva, L. M., Rogozhin, S. V., Hasaev, D. P.: Dokl. Acad. Nauk USSR 261, 669 (1981)
28. Masimov, E. A., Zaslavsky, B. Yu., Gasanov, A. A., Rogozhin, S. V.: J. Chromatogr. 284, 337 (1984)
29. Masimov, E. A., Zaslavsky, B. Yu., Gasanov, A. A., Davidovich, Y. A., Rogozhin, S. V.: J. Chromatogr. 284, 349 (1984)
30. Melander, W., Horvath, C.: Arch. Biochem. Biophys. 183, 200 (1977)
31. Ueberreiter, K.: Colloid & Polymer Sci. 260, 37 (1982)
32. Zaslavsky, B. Yu., Masimov, E. A., Gasanov, A. A., Rogozhin, S. V.: J. Chromatogr. 294, 261 (1984)
33. Tanford, C.: Science 200, 1012 (1978)
34. Zaslavsky, B. Yu., Masimov, E. A., Rogozhin, S. V.: Dokl. Acad. Nauk USSR 277, 1163 (1984)
35. Higuchi, T., Davis, S. S.: J. Pharm. Sci. 59, 1376 (1970)
36. Zaslavsky, B. Yu., Gulaeva, N. D., Rogozhin, S. V., Gasanov, A. A., Masimov, E. A.: Mol. Cell. Biochem. 65, 125 (1985)
37. Hermann, R. B.: J. Phys. Chem. 75, 363 (1971)
38. Hermann, R. B.: J. Phys. Chem. 76, 2754 (1972)
39. Hermann, R. B.: J. Phys. Chem. 79, 163 (1975)
40. Harris, S. M. J., Higuchi, T., Rytting, J. H.: J. Phys. Chem. 77, 2694 (1973)
41. Nozaki, Y., Tanford, C.: J. Biol. Chem. 246, 2211 (1971)
42. Amidon, G. L., Yalkowsky, S. H., Anik, S. T., Valvani, S. C.: J. Phys. Chem. 79, 2239 (1975)
43. Davis, S. S., Higuchi, T., Rytting, J. H.: Adv. Pharm. Sci. 4, 73 (1974)
44. Anderson, B. D., Conradi, R. A.: J. Pharm. Sci. 69, 424 (1980)
45. Hansch. C., Quinlan, J. E., Lawrence, G. L.: J. Org. Chem. 33, 347 (1968)
46. Moriguchi, J.: Chem. Pharm. Bull. 23, 247 (1975)
47. Chotia, C.: Nature 248, 338 (1974)
48. Gelles, J., Klapper, M. H.: Biochim. Biophys. Acta 533, 465 (1978)
49. Richards, F. M.: Annu. Rev. Biophys. Bioeng. 6, 151 (1977)
50. Moriguchi, J., Kanada, Y., Komatsu, K.: Chem. Pharm. Bull. 24, 1799 (1976)
51. Leo, A., Hansch, C., Jow, P. Y. C.: J. Med. Chem. 19, 611 (1976)

52. Anderson, B. D., Rytting, J. H., Higuchi, T.: J. Pharm. Sci. 69, 676 (1980)
53. Leo, A., Hansch, C., Elkins, D.: Chem. Rev. 71, 525 (1971)
54. Fendler, J. H., Nome, F., Nagyvary, J.: J. Mol. Evol. 6, 215 (1975)
55. Herskovits, T. T., Harrington, J. P.: Biochemistry, 11, 4800 (1972)
56. Scruggs, R. L., Achter, E. K., Ross, P. D.: Biopolymers 11, 1961 (1972)
57. Cratin, P. D.: Ind. Eng. Chem. 60, 14 (1968)
58. Alhaider, A. A., Selassie, C. D., Chua, S. O., Lien, E. J.: J. Pharm Sci 71, 89 (1982)
59. Martin, Y. C.: Advances in the Methodology of Quantitative Drug Design, in: Drug Design (ed.) Ariens, E. J., Vol. 8, p. 1, New York, Academic Press 1979
60. Wang, P. H., Lien, E. J.: J. Pharm. Sci, 69, 662 (1980)
61. Smith, R. N., Hansch, C., Ames, M.: J. Pharm. Sci. 64, 599 (1975)
62. Leo, A., Hansch, C.: J. Org. Chem. 36, 1539 (1971)
63. Zaslavsky, B. Yu., Miheeva, L. M., Rogozhin, S. V.: J. Chromatogr. 216, 103 (1981)
64. Tomlinson, E.: J. Chromatogr. 113, 1 (1975)
65. Baker, J. K., Ma, C. Y.: J. Chromatogr. 169, 107 (1979)
66. Kaliszan, R.: J. Chromatogr. 220, 71 (1981)
67. Valko, K.: J. Liquid Chromatogr. 7, 1405 (1984)
68. Rekker, R. F.: J. Chromatogr. 300, 109 (1984)
69. Wolfenden, R.: Biochemistry, 17, 201 (1978)
70. Cullis, P. M., Wolfenden, R.: Biochemistry 20, 3024 (1981)
71. Wolfenden, R., Andersson, L., Cullis, P. M., Southgate, C. C.: Biochemistry 20, 849 (1981)
72. Woese, C. R., Dugre, D. H., Kondo, M., Saxinger, W. C.: Cold Spring Harbor Symp. Quant. Biol. 31, 723 (1966)
73. Bina, M., Feldmann, R. J., Deeley, R. G.: Proc. Natl. Acad. Sci. USA 77, 1278 (1980)
74. Engelman, D. M., Steitz, T. A.: Cell 23, 411 (1981)
75. Heijne Von, G.: Eur. J. Biochem. 116, 419 (1981)
76. Gerson, D. F.: Biophys. J. 37, 145 (1982)
77. Capaldi, R. A., Vanderkooi, G.: Proc. Natl. Acad. Sci. USA 69, 930 (1972)
78. Daschevskii, V. G.: Mol. Biologia (Rus.) 14, 105 (1980)
79. Davis, M. A., Hauser, H., Leslie, R. B., Phillips, M. C.: Biochim. Biophys. Acta 317, 214 (1973)
80. Nandi, P. K.: Int. J. Peptide Protein Res. 8, 253 (1976)
81. Pliska, V., Schmidt, M., Fauchere, J. L.: J. Chromatogr. 216, 79 (1981)
82. Segrest, J. P., Feldman, R. J.: J. Mol. Biol. 87, 853 (1974)
83. Chothia, C., Janin, J.: Nature 256, 705 (1975)
4. Chothia, C.: J. Mol. Biol. 105, 1 (1976)
85. Janin, J., Chothia, C.: Biochemistry 17, 2943 (1978)
86. Janin, J., Wodak, S., Levitt, M., Maigret, B.: J. Mol. Biol. 125, 357 (1978)
87. Helenius, A., Simons, K.: Proc. Natl. Acad. Sci. USA 74, 529 (1977)
88. Tanford, C.: J. Mol. Biol. 67, 59 (1972)
89. Ismailova, V. N., Rebinder, P. A.: Structures Formation in Protein Systems, Moscow, Nauka 1974
90. Albertsson, P. Å.: Partition of Cell Particles and Macromolecules, New York, Wiley 1971[2]
91. Fischer, D.: Biochem. J. 196, 1 (1981)
92. Albertsson, P. Å., in: Methods of Biochemical Analysis (ed.) Glick, D., Vol. 29, p. 1, New York, Wiley 1983
93. Miheeva, L. M., Zaslavsky, B. Yu., Rogozhin, S. V.: Biochim. Biophys. Acta 542, 101 (1978)
94. Bamberger, S., Seaman, G. V. F., Brown, J. A., Brooks, D. E.: J. Colloid Interface Sci. 99, 187 (1984)
95. Laurent, T. C., in: The Chemical Physiology of Mucopolysaccharides (ed.) Quintarelli, G. I., p. 153, Boston, Little, Brown and Co. 1968
96. Ben-Sasson, S. A., Henkart, P. A.: J. Immunol. 119, 227 (1977)
97. Agostino, D.: Tumori 62, 245 (1976)
98. Middaugh, C. R., Tisel, W. A., Haire, R. N., Rosenberg, A.: J. Biol. Chem. 254, 367 (1979)
99. Evdokimov, Yu. M.: Zhur. Vsesoyus. Chim. Obshetvo 20, 259 (1975)
100. Maggio, B., Ahkong, Q. F., Lucy, J. A.: Biochem. J. 158, 647 (1976)
101. Zaslavsky, B. Yu., Miheeva, L. M., Mestechkina, N. M., Shchyukina, L. G., Chlenov, M. A., Kudrjashov, L. I., Rogozhin, S. V.: J. Chromatogr. 202, 63 (1980)

102. Zaslavsky, B. Yu., Miheeva, L. M., Rogozhin, S. V.: J. Chromatogr. 212, 13 (1981)
103. Zaslavsky, B. Yu., Mestechkina, N. M., Rogozhin, S. V.: J. Chromatogr. 260, 329 (1983)
104. Zaslavsky, B. Yu., Miheeva, L. M., Rogozhin, S. V.: Biochim. Biophys. Acta 510, 160 (1978)
105. Zaslavsky, B. Yu., Miheeva, L. M., Mestechkina, N. M., Rogozhin, S. V.: J. Chromatogr. 253, 139 (1982)
106. Zaslavsky, B. Yu., Miheeva, L. M., Mestechkina, N. M., Rogozhin, S. V.: J. Chromatogr. 253, 149 (1982)
107. Ligny De, C. L., Gelsema, W. J.: Sep. Sci. Technol. 17, 375 (1982)
108. Boguslavsky, L. I.: Bioelectrochemical Phenomena and Interface, Moscow, Nauka 1978
109. Zaslavsky, B. Yu., Mestechkina, N. M., Miheeva, L. M., Rogozhin, S. V.: J. Chromatogr. 240, 21 (1982)
110. Zaslavsky, B. Yu., Shchyukina, L. G., Rogozhin, S. V.: Mol Biologia (Rus.) 15, 1315 (1981)
111. Zaslavsky, B. Yu., Mestechkina, N. M., Miheeva, L. M., Rogozhin, S. V., Bakalkin, G. Y., Rjazhsky, G. G., Chetverina, E. V., Asmuko, A. A., Bespalova, J. D., Korobov, N. V., Chichenkov, O. N.: Biochem. Pharmacol. 31, 3757 (1982)
112. Chiou, C. T., Block, J. H., Manes, M.: J. Pharm. Sci. 71, 1307 (1981)
113. Draffehn, J., Schönecker, B., Ponsold, K.: J. Chromatogr. 216, 63 (1981)
114. Shchyukina, L. G., Zaslavsky, B. Yu., Rogozhin, S. V., Florentiev, V. L.: Mol. Biologia (Rus.) 18, 1128 (1984)
115. Zaslavsky, B. Yu., Mestechkina, N. M., Rogozhin, S. V.: Biochim. Biophys. Acta 579, 463 (1979)
116. Zaslavsky, B. Yu., Miheeva, L. M., Rogozhin, S. V.: Biochim. Biophys. Acta 588, 89 (1979)
117. Zaslavsky, B. Yu., Borovskaya, A. A., Rogozhin, S. V.: Mol. Cell. Biochem. 60, 131 (1984)
118. Zaslavsky, B. Yu., Miheeva, L. M., Rogozhin, S. V., Borsova, L. I. Kosinez, G. I.: Biochim. Biophys. Acta 597, 53 (1980)
119. Dearden, J. C., O'Hara, J. H., Townend, M. S.: J. Pharm. Pharmacol. 32, 102 P (Suppl.) (1980)

Author Index Volumes 101–146

Contents of Vols. 50–100 see Vol. 100
Author and Subject Index Vols. 26–50 see Vol. 50

The volume numbers are printed in talics

Bei Fragen zur Produktsicherheit wenden Sie sich bitte an:
If you have any questions regarding product safety,
please contact:

Walter de Gruyter GmbH
Genthiner Straße 13
10785 Berlin
productsafety@degruyterbrill.com